I0882475

WHOSE WATER? THE CONTROL AND APPROPRIATION OF WATER RESOURCES IN MEDIEVAL HUNGARY

Whose Water? The Control and Appropriation of Water Resources in Medieval Hungary addresses conflicts surrounding water management, a space where different economic and other socio-political interests meet and sometimes clash. The geographical space that it focuses on is the Kingdom of Hungary, and the time frame is the period from the foundation of the Christian state around the year 1000 to the late medieval period.

Modern politics focuses on who has legitimate claims in water-related disputes, but for historians, it is certainly more relevant to understand how such conflicts were approached and resolved in the past. These questions include what kind of disputes unfolded concerning water use; to what degree water was conceived as a private or common good; and how different interests were aligned with each other. Throughout the book's chapters, it is argued that the use of water by the societies of the Kingdom of Hungary in the Middle Ages gave rise to complex sets of customs and norms that, until the modern era, were the most important principles for settling water-use disputes.

Whose Water? The Control and Appropriation of Water Resources in Medieval Hungary will be of interest to students and scholars of environmental and landscape history, as well as those studying the legal and urban history of premodern Europe.

András Vadas is an associate professor in medieval history at Eötvös Loránd University, Budapest, Hungary. He graduated from the same institution (PhD, 2015) and the Central European University (PhD, 2020). He was awarded a Fulbright Postdoctoral Scholarship at Georgetown University (2017) and taught at Columbia University as an István Deák Visiting Professor (2025).

He is the author or editor of about a dozen books in English and Hungarian on the environmental, urban, and economic history of premodern Central Europe. His first English-language monograph, *The Environmental Legacy of War on the Hungarian-Ottoman Frontier,* was published in 2023.

WHOSE WATER? THE CONTROL AND APPROPRIATION OF WATER RESOURCES IN MEDIEVAL HUNGARY

András Vadas

Routledge
Taylor & Francis Group
NEW YORK AND LONDON

Designed cover image: Antiqua Print Gallery / Alamy

First published 2027
by Routledge
605 Third Avenue, New York, NY 10158

and by Routledge
4 Park Square, Milton Park, Abingdon, Oxon, OX14 4RN

Routledge is an imprint of the Taylor & Francis Group, an informa business

© 2027 András Vadas

The right of András Vadas to be identified as author of this work has been asserted in accordance with sections 77 and 78 of the Copyright, Designs and Patents Act 1988.

All rights reserved. No part of this book may be reprinted or reproduced or utilised in any form or by any electronic, mechanical, or other means, now known or hereafter invented, including photocopying and recording, or in any information storage or retrieval system, without permission in writing from the publishers.

For Product Safety Concerns and Information please contact our EU representative GPSR@taylorandfrancis.com. Taylor & Francis Verlag GmbH, Kaufingerstraße 24, 80331 München, Germany.

Trademark notice: Product or corporate names may be trademarks or registered trademarks, and are used only for identification and explanation without intent to infringe.

ISBN: 9781032980706 (hbk)
ISBN: 9781032980737 (pbk)
ISBN: 9781003596905 (ebk)

DOI: 10.4324/9781003596905

Typeset in Sabon
by Newgen Publishing UK

CONTENTS

List of Figures *viii*
List of Tables *x*
Acknowledgments *xi*
List of Abbreviations *xiii*
A Note on Names *xvi*

1 Introduction 1
1.1 Riverine Landscapes as Conflict Environments in Premodern Europe: Some Historiographic Reflections 3
1.1.1 New Directions in Research: Where Environmental, Social, and Technological History Meet 6
1.1.2 Hungary and Central Europe: Research Traditions and Recent Results 11
1.2 The Goals of this Work 16
1.2.1 Time and Space 21
1.2.2 About Sources 22
1.3 The Structure of the Book 24

2 The Spread of Water Mills in Medieval Hungary with a Central European Overview 43
2.1 Mills and Benedictines: Some Preliminary Considerations 44

2.2 *The First Water Mills in the Region: Historical Evidence, Linguistics, and (the Lack of) Archaeological Data 46*
2.3 *Conclusions 54*

3 Waterscapes as Conflict Environments in Late Medieval Hungary 63
3.1 *Legal Regulations and Practice of Mill Construction in Medieval Hungary 65*
3.1.1 *Locus molendini: A Place, a Right, a Claim? 67*
3.1.2 *Locus Molendini in Early Literacy: First References and Terminological Unifications 68*
3.1.3 *Terminological Uniformity: Uniformity in Meaning? 73*
3.1.4 *Mills and Mill Sites: How Do They Relate to Each Other? 77*
3.2 *Rivers in Urban Landscapes: Rivers, Streams, Moats, and Disputes 79*
3.2.1 *Major Waterways and Mills: The Example of Buda, Óbuda, and Pest 80*
3.2.2 *Streams and Mills in Urban Environments: The Cases of Zagreb and Felhévíz 84*
3.2.3 *Urban Moats and Water Use 94*
3.3 *Rivers in Mining Areas: Abundance of Water versus Lack of Access? 110*
3.3.1 *Mining Town Privileges and Water Use 113*
3.3.2 *Early Urban Privileges and Water Use 120*
3.3.3 *Kremnica and Its Water Mills 123*
3.3.4 *The Foundation of Nová Baňa and Its Water Mills 130*
3.3.5 *Mills in Mining Towns: Who Was Entitled to What? 131*
3.4 *Rivers as Borders: Another Minefield? 132*
3.4.1 *The Border between Ľubotín and Orlov: What Can a Single Case Reveal? 136*
3.4.2 *Riverbed Changes and Estate Borders: Was There a Medieval Customary Law in Hungary? 139*
3.4.3 *Riverbeds and Borders: Some Conclusions 147*
3.5 *Conclusions: Rivers as Conflict Environments 150*

4 The Men behind the Wheel... Building on Water in the Middle Ages 197
4.1 The Renovation of the Moat of Sopron in 1540: A Unique Source 198
4.1.1 The Expenditures on the Work 202
4.1.2 The Labor Organization 204
4.2 Millers 206
4.3 Ditch and Pond Diggers and Pond Masters 210
4.4 Conclusions 215

5 Conclusions and Outlook 226
5.1 The Politics of Water: An Epilogue 229

Appendices 232
Appendix A: Documents 232
Appendix B: List of Charters of Privilege studied in Chapter 3.3 242
Appendix C: List of Cases Studied in Chapter 3.4 244
Appendix D: List of the Kings of Medieval Hungary 248
Bibliography 250
Index 307

FIGURES

1.1	Areas covered by water and wetlands before river regulation in the Carpathian Basin (after Woldemár Lászlóffy)	18
1.2	Major geographical units in the Carpathian Basin (drawn by Béla Nagy)	20
3.1	Gombaszög with the location of the Pauline abbey marked with the red cross and the water mills along the artificial riverbed of the Hernád (First Military Survey)	74
3.2	The surroundings of Nová Baňa with the Büksavnica (the one running west–east) and Verence Streams (running north–south) on the First Military Survey, with the two mill sites on the latter	77
3.3	Pest, Buda, and Óbuda and their surroundings in ca. 1300 (cartography by Béla Nagy)	82
3.4	Gradec, Kaptol, and the Medveščak Stream with the mill race on the manuscript map by Nicholas Angielini	87
3.5	The mills of Zagreb in the Middle Ages (drawn by András Vadas and László Ferenczi)	88
3.6	Felhévíz and the Malomtó (Millpond) as represented in a late seventeenth-century illustration (by H. Bredekow)	90
3.7	The water-related urban infrastructure of Sopron before 1700	100
3.8	Zacharias Michel's Bird's eye view of Sopron in 1700 (after Jankó, Kücsán, and Szende, *Sopron*, Plate C.4)	101
3.9	View of the town of Prešov from 1768 by Gaspar Caspar	106
3.10	Detail of the view of the town of Prešov with the Upper Mill (no. 23) from 1768 by Gaspar Caspar	107

3.11 Topography of Prešov in the late medieval period, with the relocation of the mills and the probable direction of the western section of the urban moat before and after 1435 109
3.12 Mining towns in medieval Hungary 112
3.13 The so-called Saint Anne altar at the Virgin Mary Cathedral at Rožňava (detail, photo by author) 115
3.14 The mills and the urban topography of Kreminca in the mid-fourteenth century after György Györffy 125
3.15 The mills of Kremnica and their ownership (1331–1519) 129
3.16 The surroundings of Ľubotín and Orlov on the sheet of the First Military Survey 137
3.17 Čoltovo and Lekeňa on the sheet of the First Military Survey 142
3.18 Szentmárton and Kóród on a sheet of the First Military Survey 144
3.19 The border between Ónod and Hídvég in the mid-nineteenth century, showing the location of the former bed of the River Sajó in the center 145
3.20 Monasteries and chapters in the Kingdom of Hungary in the Late Middle Ages (after Beatrix Romhányi) 153
4.1 The incomes and expenses of the town of Sopron between 1530 and 1543 in light of its account books 203

TABLES

2.1 The pertinence lists of the Veszprém endowment charter and three imperial and royal donation charters 50
4.1 Wages according to the 1539/1540 account book of Sopron 205

ACKNOWLEDGMENTS

This book is based on a PhD dissertation defended in March 2020 during the COVID-19 pandemic at Central European University (CEU). Although I will acknowledge many others to whom I am indebted, the most significant are my two supervisors, Katalin Szende and Alice M. Choyke. Without them, neither the dissertation nor the book would be the same. They connected me with dozens of colleagues and expanded my scientific network into new realms. I received invaluable assistance from several colleagues, friends, and research institutions. Many additional faculty members and fellow students from the Department of Medieval Studies at CEU contributed to this book. József Laszlovszky continuously monitored my work, as did Balázs Nagy, to whom I owe much more than that. I am also grateful to the wonderful librarians of the unfortunately abolished ELTE-CEU Medieval Library, especially Ágnes Havasi, who, whenever she spotted a volume that might interest me, promptly notified me or showed it to me. She also overlooked my months-long delays in returning books.

In addition to individuals at the CEU Department of Medieval Studies, many others contributed to completing this project. Professor Gerrit Jasper Schenk (Technical University of Darmstadt) and Professor Richard Hoffmann (York University, Toronto) suggested numerous essential corrections that helped transform the dissertation into a book. Christoph Sonnlechner from the Municipal and Provincial Archives of Vienna was extremely helpful during my research in Vienna and by following up on my work. Bence Péterfi (Institute of History, ELTE Research Centre for the Humanities, Budapest) has shared countless data with me over the past ten years, while István Tringli (Institute of History, Research Centre for the Humanities, Budapest) consistently took the time to read different phases of this work and pointed out some of its shortcomings. I am grateful to Bruno Škreblin (Croatian Institute of History, Zagreb) for his guidance and assistance with the mills in Zagreb.

I received financial aid from several institutions that were crucial to completing this work, but the generous grant from Central European University was the most significant support. The university's travel grants enabled me to attend conferences and conduct research in Leeds, Beijing, Vienna, and Oxford. A research stay in Heidelberg was backed by a joint initiative of the University of Heidelberg and CEU, sponsored by DAAD. All these various sources of support were vital for finishing this book. The funding from the Budapest—Environmental History of an Urban Area (NKFI FK no. 142451) program was instrumental in completing this book as well as the Industrialization of the Danube as a Dynamic Socio-Natural Site (IDAN) project of the Austrian Science Fund (FWF).

The Department of Medieval History at Eötvös Loránd University has been a wonderful workplace over the past decade, which, despite the increasing political pressure on academic life in Hungary, has remained a peaceful and independent place for teaching and research. In the final stages of this work, a visiting professorship at the Harriman Institute at Columbia University in New York City was essential. I am grateful to Routledge for agreeing to publish my book, which focuses on a region and period underrepresented in environmental history research.

Finally, as always, my heartfelt thanks go to my family. Besides my father and mother, my grandmother gave me the most encouragement to pursue a career in history. I am grateful to my wife, Virág, and my daughters, Lujza and Sári, who accepted my frequent absences due to conferences and research trips and accompanied me during some of my more extended stays.

ABBREVIATIONS

ÁMTF	Györffy, György. *Az Árpád-kori Magyarország történeti földrajza*, 4 vols [A historical geography of Hungary in the Árpádian period] Budapest: Akadémiai Kiadó, 1963–1998.
AO	*Anjoukori okmánytár / Codex diplomaticus Hungaricus Andegavensis*, 7 vols. Eds. Imre Nagy and Gyula Tasnádi Nagy. Budapest: Magyar Tudományos Akadémia, 1878–1920.
AOklt	*Anjou-kori oklevéltár*, 53 vols [Angevin cartulary]. Ed. in chief, Gyula Kristó, and Tibor Almási. Szeged and Budapest: JATE – MNL OL, 1990–2025.
ÁÚO	*Árpádkori új okmánytár*, 12 vols [New Árpádian-age cartulary]. Ed. Gusztáv Wenzel. Pest–Budapest: Eggenberger Ferdinánd, 1860–1874.
BTOE, vol. 1	*Budapest történetének okleveles emlékei/ Monumenta diplomatica civitatis Budapest*, vol. 1. Ed. Albert Gárdonyi. Budapest: Budapest Székesfőváros, 1936.
CD	*Codex diplomaticus Hungariae ecclesiasticus ac civilis*, 11 vols. Ed. Georgius Fejér. Buda: Typis Regiae Universitatis Hungaricae, 1829–1844.
CDB, vol. 1	*Codex diplomaticus et epistolaris regni Bohemiae*, vol. 1. Ed. Gustavus Friedrich. Prague: Sumptibus Comitiorum Regni Bohemiae, 1904/1907.
CDES	*Codex diplomaticus et epistolaris Slovaciae*, 2 vols. Ed. Richard Marsina. Bratislava: Academia Scientiarum Slovacae, 1971–1987.

DHA, vol. 1	*Diplomata Hungariae Antiquissima*, vol. 1. 1000–1131. Ed. György Györffy. Budapest: Akadémiai, 1992.
EFHU, vol. 3/2	*Elenchus fontium historiae urbanae*, vol. 3/2. Ed in chief András Kubinyi. Budapest: Balassi, 1997.
Engel, *Archontológia*	Engel, Pál. *Magyarország világi archontológiája: 1301–1457*, 2 vols [Lay archontology of Hungary, 1301–1457]. Budapest: História, 1996.
Házi, *Sopron*	Házi, Jenő. *Sopron szabad királyi város története*, 13 vols [The history of the free royal town of Sopron]. Sopron: Székely, Szabó és Társa Könyvnyomdája, 1921–1943.
HO	*Hazai okmánytár / Codex diplomaticus patrius*, 8 vols. Ed. Imre Nagy. Győr and Budapest: Sauervein Géza – Franklin, 1865–1891.
MCZ	Tkalčić, Ivan Krstitelj. *Povjestni spomenici slob. kralj. grad. Zagreba. Monumenta historica liberae regiae civitatis Zagrabiae*, 11 vols. Zagreb: Brzotisk Albrechta, 1889–1905.
MES	*Monumenta Ecclesiae Strigoniensis*, 4 vols. Eds. Nándor Knauz et al. Esztergom and Budapest: Horák – Argumentum, 1874–1999.
MGH	*Monumenta Germaniae Historica*
MNL OL	Magyar Nemzeti Levéltár Országos Levéltára (National Archives of Hungary, State Archive), Budapest
MNL OL database	A Középkori Magyarország Levéltári Forrásai [Archival documents of medieval Hungary]. Database. https://archives.hungaricana.hu/hu/charters/ (last accessed: June 14, 2025)
MNL OL DF	Diplomatikai Fényképgyűjtemény [Photographic Collection of Diplomatics], U szekció [Section "U"]
MNL OL DL	Diplomatikai Levéltár [Archives of Diplomatics], Q szekció: Mohács előtti gyűjtemény [Section "Q": Pre-1526 Collection]
MNL OL E	Magyar Kincstári Levéltárak [Archives of the Hungarian Chambers]
MNL OL P	Családok, testületek és intézmények levéltárai [Archives of Families, Corporations, and Institutions]
MNL OL S	Térképár [Collection of Maps]

OSZK	Országos Széchényi Könyvtár (Széchényi National Library), Budapest
PMTOE	*Pest megye történetének okleveles emlékei 1002–1599-ig* [Diplomatic sources on the history of Pest County, 1002–1599]. Ed. László Bártfai Szabó. Budapest: [no publisher], 1938.
RA	*Az Árpád-házi királyok okleveleinek kritikai jegyzéke / Regesta regum stirpis Arpadianae critico-diplomatica*, 2 vols. Eds. Imre Szentpétery, and Iván Borsa. Budapest: Akadémiai, 1923–1987.
Tripartitum	*Tripartitum: The Customary Law of the Renowned Kingdom of Hungary: A Work in Three Parts Rendered by Stephen Werbőczy (The "Tripartitum")*. Ed. and transl. by János M. Bak, Péter Banyó, and Martyn Rady. Idyllwild, CA: Charles Schlacks, 2005.
UB	*Urkundenbuch des Burgenlandes und der angrenzenden Gebiete der Komitate Wieselburg, Ödenburg und Eisenburg*, 5 vols. Hans Wagner, and Irmtraut Lindeck-Pozza. Graz and Cologne: Böhlau, 1955–1999.
VMMS, vol. 1	*Výsady miest a mestečiek na Slovensku. I. 1238–1350* [Privileges of towns and cities in Slovakia, 1238–1350]. Ed. Ľubomír Juck. Bratislava: Veda – Vydavateľstvo Slovenskej akadémie vied, 1984.
Zichy	*A zichi és vásonkeői gróf Zichy-család idősb ágának okmánytára / Codex diplomaticus domus senioris comitum Zichy de Zich et Vasonkeő*, 12 vols. Eds. Imre Nagy et al. Pest – Budapest: Magyar Történelmi Társulat, 1871–1931.
ZsO	*Zsigmondkori oklevéltár*, 13 vols [Sigismundian cartulary] (Magyar Országos Levéltár kiadványai, II. Forráskiadványok 1, 3–4, 22, 25, 27, 37, 39, 41, 43, 49, 52, 55, 59, 61). Eds. Elemér Mályusz et al. Budapest: Akadémiai, 1951–2022.
Zsoldos, *Archontológia*	Zsoldos, Attila. *Magyarország világi archontológiája, 1000–1301* [Lay archontology of Hungary, 1000–1301] (História könyvtár. Kronológiák, adattárak, 11). Budapest: História – MTA Történettudományi Intézete 2011.

A NOTE ON NAMES

When the location of a settlement falls outside Hungary's present-day territory, the book always uses its current, official name. If a settlement no longer exists, the medieval or early modern name is applied. For rivers with sections in present-day Hungary, Hungarian names are used unless a common English name exists. For rivers flowing outside present-day Hungary, the names used in the relevant countries are used. Since county borders have changed significantly over time, the historical county names are provided in their Hungarian forms. Various other names for settlements and regions are included in the index.

1
INTRODUCTION

> Hungary shall cede to Czechoslovakia the villages of Horvathjarfalu [Jarovce], Oroszvar [Rusovce], and Dunacsun [Čunovo, part of present-day Bratislava], together with their cadastral territory as indicated on Map No. IA annexed to the present Treaty. Accordingly, the Czechoslovak frontier on this sector shall be fixed as follows: from the point common to the frontiers of Austria, Hungary and Czechoslovakia, as they existed on January 1, 1938, the present Hungarian-Austrian frontier shall become the frontier between Austria and Czechoslovakia as far as a point roughly 500 meters south of hill 134 (3.5 kilometers northwest of the church of Rajka), this point now becoming common to the frontiers of the three named States; thence the new frontier between Czechoslovakia and Hungary shall go eastwards along the northern cadastral boundary of the village of Rajka to the right bank of the Danube at a point approximately 2 kilometers north of hill 128 (3.5 kilometers east of the church of Rajka), where the new frontier will, in the principal channel of navigation of the Danube, join the Czechoslovak-Hungarian frontier as it existed on January 1, 1938; the dam and spillway within the village limits of Rajka will remain on Hungarian territory.[1]

The infamous Paris Peace Treaty of 1947 redrew the borders of Hungary and Czechoslovakia as described. For much of its length, as before World War II, the Danube and Ipoly Rivers delineated the boundary between the two countries. The rivers, particularly the Danube, with its substantial flow, appeared to be the most obvious border. Nonetheless, to generate energy for Communist industrialization and, no less importantly, to bring the political landscape of the countries closer together, the old concept of a major hydropower complex on the Danube resurfaced soon after World War II. The

DOI: 10.4324/9781003596905-1

Soviet Union pursued similar initiatives in other countries where communists seized power and joined the Warsaw Pact. The Iron Gate hydropower plant, located between Yugoslavia and Romania, served similar purposes. While the latter project was completed, the Hungarian-Czechoslovakian endeavor was never fully realized. The so-called Gabčíkovo–Nagymaros Dams, planned and partially constructed between the 1970s and 1990s on the Danube, became one of the most contentious industrial projects in the history of both nations. The plan was to dam the river at two points to create a hydropower complex: one near Bratislava at Gabčíkovo, and another further downstream at one of the most picturesque locations along the River Danube in Hungary, known as the Danube Bend, between Nagymaros and Visegrád. The project played a crucial role in shaping the Hungarian political opposition, which, alongside the country's unfolding financial crisis, ultimately led to the gradual cessation of the project and, on the Hungarian side, its suspension. At that stage, as the hydrological work on the Czechoslovak side neared completion, the newly formed Slovak government, emerging after the split of the Czech Republic and Slovakia, vigorously opposed terminating the project. The controversy peaked with the diversion of over 80 percent of the Danube's waters through a sluice that became operational on the Slovak side in October 1992. The project largely misfired, resulting in a prolonged lawsuit between newly established Slovakia and Hungary, as well as significant changes to the landscape in Hungary's upper section of the River Danube. The dispute caused notable political tensions between the two countries, in which Slovakia—not without justification—faced accusations of "stealing" the Danube, while Hungary—again, justifiably—was charged with violating a signed international agreement. The project's history, which curiously lacks a comprehensive study, highlights the significance of water ownership and the persistent conflicts over water use to this day.[2] Neither the damming process described above nor the protests is unusual. Numerous similar lawsuits have arisen over the past few decades concerning individuals' and communities' rights to water resources, including access to freshwater, water for hydropower, and industrial water. Recent reports on climate change indicate that this competition for water resources will pose an escalating challenge for humanity in the upcoming decades. Yet there is nothing new under the sun; similar legal battles can be traced back at least a millennium in the region examined in this book. The varying interests of people using river water for drinking, agriculture, energy production, craft industries, fishing, navigation, or organizing the defense of a settlement or a country all led to differing priorities that frequently caused conflict and compromise.

The book addresses the issues surrounding water management, an area where various economic and socio-political interests intersect and sometimes conflict. Modern politics focuses on who holds legitimate claims in such disputes. Still, for historians, it is much more relevant to understand how these conflicts were approached and resolved. These inquiries include what types of disputes arose regarding water use and the extent to which water was considered private or "common." How were different interests aligned with one another? The book explores these questions in the context of Central Europe, specifically within the Kingdom of Hungary during the Middle Ages (approximately from the tenth to the mid-sixteenth century), a period when literacy reemerged in the region following the Roman era. Throughout the various chapters, it is argued that water use by the societies of the Kingdom of Hungary in the Middle Ages led to the development of intricate sets of customs and norms that, until modern times, served as the key principles in resolving water-related disputes. Before discussing the scope and hypotheses of this work, a summary is provided on how scholarship has addressed these questions over the past century.

1.1 Riverine Landscapes as Conflict Environments in Premodern Europe: Some Historiographic Reflections

Historical scholarship over the last few decades has demonstrated the complexity of water-related disputes in modern contexts. There's strong evidence from many parts of Europe and beyond, however, that these disputes—centered on unequal access to water—began well before modern times. Due to increasing water scarcity worldwide, especially in urban areas, researchers have also increasingly addressed water-related concerns in historical contexts. The topic has been explored in scholarship from numerous research angles and studied by technological and architectural historians, political and legal historians, economic and social historians, and, finally, environmental and landscape historians and archaeologists. Recently, a completely different approach dealing with the intellectual history of water has also garnered interest among historians and literary scholars. A survey of scholarship on the history of water use and management would be worthwhile, as there is a significant gap in this area. However, exploring this field would far exceed the limitations of this book, and given the extensive nature of the scholarship, it could hardly claim completeness. Therefore, the following pages provide an overview of the main research directions and questions raised about premodern water-related conflicts in Europe. The book aims to be more comprehensive in its coverage of recent scholarship on Central Europe compared to other regions, as works produced there in recent years have only partially reached English- and German-speaking audiences despite significant growth in publications in Western languages. Primary research directions will be outlined in the following pages, with further specific works referenced and discussed in individual chapters. This survey, while referring to works on water use in general, will primarily focus on mills, as milling environments appear to have played a key role in disputes.

The study of historical water management and related conflicts is more than a century old. The monumental four-volume study by Richard Bennett and John Elton, *History of Corn Milling*, while primarily focusing on milling in England, remains a most valuable handbook on numerous aspects of milling and water use in the British Isles and continental Europe over the last millennium. This seminal work set the stage for the research directions of the twentieth century. The book was likely produced during one of the last moments when it was still feasible to study several technological aspects, as numerous mills were still standing, if not running. Not much later than the two English historians wrote their book, two significant research directions unfolded, which continue to play an important role in water history scholarship: historical topography and geography, and, more significantly from the perspective of this book, legal history. Historical topographies and regional historical geographies published in France, Germany, and Central Europe all contributed to the study of historical water management. The datasets collected within them contain numerous references to toponyms and topographic features related to human interventions in water systems, such as artificial channels, dikes, dams, sluices, fishponds, and more.[3] Nonetheless, studies on historical water rights are more relevant to this analysis since they inherently focus on the conflicts that arose in water management and use, their resolutions, and the resulting regulations in historical contexts. In this tradition, German historiography has proven to be the most influential.[4] Besides general works, issues surrounding dikes,[5] mills,[6] fishing, and fishponds,[7] along with regional variations in the regulation of water use, have been discussed in detail across many German-speaking areas.

Following in the footsteps of early legal histories that focused primarily on normative sources—law codes and other legal regulations—rather than on actual attempts at conflict resolutions, new historical approaches to waterscapes emerged during the interwar period. These works concentrated on technological, economic, and social history; however, in many instances, these three aspects are intertwined. Most significant studies from this period are as much technological and economic histories as they are social histories. It is not the number of studies from this interwar period that holds importance, but rather the influence of a few works that established the foundational questions historians would continue to revisit in the following decades. One of the most important of these works was written by the renowned French historian, Marc Bloch. His study on the spread of water mills, published in 1935, is likely one of the most frequently cited works in water history scholarship.[8]

There is virtually no comprehensive work on water history that does not mention Bloch's critical study, which is valuable from technological, economic, and social historical perspectives. His work's main argument is that the spread of watermills in the medieval West was not primarily due to population growth, but rather the power of landlords, who, by enforcing a ban on milling, generated significant income for themselves. Although Bloch also discussed the technological aspects of the spread of water mills, he primarily focused on the social organization behind their proliferation. Later, as Bloch argued, increases in mill numbers were partly attributed to the rising demand for industrial mills—especially in the textile[9] and metal[10] industries—but this trend may have had a rather limited geographical impact across Europe, particularly before the Late Middle Ages. Bloch's research raised a question that historians only revisited from the 1950s onward: to what extent mills should be regarded as urban phenomena during the High Middle Ages. Some scholars contended that, unlike most industries, mills were evenly distributed throughout the countryside and, consequently, contributed to the industrial development of rural areas.[11] In contrast, many others drew attention to the significant concentration of industrial and flour mills in urban settings. Historians repeatedly examine the role waterpower played in the intensification of medieval industry and the energy framework of the premodern (pre-coal) era.[12]

It was primarily social and technological historians in the postwar period who discussed issues of water management.[13] Georges Duby, another representative of the Annales School after Bloch, addressed water management and argued for the crucial importance of water mills in the development of the countryside in Western Europe.[14] Historians of technology, following in Duby's footsteps, published several influential—and partly popularizing—books on the role of water mills in the intensification of industry during the Middle Ages.[15] Richard Holt's monograph on English mills, in many ways, represents a critical summary of the subject until the advent of new approaches that became widespread in Europe in the 1990s and somewhat earlier in the U.S. Due to its concise and clear argumentation, this essential handbook remains influential.[16]

In addition to mills, historians of technology and social historians have shown interest in large water-regulation and canalization projects—such as the Karlsgraben of Charlemagne, which was initially conceived, in part, as a prime example of imperial power and later as a precursor to the Rhine–Main–Danube Canal.[17] However, the social context of significant medieval water-regulation works has been only partially explored in the literature, despite their potential for studying the relationship between state power and water infrastructure, on the one hand, and the social organization behind these projects, on the other.[18] The lack of research is likely due to social historians sometimes distancing themselves from concepts that are easily recognizable as Marxist. Another focal point of social history that has drawn scholars' attention is the role of mills as local centers. In rural areas, where most watermills were situated, these structures were among the most vital to locals. Due to the seigneurial ban, people from neighboring villages frequently visited the mills. Accordingly, these became gathering spots for local communities, not only because of the necessary visits but also because the miller was often the most informed and educated individual in the village.[19] This has rendered mills and millers important subjects for social historians.

In summary, the initial steps in studying historical water management emerged from various directions. First, legal history significantly influenced early research and remains one of the most vital approaches, especially in German scholarship. Additionally, historical geography and local and regional histories aided the study of water management by gathering data on waterways, fishponds, fords, and more. Research by social and economic historians, along with historians of technology from the 1930s, began to focus on premodern water management. These aspects remained crucial in research in subsequent decades, with relatively few new developments observed from the postwar period to the 1990s. Notably, there was minimal consideration of the effects of hydraulic works on their immediate and broader environments, as well as the conflicts associated with the exploitation of these water bodies across various economic spheres, by different actors in both urban and rural areas.

1.1.1 New Directions in Research: Where Environmental, Social, and Technological History Meet

The past three decades have brought significant changes to the field of historical water management. General interest in water history has grown continuously—an unsurprising development given the political and social concerns mentioned above. More importantly, the methods for approaching water history have also changed significantly. Until the 1980s, scholarly works, with very few exceptions, treated water as a passive substance that past societies exploited or modified through technology.[20] From the 1980s onward, however, several theories by philosophers, sociologists, ecologists, and, notably, historians proposed new ways of examining historical and contemporary waterscapes and landscapes.[21] Historical ecologists and environmental historians increasingly viewed rivers not as passive objects, but as dynamic actors shaping their surrounding material and social arrangements. In this regard, the research group focused on the waters of Vienna, led by Verena Winiwarter, which serves as a prime example. Just as today, riverscapes have always been among the most dynamic landscapes and must be understood as complex arrangements—socio-natural sites.[22] Building on this concept, recent scholarship has suggested studying floodplains in Central Europe as sites of a "fluvial anthroposphere."[23]

The last three decades have witnessed the expansion of environmental history alongside the growth of environmental archaeology, which introduced new sources, methods, and research questions related to rivers. While the traditional focus was on how rivers were modified for exploitation, environmental historians flipped the question, often examining rivers as active agents that influenced, collaborated with, or opposed the societies around them. As part of this recent research agenda, river histories have been written that attribute agency to waterways. Beyond the methodological studies,[24] Leona Skelton's "biography" of the Tyne is particularly important; it examines four centuries of changes in the river's social and economic organization.[25] Even though virtually none of the river environmental histories offer comprehensive accounts, the snapshots or the more extended periods they cover illustrate how various social, financial, and political contexts shape how societies perceive and use rivers. As Paula Schönach has put it:

> While rivers create opportunities and impose limits, they don't determine the eventual paths that are taken. Competing claims are made, and the various actors aspire to different kinds of rivers at specific times and places. Water is political, the riverine waters are political, and the human–river interactions are political, which brings explorations in river history down to questions of power.[26]

In many recent studies, the focus has been on the contest between water as a resource and its impact on power relations in efforts to harness waterscapes. However, as recent historiographic overviews show, this problem has primarily been studied in works addressing the modern period.[27] Nonetheless, as demonstrated by Skelton and other scholars, this approach can also be applied to the premodern period. Skelton's book is an exception, as it covers a more extended period to trace how the different functions necessitated varying arrangements. The book discusses the "Tyne as an industrial river, a port, a liquid highway, a salmon river, an open sewer, a source of energy for mills, a regional icon, a recreational facility, a tourist attraction, and a source of regional pride and artistic inspiration."[28] Although not all aspects are relevant in premodern Central European contexts, the primary sources of conflict, the different interests of the actors living along the river, are succinctly summarized within these lines. The Tyne, like other rivers in England and, of course, in Central Europe as well, was exploited for fishing, well before the Early Modern period that Skelton primarily focused on. Although the catchment of the Tyne is not represented in the *Domesday Book*, mills were constructed along the river by the High Middle Ages.[29] The growing industrial activity necessitated different installations and arrangements along its course, as did the construction of harbors. This also occurred throughout Central Europe in the medieval period. The Tyne was rich in salmon, just as the Danube and other major rivers in Central Europe were rich in sturgeon and other species of fish. Still, many obstacles—especially weirs and dams—were introduced into the river for milling and industrial activities, hindering their migration, a phenomenon that occurred along many rivers in the British Isles and on the continent.[30] In England, research has particularly emphasized the impact of mills and fishing (weirs) on inland transportation.[31] Changes in the ecology of riverine environments, primarily due to navigation, the disposal of industrial and domestic waste, and physical interventions in river flow, have also sparked controversies.[32] This already led to changes in fishing conditions in the Middle Ages, as evidenced by various sources.[33] Most relevant studies, however, have focused on major water flows, often overlooking smaller tributaries, as is the case in modern contexts. While struggles for the use of major waterways were certainly sources of conflict, especially along urban riverscapes, this book will argue that small catchments may have been just as frequent sources of disputes between competing users.[34] That said, a few studies have attempted to address the problem in recent years, notably many papers from two conferences: one held at the Abbey of Flaran in France and another at Lorsch in Germany.[35] As the studies in these conference proceedings volumes and the upcoming chapters of this book demonstrate, in small catchments, the involved parties were often engaged in similar economic activities. Still, due to limited water resources, the amount of water each person used became critical. Even minor interventions in these rivers and streams could have significant impacts upstream and downstream, often leading to lawsuits.

Scholarship has also shown interest in the role and construction of artificial water systems, such as canals, irrigation channels, and mill races. However, until the last few decades, most of these works emphasized the despotic nature of the ruling authorities who oversaw their creation. The concept of hydraulic societies—as many studies referred to them—was used to describe how individual interests were often overshadowed by "empire building."[36] In recent decades, the focus has shifted from revealing despotism to examining the role of cooperation across different historical scales in the construction and maintenance of water management systems—including irrigation and flood mitigation. Several recent studies argue that rulers or ruling groups often entrusted local societies with power and authority over water regulation.[37] In one prime example, Alan Mikhail demonstrated how different peasant communities in Ottoman Egypt were granted full authority to govern irrigation networks and had the power to compel local elites to defer to their will.[38]

While the above new research trends have flourished in the past two decades, earlier fields of scholarship have not, of course, disappeared. Beyond the social aspects of historical water management discussed in the previous paragraphs, technological and economic history, as well as to some extent legal history, continue to be represented in the scholarship.

In history of technology, several essential volumes have been published, many of which utilize methods and results from another relatively recently emerging research field mentioned above, namely, environmental archaeology. This combination has significantly contributed to the study of medieval fishing, milling, irrigation networks, and navigation, among other aspects.

Among recent summarizing works, a volume edited by Paolo Squatriti deserves mention for its comprehensiveness and for integrating technological, social, and landscape history aspects of historical water management, making it one of the most frequently cited works in the field.[39] In addition to this edited volume, historians of technology have increasingly focused on the survival of ancient water technologies, including water wheels, during Late Antiquity and the Early Middle Ages. They have also examined their subsequent spread through the movement of monastic orders in "New Europe" and other means of knowledge transfer. This topic will be revisited in Chapter 2 of this book, focusing on Central Europe.[40] Furthermore, archaeology, influenced by the growth of environmental archaeology and new prospecting techniques, has contributed significantly to understanding the survival of ancient technologies by uncovering several early medieval mill complexes.[41]

Water-related economies have remained an important research topic, evidenced by numerous conferences and proceedings volumes over the past decades.[42] In these studies, water mills have remained a significant focus. As before the 1990s, key works addressed the milling industry in England, but growing scholarship has been dedicated to mills in Iberia and to their roles in the Italian,[43] German,[44] Austrian,[45] and French[46] economies. Two books stand out from the many that focus on milling industry in England. The first, authored by John Langdon, whose work, despite its title, *Mills in the Medieval Economy*, serves as both a technological and economic history of the medieval milling industry. His main argument, however, centers on a financial question. He contends that while the late medieval period is traditionally associated with significant economic restructuring and the birth of capitalism, this does not apply to the English milling industry. From the perspective of this book, though, his summary of the challenges of maintaining and rebuilding mills in the late medieval period is at least as important as his overall conclusion.[47] The other significant work is a relatively recent monograph by Adam Lucas, who studied ecclesiastical involvement in milling in medieval England, a topic highly relevant to the present book because, as will be demonstrated in the coming chapters, church institutions were especially active in water-related economies in medieval Central Europe, including Hungary.[48] Lucas demonstrates that in England, primarily with the Benedictines and partially with episcopal lords, great emphasis was placed on milling incomes (which for the Benedictines amounted to approximately 10 percent of their manorial incomes, a figure similar to that of some church institutions in Central Europe). Thus, they took great care to maintain their related privileges.[49]

Of the various spheres of water use in premodern societies, very little attention has been paid to fishing in this relatively brief survey. This lack of focus can be attributed to limited interest in the topic before World War II, and even the subsequent increase in interest was only noticeable from the 1980s onward.[50] Since then, the rapidly growing body of work—thanks in large part to the rise of ichthyoarchaeology and historical ecology—indicates that this field has emerged as a distinct research area.[51] In these studies, as in recent research on milling and historical navigation, the issue of differing interests among those accessing water bodies has been repeatedly addressed. Finally, one notable recent development should be acknowledged, which examines water infrastructure through the lens of critical infrastructure and evaluates the evolution of hydraulic works from a network perspective. These studies, primarily originating in Germany thus far, contribute important methodological insights by incorporating concepts from the social sciences, such as resilience, vulnerability, and criticality.[52]

In addition to the specific works mentioned in the previous pages, the growing general interest in water and society in the past is reflected in new publication forums. Several relevant publication forums were established in the 1990s and 2000s that focus on water history. As environmental history developed in both the U.S. and Western Europe, two significant journals were founded: *Environment and History* (founded in 1995), which focuses more on Europe but is increasingly opening up to a global perspective, and *Environmental History* (founded in 1996), which, from the start, has centered heavily on the Americas, Africa, and Asia, primarily on the modern times.[53] A more recent but promising undertaking is the International Water History Association's journal, *Water History*, founded in 2009. This journal focuses on water management and water as an energy resource across various historical contexts. Articles from these journals, along with recent decades' conferences on the topic, illustrate the diversity of issues that can be addressed within this broad framework and the sources available to address relevant questions.

These previous paragraphs aimed to illustrate that after initially being studied through the lens of technological and legal history, other approaches to historical water management emerged, particularly environmental history, which became predominant from the 1990s onward. The introduction of new sources, research questions, and methodologies, along with an interdisciplinary approach to related research issues, has led to significant new findings in the social history of water and an increasing interest in historical water management among scholars in recent decades. Nonetheless, the primary focus so far has been on scholarly works from England, the U.S., and German-speaking regions. However, from this book's perspective, the most relevant research—specifically concerning Central Europe—has not yet been addressed.

1.1.2 Hungary and Central Europe: Research Traditions and Recent Results

Scholarship on historical water management in Central European countries has developed very differently. Polish, Czech, Slovak, and Hungarian historians have focused on somewhat different questions, and despite similar political environments, scholars have researched distinct aspects of historical water management. This is slightly surprising, as the questions raised by social historians in the interwar period and after World War II regarding the social constructions behind water management projects would have been a topic welcomed in Marxist historiography. Nonetheless, in all three countries, significant results have emerged from the study of historical waterscapes and management. The varying scholarly interests naturally stem from the distinct geographical characteristics of these countries. The Czech lands and Hungary share similarities, as both are centered in basins. However, the dominant lowland features of Poland, with the River Vistula flowing along its long axis from the Silesian Beskids in the south to the Baltic Sea in the north, tell a different story altogether. While in Poland, the northern areas, similar to the central lowlands of the Carpathian Basin, possessed extensive wetlands, this was not typical of the Czech and Moravian landscapes.[54]

In Hungary, after the initial steps taken in the first third of the twentieth century to study historical water management from the perspectives of legal history and, to some extent, technological history, minimal research attention was dedicated to this question between the 1940s and the 1990s.[55] Aside from primarily ethnographic studies on premodern water management, virtually no research based on primary sources was conducted during these five decades. When the issue was considered, it was generally part of a broader economic history of individual settlements or institutions.[56] Despite the lack of innovative research on the role of water in the Middle Ages and the Early Modern period, some scholars produced works on the topic. However, when writing these, scholars seemed not to have engaged with primary sources predating the eighteenth century at all.[57]

In recent decades, as in Western Europe, Hungarian historians and archaeologists have addressed issues of water supply and wastewater disposal, albeit only to a limited extent due to the problematic source situation. The technological aspects of water supply have been studied at specific urban sites, including Buda, Fehérvár, Sopron, and Esztergom. In the past few years, an increasing number of articles have focused on wells and cisterns in rural areas and market towns.[58] The disposal of wastewater and human waste has also been examined and identified in archaeological studies of medieval Buda.[59]

Fishing, unlike in many countries, has been extensively studied in Hungary. An archivist at the turn of the nineteenth century, Sándor Takáts—whose works, in the absence of more modern research, are still frequently cited wrote studies on fishing along the Danube, with particular regard to sturgeon catching.[60] Somewhat later, Alajos Degré, using diplomatic sources, published a fundamental analysis of fishing rights in Hungary.[61] He traced the roots of fishing rights in the country to the German legal tradition, though he often failed to demonstrate its actual presence. Nonetheless, he uncovered numerous unedited documents discussing various legal disputes related not only to fishing but also to the usage of both navigable rivers and streams. Since then, scholars have highlighted the importance of fishing to the population's food supply, especially during Lent. Fishing along major rivers—as in most parts of Europe—was a significant privilege in Hungary, which kings gradually transferred to groups such as settlers and church institutions, sparking local controversies.[62] The economic role of medieval fishing is evident in the construction of numerous and sometimes extensive fishpond complexes on the estates of both laypeople and the church. Studies have focused on fishponds at various locations, but never on the scale of the Kingdom of Hungary as a whole. In comparison to these works, similar to Western Europe, both archaeozoology[63] and landscape archaeology[64] have introduced significant developments.

Like in Western Europe, historians of technology have also focused on the history of water management. Following several studies on the technological aspects of mills and, most notably, dam construction by the previously mentioned archivist, Sándor Takáts, in the early twentieth century,[65] interest in the subject gradually increased after World War II. However, this interest primarily centered on the modern transformations of waterscapes, which were not unaffected by the visible impact of postwar industrialization on the environment. Zsigmond Károlyi, a historian of technology, authored the first work that addressed significant aspects of water management and the role of water in the lowland areas of the Carpathian Basin during the early modern period.[66] As a founding figure of the Hydrology—later Danube—Museum, he was among the few Hungarian scholars to explore the history of water in a monograph. Due to his expertise, he focused on the modern period, particularly the era of extensive river control projects in the nineteenth century. Nonetheless, he was one of the first to highlight the extensive farming of the lower floodplains during the Middle Ages and the early modern period, along with the increased use of waterpower.[67] He examined the challenges of fish farming and the proliferation of various types of industrial mills. Further historians have explored the diverse uses of mills, contributing to the discussion on industrial mills, their significance in the country's economic intensification, and their spread in rural areas.[68]

Ethnographers and later archaeologists also made significant discoveries about the technological and social-historical aspects of water management. Drawing on entirely different sources and methods, scholars have hypothesized the existence of diverse hydraulic systems. Ethnographer Bertalan Andrásfalvy argued for the existence of a complex flood mitigation system in the Great Hungarian Plain during the premodern period, which he called the "fok" system. The dating and extent of human interventions to create this system are subjects of considerable debate among scholars. Andrásfalvy's seminal book highlighted effective flood mitigation techniques that predated modern drainage and regulation efforts in the eighteenth and nineteenth centuries.[69]

The beginning of archaeological research into water management systems in the dense network of waterways covering the lowland areas of the Carpathian Basin had to wait until the 1990s. An archaeologist led this research. Based on fieldwork in the Rábaköz (a micro-region in northwest Hungary), Károly Takács assumed the existence of a complex system of artificial channels in the plains of the Carpathian Basin.[70] Maintaining this channel system must have required significant labor expenditure. Takács argued that these channels—according to him, functioning in the early centuries of the Kingdom of Hungary—were maintained by a service population serving the ispánok (*comites*) in their castles.[71] He posited that maintenance of the system ended in the thirteenth century, a period when Hungarian society underwent major social transformations, including the dissolution of the royal castle system, whose population, according to his theory, served as the workforce for maintenance. It is not the place to discuss the questions surrounding the validity of Takács's otherwise fascinating idea. Still, it is important to note that he was among the first to ask who oversaw the maintenance of these water systems and who were the individuals interested in or involved in managing waters in the Carpathian Basin. Although the hydrography in this region was very dense, most rivers had a slow current, allowing only limited sections for exploiting waterpower. Thus, those river sections were just as valuable as rivers in areas where less dense river networks served local societies. To maintain this system, dams, sluices, and races were constructed in hilly areas to ensure a more even distribution of water, necessary for milling, fish cultivation, and irrigation (the latter probably playing a limited role in premodern Hungary). Building dams and sluices on streams required more limited expertise compared to interventions into major rivers, nevertheless partial dams and water conduits were also built on larger waterways. Consequently, certain landed properties would have received more water while others would have lost their water supply. This led to several legal controversies throughout Europe as well, although, as mentioned, research has thus far focused on major rivers.[72] Interventions into rivers and streams were a hotbed for legal disputes in the Middle Ages, just as they are today. The long-lasting debate mentioned in the introduction to this book, which will be followed up in the epilogue, over the Gabčíkovo–Nagymaros Dam in the 1980s and 1990s, connected to the diversion of the River Danube, serves as a modern example of an ancient trend. While the proposed Gabčíkovo–Nagymaros Dam was, of course, a larger-scale project, it closely parallels the jurisdictional problems that arose when medieval rivers and moreover streams were dammed, and part of their water was diverted from the original riverbeds to create new channels, mill races, fishponds, etc. It also highlights the challenges of adjudicating the different economic interests of various actors. Even if the rivers were smaller, the problem of diverting rivers along country borders, as recently demonstrated from a political history perspective, was recognized in medieval Hungary, particularly along its borders with the Austrian duchies and the Kingdom of Bohemia.[73]

Over the past three decades, historians have shown a notable resurgence of interest in premodern water management systems in Hungary. Beyond the undeniable influence of expanding research in Western Europe, a crucial factor in this revival has been the publication of several new sources, particularly charter summary collections, DVDs, and subsequently, the online publication of the vast majority of medieval Hungarian documentary evidence (charters).[74] Regesta collections have enabled quantitative analyses and attracted scholars' attention to numerous relevant cases that deserve more nuanced exploration. The online databases have facilitated research by providing easier access to tens of thousands of unedited legal sources. This ready access has led to the creation of several collections focused on water management, utilizing medieval documentary evidence.[75]

The impact of water mills on riverine environments has also been addressed, including their effects on river water regimes and the artificial floods they frequently caused.[76] A brief but significant study by István Tringli merits mention here, as it was the first after Degré's work in the 1930s to tackle the problem of the customary law governing mill construction.[77] He had a considerably broader dataset than Degré, partly because he published the cartulary of an influential Hungarian noble family, the Perényis.[78] An urban historian, Katalin Szende, recently revisited the question of the conflicting interests of mill owners operating along the same waterway in certain Hungarian towns, using a topographical approach largely overlooked in prior research.[79]

In recent years, two research directions have emerged, mirroring what has dominated Western European research: environmental and landscape history on one hand and the social history of water management on the other. Archaeologists and historians have combined written evidence, cartographic data, and field surveys to illustrate how water management systems operated at the local scale in premodern times. László Ferenczi examined mills in the immediate vicinity of present-day Budapest,[80] Zsuzsa Pető analyzed monastic fishponds in the Pilis Mountains,[81] and Elek Benkő, Ünige Bencze, along with Gusztáv Jakab and his colleagues, studied water management systems in Eastern Transylvania.[82] Meanwhile, István Viczián and Csilla Zatykó reconstructed the water management system along a smaller river, the Dráva, and one of its tributary streams in southern Transdanubia (in Hungarian Dunántúl, referring to the areas west and south of the River Danube in Hungary) based on legal evidence and field surveys.[83] The water supply of individual monasteries has also been explored recently.[84] Another issue that has been addressed in numerous studies is the social organization of various water management systems. A recent volume was dedicated to this topic, presenting premodern and modern case studies in a balanced manner.[85] In the volume, a survey by Tamás Vajda addressed the seigniorial rights related to mills and the milling business opportunities available to different social groups. At the same time, László Gálffy focused on how societies along the Danube depended on water. The ways wars influenced water management systems in the frontier zones of Hungary and the Ottoman Empire, as well as their impact on riverine societies, have also been explored.[86]

Even though there has been a slow but steady increase in interest in Hungary in recent years, the methods and results of Anglo-Saxon, German, French, and Italian research have only gradually influenced Hungarian scholarship. Research remains somewhat fragmented, and despite some interdisciplinary projects initiated in the past decade, most historians focusing on water and related topics still work in isolation from archaeologists, let alone ecologists, botanists, or palynologists.

Research in Poland over the last two decades has likely been even less active in the study of historical water management until recently than that of Hungarian historians.[87] Aside from a few individual studies primarily focusing on the technological and economic aspects of water exploitation, little research on these issues was conducted until around 2000.[88] In recent years, there has been a noticeable increase in research activity, with innovative methods employed to understand premodern water management systems. Scholars have explored the potential for reconstructing these systems using geographical information systems (GIS) and environmental archaeology, alongside more traditional sources such as urban statutes and tax registers.[89] Examining water supply systems in medieval Polish towns and those of the Teutonic Order has been particularly significant.[90] One of the most important outcomes of these recent efforts is Urszula Sowina's monograph, which draws on a wide range of written evidence to examine how towns in Poland and Silesia, such as Kraków and Wrocław, managed to supply water for their industries and populations.[91] In addition to studying written sources, water supply systems in Poland have also been analyzed from an archaeological perspective.[92] While economic history remains essential to research, the social history of water management remains marginal in Polish studies.[93]

Research by Czech and Slovak scholars on water history has been relatively limited until very recently. Even in recent years, only limited research interest is evident, which is somewhat surprising given two factors: first, the significant output of Czech forest and climate historians and historical ecology; and second, the high relevance of fish culture in premodern Bohemia. Milling generated some interest,[94] as did fishponds and fish cultures, but there is a gap in basic handbooks and overviews of the subject.[95] Mills and fishponds have primarily been studied by archaeologists,[96] while, despite the research potential of written sources, historians have paid less attention to these.[97] Nonetheless, a recently published volume indicates that medieval waterscape research is becoming more active in the Czech Republic (while remaining completely marginal in Slovakia).[98] This volume includes a summary of the technological aspects of water mills in Bohemia[99] and addresses the issue of urban water supply.[100]

Even though promising research has been initiated in recent years across Central Europe, including Hungary, the dissemination of research results to international audiences is only slowly gaining traction, while findings from the region rarely appear in Western scholarly work. Conferences are showing growing attendance from these countries, and the increasing number of English-language publications in Hungary and Poland, along with German publications in Bohemia, is likely to change this situation gradually.

Not every single study dedicated to the topic has been cited in the previous pages. Several additional monographs or collective volumes on different aspects of water management will be discussed in the forthcoming relevant subchapters. Nonetheless, those presented indicate a wide range of research directions in historical water management. In past decades, there has been a clear trend toward examining how societies organized themselves, either top-down or increasingly from bottom-up, in managing water bodies and systems of various sizes. Numerous studies have addressed how different conflicts were resolved; what legal regulations, customs, and social norms underpinned the organization of the network of waterways in certain regions, and not least, how these social arrangements affected the environment and how environmental processes influenced the surrounding social structure. These issues will be the focus of this book.

1.2 The Goals of this Work

Looking at the map of the medieval Kingdom of Hungary (see Figure 1.1), it becomes clear that water has always been a vital resource. The kingdom's geographical setting, the Carpathian or Pannonian Basin, is defined by a basin formed by an ancient sea and bordered by the towering Carpathian Mountains. The major flat areas in the central part—the so-called Great and Little Hungarian Plain (Alföld or Nagyalföld and Kisalföld in Hungarian, respectively)—cover a considerable portion of the inhabited area during the Middle Ages (*see* Figure 1.2). In these areas, managing water from the early stages of Hungarian state formation (ca. 900-) was essential as vast regions lay within the floodplains of the rivers flowing through the Basin. Additionally, water was a valuable resource in the hilly and more mountainous areas. These waters, as elsewhere in Europe, were used to generate waterpower, sustain fishponds, irrigate farmland, supply drinking water for people, provide water and winter pasturage for livestock, and support various industrial activities, among other uses.

This book argues that a system of water use norms developed from the tenth century onward, specifically due to the dense network of rivers and their uneven water regimes. By the Late Middle Ages, water systems became increasingly complex; at the same time, unwritten norms stabilized. In these water systems, despite the varying interests, social status, and political power of the parties involved, long-term solutions to water management could be found—a task that was a top priority for the various actors. Ultimately, landowners' access to water depended on the cooperation of upstream and downstream landowners along the same waterway.

The ways waters were manipulated for exploitation are anything but obvious. In the Kingdom of Hungary, substantial legal evidence indicates that disputes over water use have been dominated by questions of the utilization rights held by individuals and institutions since the earliest days of the Hungarian state. The key issues that arose included: What rights did landowners possess regarding a river that flowed through their estate? Did someone have the right to restrict the flow of a river when its source originated on their property? How did various members of society perceive water as a resource? Was it regarded as any other type of property or natural resource on an estate? Was it regarded differently from mineral resources or forests? What landscape changes resulted from the manipulation of water bodies? How were these multifunctional waterways managed, and who were the primary actors in these water systems? How did rivers influence social relations, arrangements in floodplains, and their broader environments?

The questions arising from the legal evidence concern the environment in which the debated actions occurred. Since legal evidence is the most important category of sources, it is also true that the environmental processes behind the water constructions can only be studied through the analysis of these written records—along with the other evidence discussed below (to be discussed in Chapter 1.2.2). The production of such texts requires two elements: on the one hand, the existence of water systems in which various actors—landowners—located near a body of water had differing economic interests; and on the other, a legal system that permitted the players to negotiate their way through the issues that inevitably arose. The evolution of these water systems across different parts of the Carpathian Basin, along with

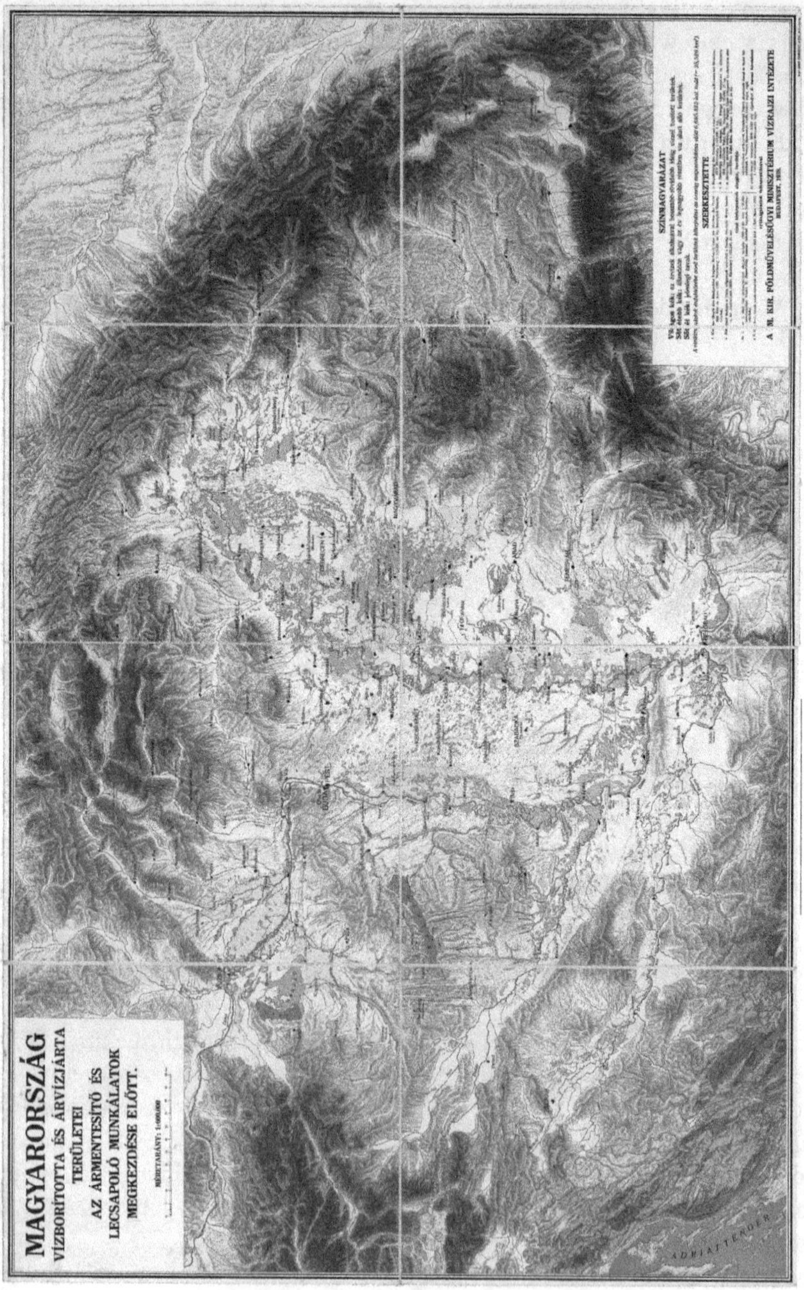
MAGYARORSZÁG
VÍZBORÍTOTTA ÉS ÁRVÍZJÁRTA
TERÜLETEI
AZ ÁRMENTESÍTŐ ÉS
LECSAPOLÓ MUNKÁLATOK
MEGKEZDÉSE ELŐTT.
SZÍNMAGYARÁZAT
SZERKESZTETTE
A M. KIR. FÖLDMŰVELÉSÜGYI MINISZTÉRIUM VÍZRAJZI INTÉZETE
ADRIATENGER

FIGURE 1.1 Areas covered by water and wetlands before river regulation in the Carpathian Basin (after Woldemár Lászlóffy). *A Kárpát-medence vízborította es vízjárta területei az ármentesitő es lecsapoló munkálatok megkezdése előtt*. [Map] [Water-Covered Areas and Wetlands in the Carpathian Basin Before the Beginning of Flood Protection and Drainage Works], ed. Woldemár Lászlóffy (Budapest: Vízrajzi Intézet, 1938). Though it has been criticized, the map can be used to understand the main characteristics of the waterscape. See: Zsolt Pinke, and Beatrix Szabó, "Analysis of the Map of the Ministry of Agriculture: Water Covered Areas and Wetlands in the Carpathian Basin Before the Commencement of Flood Protection and Draining," in *8. országos interdiszciplináris Grastyán konferencia előadásai* [Papers presented at the 8th Grastyán interdisciplinary conference], eds. Virág Rab, and Melinda Szappanyos (Pécs: PTE Grastyán Endre Szakkollégium, 2010), 207–217.

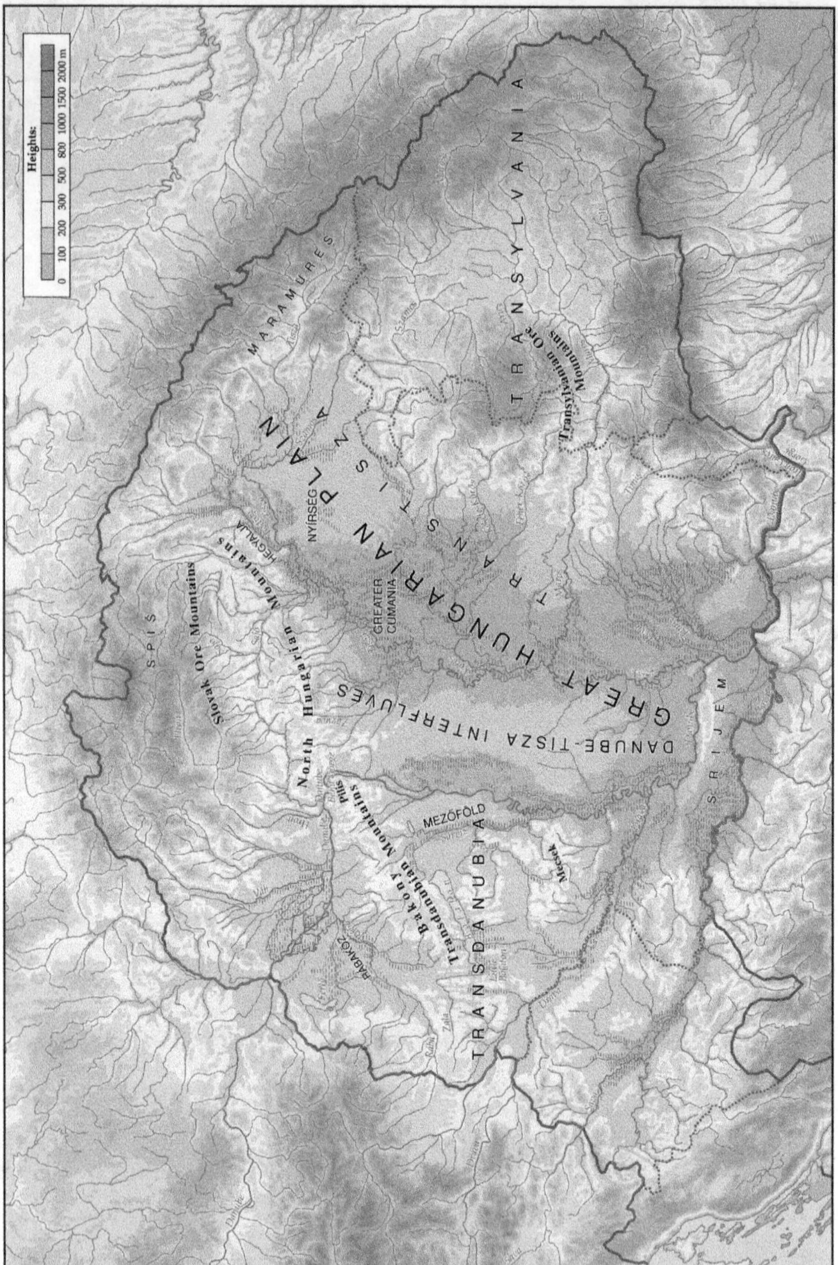

FIGURE 1.2 Major geographical units in the Carpathian Basin (drawn by Béla Nagy).

the legal evidence documenting these disputes, delineates the space and time frame addressed in the forthcoming chapters of the book.

1.2.1 Time and Space

Although, as mentioned above, there are signs that hydraulic systems already functioned in the Carpathian Basin from the period of early statehood onwards, it is pretty difficult to understand the conflicts surrounding the use of various water bodies in the first centuries of the rule of the Árpádian dynasty (1000 to 1301) due to the scarcity of available written evidence. In previous decades, research has made tremendous strides in reconstructing settlement patterns in the Hungarian Kingdom during the Árpádian period. The monumental, yet-to-be-completed work of György Györffy (1917–2000) on the historical geography of Hungary up to the early fourteenth century clearly illustrates the limitations in gathering information on individual settlements, not to mention water systems.[101] Written sources up to the second half of the thirteenth century, aside from registering the existence of some elements of water-related infrastructure, seldom refer to any of their features. One can presume that these constructions frequently caused controversy during this early period as well, but these disputes left little trace in the surviving legal evidence. Thus, it would be premature to draw conclusions based on this body of evidence. Chapter 2 addresses this early period of literacy to understand how the most widespread element in medieval Hungary's waterscapes, water mills, began to spread. Given the nature of the sources and the likely growing number of disputes, the later chapters of the book focus on the period usually referred to as late medieval in the context of the Kingdom of Hungary, specifically from the beginning of Angevin rule (early fourteenth century) to the fall of the medieval state to the Ottomans in 1526. Of course, neither the death of the last Árpádian ruler, King Andrew III, in 1301, nor the loss at Mohács to the Ottomans and the death of King Louis II in 1526 can be viewed as turning points in water management. Nonetheless, the well over two-hundred-year period allows for pursuing processes, observing how different customs and norms remained stable or changed, how the privileges of various social groups related to water use were negotiated and practiced, and how water systems were managed over the long term. From the fourteenth century onward, there was a clear increase in surviving written evidence produced in the Kingdom of Hungary, particularly in legal documents issued each year rather than in narrative sources. From this period onward, in at least some regions and social contexts, there was an intention to create a written record of how a dispute was settled. There is no point in limiting the book's scope to 1301. However, it will clearly focus on the period from the Angevin rule onward, as from then the sources allow a relatively nuanced view of some water-related legal arrangements. Not only can the existence of water-related infrastructure be confirmed, but some legal and non-legal evidence also allows for analyzing the social and environmental context of the recorded disputes. Given the logic of the archival sources, and partly the edited ones, as well as the aims of this book, however arbitrary, the primary emphasis will be on the later medieval period. To study the material from after 1526, Hungary has an entirely different archival system, and gradually, the area the book focuses on becomes part of three different polities with distinct legal norms, practices, and language use, making it impossible to follow some processes.

It is therefore easier to limit the geographic space that will serve as the basis for the book. The focus will be on the medieval Kingdom of Hungary. Although the borders of this political entity changed significantly during the studied period, the area under the authority of the kings of Hungary clearly defines the boundaries to be considered. The book does not discuss Dalmatia, even though for specific periods it fell under the authority of the kings of Hungary; however, its legal system was never aligned with that of Hungary, and only a limited number of case studies concern Croatia and Slavonia. The overwhelming majority of the cases discussed, both urban and rural, were under the aegis of the Kingdom of Hungary. Wherever relevant, the book continually references similarities with other parts of Central Europe, particularly regarding urban water use, as in these settings, practices show remarkable parallels with those of Polish, Silesian, and Czech towns.

1.2.2 About Sources

As shown above, various types of written evidence have been utilized to examine historical water management in Western Europe. However, to fully understand water-related disputes, sources beyond written records will also be considered. Archaeology enhances scholarly understanding of how different types of interventions to water affect other water-related infrastructural elements, both upstream and downstream. Environmental archaeology has been crucial in analyzing the effects of urbanization and industrialization on water systems and water quality. Nevertheless, at this stage of research, the primary focus will remain on written evidence. However, where available, visual sources, along with archaeological data and environmental and geographic information, will also be included to enrich our understanding of the hydrological processes and legal conflicts that emerged around water use.

For the medieval period, the written sources that provide good insight into the above problems are mostly referred to in Hungarian scholarship as *oklevelek* (somewhat similar to charters in English or *litterae* in Latin), as well as laws and statutes. Both types of sources constitute legal evidence, as do most sources from medieval Hungary that have survived. This legal focus has inevitable consequences. First and foremost, water-use-related issues appear as written evidence when disputes arose, or at least when the customs and rights concerning water use were unclear. It also means that, as long as water was not used intensively or systematically, there is little chance of finding written evidence of contemporary understandings of water as a resource or property.

Several charters address water control, such as the construction of dams, mills, sluices, fishponds, and similar structures. This body of evidence—comprising various legal documents primarily designed to grant individuals or specific groups rights, including but not limited to property rights, and to resolve disputes among the parties named in the texts—constitutes the most significant corpus for research. The disputes themselves highlight the role of water in agriculture and how certain groups perceived their rights to it. Hundreds of cases arose concerning the establishment of mills and other water-related works, frequently involving conflicts between local nobility and townspeople. Disputes also arose with various church institutions. Typically, only brief snapshots of disputes, sometimes spanning decades, are available. Although few negotiations have extensive documentation, many of these existing records do contain texts of earlier charters related to the same issue.

The charters relevant to the different chapters were gathered from existing secondary literature and three distinct sources: regesta (document summary) collections, primarily the Angevin and Sigismundian Cartularies,[102] along with the online database of the regestas of the Hungarian National Archives; published charters found in major cartularies,[103] parts of which are included in the Digital Library of Medieval Hungary[104]; and, in the most significant cases, unpublished charters held in the digital collection of the Hungarian National Archives.[105] The indexing systems of the Angevin and Sigismundian Cartularies, despite occasional brief charter summaries and the oversight of numerous archives in earlier editions of both series, enabled the collection of charters referencing dams, dikes, water mills, channels, and fishponds. Additionally, the *Lexicon latinitatis medii aevi Hungariae* (and its recent supplement by Kornél Szovák) clarifies some of the terminology used in medieval charters related to water constructions.[106] The sources from the fifteenth to the sixteenth centuries are less systematically published or summarized than those from before 1428.[107] In most instances, fifteenth- to sixteenth-century charters are published in the cartularies of noble families (Zichy, Perényi, Károlyi, Frangepán, Héderváry, Teleki, Szapolyai, etc.). In addition to these charters, many late medieval document summaries, which vary significantly in quality, are accessible in the online Database of Archival Documents of Medieval Hungary. These have proven essential for the study of late medieval disputes over water use.

The other type of written evidence comprises laws and regulations. A few late medieval regulations preserved in the *Corpus Juris Hungarici* refer to the use of water.[108] The early sixteenth-century summary of customary law, the

Tripartitum of Stephen Werbőczy, extensively addresses the legal questions connected to the construction of mills and dams. Additionally, sixteenth-century statutes and town council registers preserved water-related decisions. Comparing the legislative documents with the disputes reflected in charters will provide insight into how water was understood as property, both in theory and practice.[109]

In the last decades of the Middle Ages, as literacy spread, new forms of evidence emerged, including terriers (*urbaria*), account books, minutes from noble gatherings across the counties, and, somewhat later, maps. These sources are significant and will be used throughout the chapters and subchapters. However, they constitute only a relatively small portion of the book's source base. Maps merit a brief mention here. In many instances in Western Europe from the medieval period onward, resolutions following border disputes—including those along waterways—were documented by depicting the area in a sketch or map.[110] This documentation can take various forms. In England, maps were often used in ways like our contemporary understanding, illustrating landscape features crucial to understanding the situation. Conversely, in regions like the Kingdom of Hungary, it was extremely rare to produce a map before modern times. Instead, borders were described in prose, akin to what has been shown for Ottoman Egypt.[111] Nonetheless, given the strong continuity in the locations of mills, dams, fords, and similar features between the medieval period and the seventeenth or eighteenth centuries, even modern maps can be helpful in some contexts.[112] The detailed maps from the so-called First Military Survey (*Josephinische Landesaufnahme*) of the Kingdom of Hungary, conducted in the 1780s, depict all water constructions in the landscape, including mills and fishponds.[113]

The book's source base is diverse, though it primarily focuses on the extant written—predominantly legal—evidence from the medieval Kingdom of Hungary: charters. The upcoming chapters will demonstrate that these sources provide a detailed account of how local societies interacted with water for their benefit. Additionally, they may offer insight into how past societies resolved conflicts when these benefits were not mutually satisfactory for all parties involved in the life of a waterway.

1.3 The Structure of the Book

The book examines how human interventions for economic gain transformed riverscapes and how the resulting socio-natural sites affected the diverse people who lived near and far from these water bodies. To achieve this, using traditional "Rechtsgeschichte", it will first discuss the formation of various means by which local economies exploited different water bodies. Chapter 2 examines the spread of the key infrastructural elements that modified riverscapes in the Kingdom of Hungary: water mills.

This development has traditionally been linked to the establishment of the Christian state around the year 1000 and the settlement of monastic orders, particularly the Benedictines, within the kingdom. The chapter suggests that the significance of monastic orders in introducing this technology to Central Europe has been overstated in research. Nevertheless, from around the eleventh to early twelfth century, there appears to have been a slight increase in the number of water mills. By the thirteenth century, this rise gave rise to a variety of legal cases, which are explored in the subsequent chapters.

Chapter 3 focuses on the rural and urban waterscapes of the Kingdom of Hungary from the fourteenth to the sixteenth centuries. It examines how dams, weirs, and other stream structures steered legal controversies, as well as the arrangements made to avoid or resolve them. The first subchapter (Chapter 3.1) primarily investigates rural environments. While studying the legal term *locus molendini*, this section argues that by the early fourteenth century, rights related to water use for milling had already acquired well-defined legal norms. The subsequent subchapters analyze the areas in which the issue of water rights and access most frequently arises, particularly in urban spaces. Here, a relatively large number of people living in a confined area seek to use water resources for a broader range of activities compared to rural settings. Thus, urban spaces serve as excellent laboratories to examine water-related legal disputes. The various subchapters examine different urban environments from two perspectives: first, the water bodies that supported (or sometimes hindered) the well-being of these settlements, and second, the function and legal status of these settlements. The discussed water bodies include major waterways, small rivers, streams, rivers fed by hot springs, and urban moats (Chapter 3.2). Following this, the book examines water access in a unique type of town in the Kingdom of Hungary: mining towns (Chapter 3.3). These settlements received special privileges from the kings to encourage miners to settle there, thereby facilitating urban development and increasing royal revenues from gold, silver, and other metals. Waterpower is a vital resource for mining settlements; therefore, the privileges granted to these towns sometimes indicate different groups' access to water. A specific type of water-related conflict that has been poorly studied, both in Hungarian and foreign scholarship, is the issue of rivers as borders, addressed in the following subchapter (Chapter 3.4). As will be demonstrated, this was a source of numerous conflicts, where environmental changes—whether natural shifts in the riverbed or artificial alterations to rivers—led to enduring disputes over riverbeds, adjacent floodplains, and other lands.

Chapter 4 focuses on the social groups involved in various water control and regulation projects. Drawing on late medieval and, sometimes, later sixteenth-century sources, it is argued that both millers and pondmasters had intricate responsibilities in managing water bodies. Their duties went far beyond merely grinding grain and overseeing lake levels. In Chapter 5, in addition to revisiting the work's main findings, modern parallels to water-related issues and their potential significance for understanding contemporary contests over waterscapes are discussed.

Notes

1 *Treaties of Peace with Italy, Bulgaria, Hungary, Roumania, and Finland* (Washington, D.C.: U.S. G.P.O., 1947), 1–2. (part I.4.c).

2 For the case, see: Judit Galambos, "An International Environmental Conflict on the Danube: the Gabcicovo – Nagymaros Dams," in *Environment and Democratic Transition in Central and Eastern Europe. Policy and Politics in Central and Eastern Europe* (Technology, Risk and Society, 7), eds. Anna Vári, and Pál Tamás (Dordrecht, Boston, and London: Springer, 1993), 146–156, and John Fitzmaurice, *Damming the Danube: Gabčikovo and Post-Communist Politics in Europe* (Boulder, CO and Oxford: Westview, 1996).

3 For France, see the regional monographs written based on the principles laid down by Paul Vidal de la Blache (Paul Vidal de la Blache, *Tableau géographique de la France* [Paris: Hachette, 1908]). For a critical analysis of these monographs: André Thibault, "L'analyse des espaces régionaux en France depuis le début du siècle," *Annales de Géographie* 81, no. 444 (1972): 129–170. For Hungary, see: Tivadar Ortvay, *Magyarország régi vízrajza a XIII. század végéig*, 2 vols [Hydrography of ancient Hungary until the end of the 13th century] (Budapest: MTA, 1892).

4 Heinrich Geffcken, "Zur Geschichte des deutschen Wasserrechts," *Zeitschrift der Savigny-Stiftung für Rechtsgeschichte. Germanistische Abteilung* 21, no. 1 (1900): 173–217. https://doi.org/10.7767/zrgga.1900.21.1.173, Reinhard Härtel, "Von Wasser- und Mühlenrecht im Hochmittelalter." in *Recht und Geschichte. Festschrift Hermann Baltl zum 70. Geburtstag*, ed. Helfried Valentinitsch (Graz: Leykam, 1988), 219–236, J. Getzler, *A History of Water Rights at Common Law* (Oxford: Oxford University Press, 2004). https://doi.org/10.1093/acprof:oso/9780199207602.001.0001. For a critical survey of the German legal history research into water use and milling see Niels Petersen, and Arnd Reitemeier, "Die Mühle und der Fluss. Juristische Wechselwirkungen," in *Wasser in der mittelalterlichen Kultur: Gebrauch – Wahrnehmung – Symbolik = Water in Medieval Culture: Uses, Perceptions, and Symbolism* (Das Mittelalter. Beihefte, 4), eds. Gerlinde Huber-Rebenich, Christian Rohr, and Michael Stolz (Berlin: de Gruyter, 2017), 276–290.

5 Julius von Gierke, *Geschichte des deutschen Deichrechts*, 2 vols (Deutschen Staats- und Rechtsgeschichte, 63 and 128) (Breslau: M. & H. Marcus, 1901–1917), esp. vol. 1, 16–183.

6 Carl Koehne, *Das Recht der Mühlen bis Zum Ende der Karolingerzeit. Ein Beitrag znr Geschichte des Deutschen Gewerberechts* (Deutschen Staats- und Rechtsgeschichte, 71) (Breslau: M. & H. Marcus, 1904), Carl Koehne, "Die Mühle im Rechte der Völker," *Beiträge zur Geschichte der Technik* 5 (1913): 27–53, Harm Wiemann, "Beiträge zur Geschichte des Mühlenrechts. Dargestellt an Mühlen der Herrschaft Crimmitschau vom 14.–17. Jahrhundert," *Zeitschrift der Savigny-Stiftung für Rechtsgeschichte, Germanistische Abteilung* 66 (1948): 477–500. http://doi.org/10.7767/zrgga.1948.66.1.477, Rüdiger Moldenhauer, "Mühlen und Mühlenrecht in Mecklenburg," *Zeitschrift der Savigny-Stiftung zur*

Rechtsgeschichte, Germanistische Abteilung 79 (1962): 195–236. https://doi.org/10.7767/zrgga.1962.79.1.195, Werner Kohl, *Recht und Geschichte der alten Münchner Mühlen* (Miscellanea Bavarica Monacensia, 15 = Neue Schriftenreihe des Stadtarchivs München, 32) (Munich: Stadtarchiv, 1969), Ingo Herbert Kropač, *Mühlen und Mühlenrecht in der Steiermark während des Mittelalters* (Dissertationen der Karl-Franzens-Universität Graz, 61) (Graz: Verlag für d. Techn. Univ. Graz, 1982), esp. 40–65, Christoph Bachmann, "Altbayerns erste bisher bekannte Mühlordnung. Die Ordnung der Ingolstädter Müller aus dem Jahr 1437," *Sammelblatt des Historischen Vereins Ingolstadt* 102/103 (1993): 309–316.

7 Alajos Degré, *Magyar halászati jog a középkorban* [Fishing law in medieval Hungary] (Budapest: Sárkány Nyomda, 1939), Hermann Heimpel, "Die Federschnur. Wasserrecht und Fischrecht in der 'Reformation Kaiser Siegmunds'," *Deutsches Archiv für Erforschung des Mittelalters* 19 (1963): 451–488 and Guido Kisch, *Das Fischereirecht im Deutschordensgebiet* (Forschungen und Quellen zur Rechts- und Sozialgeschichte des Deutschordenslandes, 3) (2nd ed. Sigmaringen: Thorbecke, 1978).

8 Marc Bloch, "Avènement et conquêtes du moulin à eau," *Annales d'histoire économique et sociale* 7 (1935): 538–563. https://doi.org/10.1017/S0003441X00017556.

9 E[leonora] M[ay] Carus-Wilson, "An Industrial Revolution of the Thirteenth Century," *The Economic History Review* 11, no. 1 (1941): 39–60. https://doi.org/10.1111/j.1468-0289.1941.tb01589.x (published again in: *Internal Colonization in Medieval Europe* [The expansion of Latin Europe, 1000–1500, 2], eds. Felipe Fernández-Armesto, and James Muldoon [Farnham: Ashgate, 2008], 301–322) and R[eginald] A[rthur] Pelham, *Fulling Mills: A Study in the Application of Water Power to the Woollen Industry* (London: Society for the Protection of Ancient Buildings, 1958). For fulling mills, see recently: Adam Lucas, "Fulling Mills in Medieval Europe. Comparing the Manuscript and Archaeological Evidence," in *Archéologie des moulins hydrauliques, à traction animale et à vent, des origines à l'époque médiévale et moderne en Europe et dans le monde Méditerranéen: Actes du colloque international, Lons-le-Saunier du 2 au 5 novembre 2011* (Annales littéraires de l'Université de Franche-Comté, 959), eds. Luc Jaccottey, and Gilles Rollier (Besançon: Presses Universitaires de Franche-Comté, 2016), 581–594.

10 For a pioneering work, see: B. Gille, "Le moulin à eau, une révolution médiévale," *Techniques et civilizations* 3 (1954): 1–15. See the more recent works of Mathieu Arnoux in the topic, most importantly: "Moulins à fer et procédé indirect. Innovation technique et conditions géographiques dans la sidérurgie européenne (XIIIe–XVIe siècles)," in *I mulini nell'Europa medievale: Atti del Convegno di San Quirico d'Orcia, 21–23 settembre 2000*, eds. Paola Galetti, and Pierre Racine (Biblioteca di storia agraria medievale, 21) (Bologna: Clueb, 2003), 317–328.

11 Bloch, "Avènement et conquêtes," André Guillerme, *Les temps de l'eau: la cité, l'eau et les techniques: nord de la France: fin IIIe–début XIXe siècle* (Seyssel: Champ Vallon, 1983) and more importantly Germain Sicard, *Aux origines des sociétés anonymes: les moulins de Toulouse au moyen âge* (Paris: Édition de l'EHESS, 1953). Both books have been translated into English: André Guillerme, *The Age of Water: The Urban Environment in the North of France, A. D. 300–1800* (Environmental History Series, 9) (College Station, TX: A & M University Press, 1988) and Germain Sicard, *The Origins of Corporations: The Mills of Toulouse*

in the Middle Ages (New Haven, CT and London: Yale University Press, 2015). The latter has oddly been published in English more than 60 years after its original publication.

12 John Langdon, "The Economics of Horses and Oxen in Medieval England," *Agricultural History Review* 30 (1982): 31–40; John Langdon, "Water-Mills and Windmills in the West Midlands, 1086–1500," *Economic History Review* 2nd series 44 (1991): 424–444 and more recently: John H. Munro, "Industrial Energy from Water-Mills in the European Economy, 5th to 18th Centuries: The Limitations of Power," in *Economia e energia, secc. XIII–XVIII: atti della trentaquattresima Settimana di studi, 15–19 aprile 2002* (Istituto Internazionale di Storia Economica F. Datini, Prato. Serie 2. Atti delle settimane di studio e altri convegni, 34), ed. Simonetta Cavaciocchi ([Florence]: Le Monnier, 2003), 223–269.

13 E.g., Lewis Mumford, *Technics and Civilization* (New York: Harcourt, Brace and Co., 1934).

14 Georges Duby, *L'économie rurale et la vie des campagnes dans l'Occident médiéval*, 2 vols (Paris: Aubier, 1962).

15 Lynn White Jr., *Medieval Technology and Social Change* (London: Oxford University Press, 1962), Jean Gimpel, *La Révolution industrielle du Moyen âge* (Paris: Seuil, 1975). Published in English as: *The Medieval Machine: The Industrial Revolution of the Middle Ages* (New York: Holt, Rinehart & Winston, 1976). For its impact, see: Shana Worthen, "The Influence of Lynn White, Jr's Medieval Technology and Social Change," *History Compass* 7, no. 4 (2009): 1201–1217. https://doi.org/10.1111/j.1478-0542.2009.00615.x and moreover, *Fifty Years of Medieval Technology and Social Change*, ed. Steven A. Walton (London and New York: Routledge, 2020) https://doi.org/10.4324/9781315582313.

16 Richard Holt, *The Mills of Medieval England* (Oxford and New York: Basil Blackwell, 1988).

17 Ernst Eichhorn, "Ein Kaiser an der Altmühl. Zwischen Solagrab und Fossa Carolina," in *Karolingisches Franken*, ed. Wolfgang Buhl (Würzburg: Echter, 1973), 107–143, Klaus Grewe, "Der Karlsgraben bei Weißenburg i.B. und der Fulbert-Stollen von Maria Laach: zwei große Wasserbauprojekte des Mittelalters in Deutschland," in *Europäische Technik im Mittelalter, 800 bis 1400: Tradition und Innovation*, ed. Uta Lindgren (Berlin: Gebr. Mann, 1996), 111–115, Theodor Fischer, and Markus Strobel, "Von der Fossa Carolina zur Main-Donau-Überleitung," in *Historische Wassernutzung an Donau und Hochrhein sowie zwischen Schwarzwald und Vogesen* (Schriften der Deutschen Wasserhistorischen Gesellschaft, 10), ed. Christoph Ohlig (Norderstedt: Books on Demand, 2008), 61–71. Cf. Paolo Squatriti, "Digging Ditches in Early Medieval Europe," *Past and Present* 176, no. 1 (2002): 11–65. https://doi.org/10.1093/past/176.1.11. Most recently, see: *Großbaustelle 793: das Kanalprojekt Karls des Großen zwischen Rhein und Donau; Begleitbuch zur Ausstellung im Museum für Antike Schiffahrt des Röömisch-Germanischen Zentralmuseums in Mainz, 30. April 2014 bis 10. August 2014* (Mosaiksteine, 11), eds. Stefanie Berg-Hobohm, and Peter Ettel (Mainz: Verl. des Röm.-German. Zentralmuseums, 2014).

18 Karl A. Wittfogel, *Oriental Despotism: A Comparative Study of Total Power* (New Haven, CT: Yale University Press, 1957), passim. Cf. Terje Tvedt, and Eva Jakobsson, "Introduction," in *Water History is World History*, vol. 1/1. xvii and Squatriti, "Digging Ditches."

19 See Chapter 4 in detail on millers with the relevant literature. See also: Carlo Ginzburg, *The Cheese and the Worms. The Cosmos of a Sixteenth-Century Miller* (Baltimore, MD: Johns Hopkins University Press, 1992), Jean Chapelot, and Robert Fossier, *Le village et la maison au moyen âge* (Paris: Hachette, 1985). Published in English: *The Village and House in the Middle Ages* (Berkeley, CA: University of California Press, 1986), 150ff. (I consulted the latter when writing this book).
20 Cf. Tvedt, and Jakobsson, "Introduction."
21 Verena Winiwarter, Martin Schmid, and Gert Dressel, "Looking at Half a Millennium of Co-Existence: The Danube in Vienna as a Socio-Natural Site," *Water History* 5 (2013): 108. https://doi.org/10.1007/s12685-013-0079-x.
22 See as historiographical introductions to the topic: Christof Mauch and Thomas Zeller, "Rivers in History and Historiography: An Introduction," in *Rivers in History: Perspectives on Waterways in Europe and North America*, eds. Christof Mauch and Thomas Zeller (Pittsburgh: University of Pittsburgh Press, 2008), 1–10, Winiwarter, Schmid, and Dressel, "Looking at Half a Millennium," Paula Schönach, "River Histories: A Thematic Review," *Water History* 9 (2017): 233–257. https://doi.org/10.1007/s12685-016-0188-4, Giacomo Bonan, "Riflessi sull'acqua. Ricerca storica e biografie fluviali," *Contemporanea: Rivista di storia dell'800 e del '900* 22, no. 2 (2019): 317–328. https://doi.org/10.1409/93534, Matthew Evenden, "Beyond the Organic Machine? New Approaches in River Historiography," *Environmental History* 23 (2018): 698–720. https://doi.org/10.1093/envhis/emy054. Very recently, with a focus on smaller waters and water-related infrastructure, Raphael Longoni, and Oliver Wetter, "Urban Stream Works in Central Europe 1200–1700," *Water History* 11 (2019): 31–57. https://doi.org/10.1007/s12685-019-00231-6. See furthermore: Theodore R. Schatzki, "Nature and Technology in History," *History and Theory* 42 (2003): 82–93. https://doi.org/10.1046/j.1468-2303.2003.00259.x and Martin Knoll, "Fluss und Flusslandschaft. Eine Kultur- und Umweltgeschichte," *Mitteilungen des Verbandes bayerischer Geschichtsvereine* 26 (2014): 65–86.
23 Lukas Werther et al. "On the Way to the Fluvial Anthroposphere—Current Limitations and Perspectives of Multidisciplinary Research," *Water* 13, no. 16 (2021): no. 2188. https://doi.org/10.3390/w13162188
24 Although not focusing on the medieval period, Mark Cioc's book on the Rhine is pioneering in this respect: Mark Cioc, *Rhine: An Eco-Biography, 1815–2000* (Seattle, WA: Seattle: University of Washington Press, 2002).
25 Leona J. Skelton. *Tyne after Tyne: An Environmental History of a River's Battle for Protection, 1529–2015* (Winwick: White Horse Press, 2017). For the problem of river biographies as labels, see the excellent review of Schönach, "River Histories," 248–252.
26 Schönach, "River Histories," 244.
27 Patrick Fournier, and Sandrine Lavaud, "Rapport historiographique," in *Eaux et conflits dans l'Europe médiévale et moderne*, eds. Sandrine Lavaud, and Roland Viader (Toulouse-Le Mirail: Université de Toulouse-Le Mirai, 2012), 7–31.
28 Skelton, *Tyne after Tyne*, 25.
29 For the areas covered, see https://opendomesday.org/.
30 Richard C. Hoffmann, "A Brief History of Aquatic Resource Use in Medieval Europe," *Helgoland Marine Research* 59 (2005): 25–26. https://doi.org/10.1007/s10152-004-0203-5 and H. J. R. Lenders et al. "Historical Rise of Waterpower

Initiated the Collapse of Salmon Stocks," *Nature. Scientific Reports* 6, art. no. 29269 (2016): 9–18. https://doi.org/10.1038/srep29269.

31 See for the example of the Thames: R. H. C. Davis, "The Ford, the River and the City," *Oxoniensis* 38 (1973) 258–267, R. B. Peberdy, "Navigation on the River Thames between London and Oxford in the Late Middle Ages: A Reconsideration," *Oxoniensis* 61 (1996): 311–340, esp. the highly informative Fig. 2. on 318, and Derek Keene, "Issues of Water in Medieval London to c. 1300," *Urban History* 28 (2001): 161–179. https://doi.org/10.1017/S0963926801002012. See also the following polemic articles for the problem in a broader context: J. Edwards and B. Hindle, "The Transportation System in Medieval England and Wales," *Journal of Historical Geography* 17 (1991) 123–134. https://doi.org/10.1016/S0305-7488(05)80070-2, J. Langdon, "Inland Water Transport in Medieval England," *Journal of Historical Geography* 19 (1993): 1–11. https://doi.org/10.1006/jhge.1993.1001, J. Edwards, and B. Hindle, "Comment: Inland Water Transportation in Medieval England," *Journal of Historical Geography* 19 (1993): 12–14. https://doi.org/10.1006/jhge.1993.1002, E. T. Jones, "River Navigation in Medieval England," *Journal of Historical Geography* 26 (2000): 60–75. https://doi.org/10.1006/jhge.1999.0191, and J. Langdon, "Inland Water Transport in Medieval England: The View from the Mills: A Response to Jones," *Journal of Historical Geography* 26 (2000): 75–82. https://doi.org/10.1006/jhge.1999.0195. See more recently in a German context: Sascha Bütow, *Straßen im Fluss. Schifffahrt, Flussnutzung und der lange Wandel der Verkehrsinfrastruktur in der Mark Brandenburg und der Niederlausitz vom 13. bis zum 16. Jahrhundert* (Studien zur brandenburgischen und vergleichenden Landesgeschichte, 18) (Berlin: Lukas, 2015), Gerrit Jasper Schenk, "»der straßen halb uff der Yll«. Wasserwege im Elsass als »kritische Infrastruktur« für Wirtschaft und Gesellschaft (1350–1550)," in *Weg und Steg. Aspekte des Verkehrswesens von der Spätantike bis zum Ende des Alten Reiches* (Kraichtaler Koll, 11), eds. Kurt Andermann, and Nina Gallion (Ostfildern: Thorbecke, 2018), 121–152 and Gerrit Jasper Schenk, "»Knoten« im »Netz«? Überlegungen zur Hafenstadt als »kritischer Infrastruktur«," in *Flusstäler, Flussschifffahrt, Flusshäfen: Befunde aus Antike und Mittelalter Der Tagungsband fasst in 15 Beiträgen die Ergebnisse der gleichnamigen, internationalen Tagung zusammen* (RGZM – Tagungen, 39 = Interdisziplinäre Forschungen zu den Häfen von der Römischen Kaiserzeit bis zum Mittelalter in Europa, 7), eds. Peter Ettel, and Achim Thomas Hack (Mainz: Verlag des Römisch-Germanischen Zentralmuseums, 2019), 207–236.

32 E.g., Dolly Jørgensen, "Local Government Responses to Urban River Pollution in Late Medieval England," *Water History* 2, no. 1 (2010): 35–52. https://doi.org/10.1007/s12685-010-0016-1, Roberta J. Magnusson, "Water and Wastes in Medieval London," in *A History of Water*, vol. 1/1, 299–313, and Ronald Edward Zupko, and Robert Anthony Laures. *Straws in the Wind: Medieval Urban Environmental Law. The Case of Northern Italy* (Boulder, CO: Westview Press, 1996).

33 See most importantly: Richard C. Hoffmann and Christoph Sonnlechner, "Vom Archivobjekt zum Umweltschutz: Maximilians Patent von 1506," *Studien zur Wiener Geschichte. Jahrbuch des Vereins für Geschichte der Stadt Wien* 62–63 (2006–2007) [2011]: 79–133.

34 Jean-Pierre Leguay, *L'eau dans la ville au moyen âge* (Rennes: Presses universitaires de Rennes, 2002) and *La città e il fiume (secoli XIII–XIX)*, ed. Carlo M. Travaglini (Rome: École Française de Rome, 2008). For the modern period in general, see: *Rivers Lost, Rivers Regained: Rethinking City-River Relations* (History of the Urban Environment), eds. Martin Knoll, Uwe Lübken, and Dieter Schott (Pittsburgh, PA: University of Pittsburgh Pres, 2017) https://doi.org/10.2307/j.ctt1qnw8gv. For individual rivers, see for instance: Pierre-Henri Guittonneau, "Petites villes et espace fluvial: pratiques sociales et conflits d'usage autour de Paris au XVe siècle," in *The Power of Space in Late Medieval and Early Modern Europe: The Cities of Italy, Northern France and the Low Countries* (Studies in European Urban History, 30), eds. Marc Boone, and Martha C. Howell (Turnhout: Brepols, 2013), 57–76. https://doi.org/10.1484/M.SEUH.1.101407 (for the Loire), the special issue of *Water History* 5, no. 2 (2013) (for the Danube), Maria Grazia Nico Ottaviani, *Statuti, territorio e acque nel Medioevo: Perugia e Marsciano, Tevere e Nestóre* (Spoleto: Fondazione Centro italiano di studi sull'alto Medioevo, 2008) (for the Tiber). See also the excellent case studies on competing interests in urban spaces by Marco Leonardi, "Die Wassernutzung in Freiburg im späten Mitelalter und der frühen Neuzeit," *Zeitschrift des Breisgau-Geschichtsvereins "Schau-ins-Land"* 129 (2010): 55–75. https://doi.org/10.57962/regionalia-17831 and Marco Leonardi, *Aqua curanda est: le acque e il loro utilizzo nei territori di Friburgo in Brisgovia e Catania dal XIII al XVI secolo* (Aquae. Studi e testi sulle terme, 9) ([Florence]: Leo S. Olschki, 2017). For the problem of water use in major towns, see the example of medieval Pest and Buda in Chapter 3.2.1.

35 For the Flaran papers, see: Mark F. Gardiner, "Bateliers, pêcheurs et habitants des ports: les conflits sur les droits des rivières dans l'Angleterre medieval,", Marc Suttor, "Les conflits pour l'usage et le contrôle de l'eau sur les rivières entre Seine et Meuse à la fin du Moyen Âge et au début de l'Époque modern," Emmanuel Grélois, "Les usages concurrents de la rivière: l'Allier en Basse-Auvergne (XIIIe–XVIIIe siècle)," Denis Reynard, "Conflits et procès autour de l'utilisation de l'eau de la Sionne (Valais, Suisse), du XVe au début du XXe siècle," in *Eaux et conflits*, 59–70, 71–86, 107–118, and 189–208, respectively. See also: *Fleuves et marais, une histoire au croisement de la nature et de la culture: sociétés préindustrielles et milieux fluviaux, lacustres et palustres: pratiques sociales et hydrosystèmes* (Archéologie et histoire de l'art, 19), eds. Joëlle Burnouf, and Philippe Leveau (Paris: Ministère de l'éducation nationale, de l'enseignement supérieur et de la recherche, 2004), esp. 285–340. For the Lorsch conference papers, see *Conflicts over Water Management and Water Rights from the End of Antiquity to Industrialisation*, eds. Michele Campopiano, and Gerrit Jasper Schenk (Vierteljahrschrift für Sozial- und Wirtschaftsgeschichte – Beihefte, 260) (Stuttgart: Franz Steiner, 2024) https://doi.org/10.25162/9783515137539.

36 See e.g., Wittfogel, *Oriental Despotism.* See its criticism: Robert L. Carneiro, "A Theory of the Origin of the State," *Science* 169, no. 3947 (1970): 733–738. https://doi.org/10.1126/science.169.3947.733.

37 E.g., Paolo Squatriti, *Water and Society in Early Medieval Italy, 400–1000* (Cambridge: Cambridge University Press, 1998) https://doi.org/10.1017/CBO9780511583094, Alan Mikhail, "Oriental Democracy," *Global Environment* 7, no. 2 (2014): 381–404. https://doi.org/10.3197/ge.2014.070206, Michele Campopiano, "Rural Communities, Land Clearance and Water Management in

the Po Valley in the Central and Late Middle Ages," *Journal of Medieval History* 39 (2013): 377-393. https://doi.org/10.1080/03044181.2013.834644, Daniel R. Curtis, and Michele Campopiano, "Medieval Land Reclamation and the Creation of New Societies: Comparing Holland and the Po Valley, c.800-c.1500," *Journal of Historical Geography* 44 (2014): 93–108. https://doi.org/10.1016/j.jhg.2013.10.004, Greg Bankoff, "The 'English Lowlands' and the North Sea Basin System: A History of Shared Risk," *Environment and History* 19, no. 1 (2013): 3–37. https://doi.org/10.3197/096734013X1352832843899, John Emrys Morgan, "The Micro-Politics of Water Management in Early Modern England: Regulation and Representation in Commissions of Sewers," *Environment and History* 23 (2017): 409–430. https://doi.org/10.3197/096734017X1497947387387, and *Fleuves et marais*, esp. the part: Gestion des espaces hydrauliques at 341–393. See also several works dedicated to the topic by Petra van Dam.

38 Mikhail, "Oriental Democracy."

39 *Working with Water in Medieval Europe* (Technology and Change in History, 3), ed. Paolo Squatriti (Leiden, Boston, and Cologne: Brill, 2000). https://doi.org/10.1163/9789047400110

40 Örjan Wikander "The Water-Mill," in *Handbook of Ancient Water Technology* (Technology and Change in History, 2), ed. Örjan Wikander (Leiden: Brill, 2002), 371–400, Roberta J. Magnusson, *Water Technology in the Middle Ages. Cities, Monasteries, and Waterworks after the Roman Empire* (Baltimore, MD: The Johns Hopkins University Press, 2001), and Adam Lucas, *Wind, Water, Work: Ancient and Medieval Milling Technology* (Technology and Change in History, 8) (Leiden: Brill, 2005). https://doi.org/10.1163/9789047417224. See also: *Wind and Water in the Middle Ages: Fluid Technologies from Antiquity to the Renaissance* (Penn State Medieval Studies, 2 = Medieval and Renaissance Texts and Studies, 322), ed. Steven A. Walton (Tempe, AZ: Arizona Center for Medieval and Renaissance Studies, 2006).

41 See most recently: *Römische und frühmittelalterliche Wassermühlen im Paartal bei Dasing Studien zur Landwirtschaft des 1. Jahrtausends* (Materialhefte zur bayerischen Vorgeschichte, 103), ed. Wolfgang Czysz (Kallmünz: Verlag Michael Laßleben, 2016). For archaeological research into milling and water management, see also: *Water Management in Medieval Rural Economy* (Ruralia, 5 = Památky archeologické, Supplementum, 17) (Prague: Institute of Archaeology, Academy of Sciences of the Czech Republic, 2005), *Archéologie des moulins hydrauliques, à traction animale et à vent, des origines à l'époque médiévale et moderne en Europe et dans le monde Méditerranéen: Actes du colloque international, Lons-le-Saunier du 2 au 5 novembre 2011*, 2 vols, eds. Luc Jaccottey, and Gilles Rollier (Besançon: Presses Universitaires de Franche-Comté, 2016) and Jens Berthold, "Mühlen im Befund – Eine Übersicht zu archäologischen Erscheinungsformen von Wassermühlen," in *Wassermühlen und Wassernutzung im mittelalterlichen Ostmitteleuropa* (Forschungen zur Geschichte und Kultur des östlichen Mitteleuropa, 50), eds. Martina Maříková, and Christian Zschieschang (Stuttgart: Steiner, 2015), 235–268.

42 See for the most important volumes: *Die Wasserversorgung im Mittelalter* (Mainz am Rhein: von Zabern, 1991), *I mulini nell'Europa medievale, L'acqua nei secoli altomedievali: Spoleto, 12–17 aprile 2007*, 2 vols (Settimane di Studio del Centro Italiano di Studi sull'Alto Medioevo, 55) (Spoleto: Presso La sede della Fondazione,

2008), *La città e il fiume, Historische Perspektiven auf Wasserhaushalt und Wassernutzung in Mitteleuropa* (Cottbuser Studien zur Geschichte von Technik, Arbeit und Umwelt, 38), eds. Knut Kaiser et al. (Münster:Waxmann, 2012), *L'eaux et conflits, Wassermühlen und Wassernutzung, Wasserinfrastrukturen und Macht von der Antike bis zur Gegenwart* (Historische Zeitschrift. Beiheft, 63), eds. Birte Förster, and Martin Bauch (Berlin: de Gruyter Oldenbourg, 2015). https://doi.org/10.1515/9783486781052, *Gestione dell'acqua in Europa (XII–XVIII secc.): selezione di ricerche = Water management in Europe (12th–18th centuries): Selection of Essays*) (Serie II, Atti delle 'settimane di studio' e altri convegni, 49) (Florence: Firenze University Press, 2018) https://doi.org/10.36253/978-88-6453-700-9, *Usus aquarum: Interdisziplinäre Studien zur Nutzung und Bedeutung von Gewässern im Mittelalter* (Forschungen zur Geschichte und Kultur des östlichen Mitteleuropa, 54), eds. Christoph Mielzarek, and Christian Zschieschang (Vienna: Böhlau, 2019). https://doi.org/10.7788/9783412500894.

43 Milan is overrepresented in the studies, because of the lack of navigable rivers within the town, and the efforts made by the urban government to provide canals to allow navigation in the water system of the River Po: Luisa Chiappa Mauri, *I mulini ad acqua nel milanese (secoli X-XV)* ([Milan]: Societa' Editrice Dante Alighieri, 1984), Giuliana Fantoni, *L'acqua a Milano. Uso e gestione del basso medioevo* (1385–1535) (Bologna: Cappelli, 1990), Patrick Boucheron, "Water and Power in Milan, c. 1200–1500," *Urban History* 28, no. 2 (2001): 180–193. https://doi.org/10.1017/S0963926801002024, Maria Paola Zanoboni, "L'acqua come spazio economico: attività commerciali e manifatturiere lungo i navigli milanesi (sec. XV)," *Storia economica* 16 (2013): 143–193. For the Italian Peninsula in general, see a number of studies in the above mentioned three conference proceedings: *I mulini nell'Europa medievale, L'acqua nei secoli altomedievali,* and *Gestione dell'acqua in Europa.* From a landscape historical point of view, see also: *Este, l'Adige e i Colli Euganei: storie di paesaggi*, ed. Gian Pietro Brogiolo (Sesto Fiorentino: All'Insegna del Giglio, 2017). See also: *L'acqua nemica: fiumi, inondazioni e città storiche dall'antichità al contemporaneo: Atti del Convegno di studio a cinquant'anni dall'alluvione di Firenze (1966–2016), Firenze, 29–30 gennaio 2015* (Biblioteca del Centro per il collegamento degli studi medievali e umanistici in Umbria, 37), eds. Concetta Bianca, and Francesco Salvestrini (Spoleto: Fondazione Centro italiano di studi sull'alto Medioevo, 2017), esp. the contribution of: Francesco Salvestrini, "Le inondazioni a Firenze e nella valle dell'Arno dal XII al XVI secolo," at 31–60.

44 See the studies in the following volumes: *Wasserversorgung im Mittelalter, Historische Perspektiven auf Wasserhaushalt, Wassermühlen und Wassernuztung*, and *Usus aquarum.* Apart from these, mostly comprehensive studies, about a dozens of local histories and regional histories of milling were published in the past decades. To name but a few, more important ones: Ilka Göbel, *Die Mühle in der Stadt: Müllerhandwerk in Göttingen, Hameln und Hildesheim vom Mittelalter bis ins 18. Jahrhundert* (Veröffentlichungen des Instituts für historische Landesforschung der Universität Gööttingen, 31) (Bielefeld: Verl. für Regionalgeschichte, 1993), Peter Theissen, *Mühlen im Münsterland: der Einsatz von Wasser- und Windmühlen im Oberstift Münster vom Ausgang des Mittelalters bis zur Säkularisation (1803)* (Veröffentlichungen der Historischen Kommission für Westfalen, 22 A = Geschichtliche Arbeiten zur westfälischen Landesforschung,

wirtschafts- und sozialgeschichtliche Gruppe, 13 = Beiträge zur Volkskultur in Nordwestdeutschland, 101) (Münster and New York: Waxmann, 2001), Willi Klein, *Zur Geschichte des Mühlenwesens im Main-Kinzig-Kreis* (Hanauer Geschichtsblätter, 40) (Hanau: Selbstverl. d. Hanauer Geschichtsvereins u. d. Wetterauischen Gesellschaft für die Gesamte Naturkunde zu Hanau, 2003), and Peter Johannes Droste, *Wasserbau und Wassermühlen an der Mittleren Rur: Die Kernlande des Herzogtums Jülich, 8.–18. Jahrhundert* (Aachener Studien zur alteren Energiegeschichte, 9) (Aachen: Shaker Verlag, 2003).

45 *Wasser – Fluch und Segen: Schwazer Silber*, ed. Wolfgang Ingenhaeff ([Innsbruck]: Berenkamp, 2004), and especially *Verwaltetes Wasser im Österreich des Spätmittelalters und der Frühen Neuzeit* (Forschungen zur Landeskunde von Niederösterreich, 37), eds. Martin Scheutz, and Herwig Weigl (St. Pölten: Verein für Landeskunde von Niederösterreich, 2016).

46 Leguay, *L'eau dans la ville*, *L'eaux et conflits*, the special issue of the journal, *Médiévales* no. 36 (1999): Le fleuve, eds. Odile Kammerer, and Odile Redon, and numerous works of Mathieu Arnoux in the topic. Most importantly: Mathieu Arnoux, "Les moulins à eau en Europe occidentale (IXe–XIIe siècle). Aux origines d'une économie institutionnelle de l'énergie hydraulique," in *L'acqua nei secoli altomedievali*, vol. 2, 693–746.

47 John Langdon, *Mills in the Medieval Economy: England 1300–1540* (Oxford: Oxford University Press, 2004).

48 Adam Lucas, *Ecclesiastical Lordship, Seigneurial Power and the Commercialization of Milling in Medieval England* (Farnham and Burlington, VT: Ashgate, 2014). https://doi.org/10.4324/9781315578620.

49 Cf. Erik Fügedi, "Az esztergomi érsekség gazdálkodása a XV. század végén," [The farming of the archbishopric of Esztergom at the end of the 15th century] *Századok* 107 (1972): 536 and Hajnalka Kuffart, "Az esztergomi érsek *pisetum*-jövedelme Battyányi István udvarbíró idején," [The *pisetum*-income of the archbishop of Eszterom in the period of the provisorship of István Batthyányi] in *Veretek, utak, katonák. Gazdaságtörténeti tanulmányok a magyar középkorról* [Mints, routes, soldiers. Studies in the economic history of medieval Hungary], eds. István Kádas, Renáta Skorka, and Boglárka Weisz (Budapest: MTA BTK Történettudományi Intézet, 2018), 93–109.

50 Probably the earliest of these works is: Simon Barthélemy Joseph Noel de la Moriniere, *Histoire generale des peches anciennes et modernes, dans les mers et les fleuves des deux continens,* vol. 1 (Paris: Impr. royale, 1815). See from the early works for Italy: Giuseppe Mira, *La pesca nel medioevo: nelle acque interne italiane* (Milan: A. Giuffre, 1937). For a more comprehensive survey of the research tradition into freshwater fisheries with a focus on Germany, see: Angelika Lampen, *Fischerei und Fischhandel im Mittelalter: Wirtschafts- und sozialgeschichtliche Untersuchungen nach urkundlichen und archäologischen Quellen des 6. bis 14. Jahrhunderts im Gebiet des Deutschen Reiches* (Historische Studien, 461) (Husum: Matthiesen, 2000), 14–18. See for comprehensive surveys on a European scale: Richard C. Hoffmann's seminal work on the topic, "Development of Aquatic Ecosystems in Medieval Europe," *The American Historical Review* 101 (1996): 631–669. https://doi.org/10.1086/ahr/101.3.631 and Richard C. Hoffmann, "Medieval Fishing," in *Working with Water*, 331–393. https://doi.org/10.1163/9789047400110_012.

51 *Fish Exploitation in the Past: Proceedings of the 7th Meeting of the ICAZ Fish Remains Working Group* (Tervuren: Koninklijk Museum voor Midden-Afrika, 1994), Archéologie *du Poisson. 30 ans d'archéo-ichtyologie au CNRS*, eds. Philippe Béarez, Sandrine Grouard, and Benoit Clavel (Antibes: Éditions APDCA, 2008), and for the role of historical ecology in the study of medieval fishing and fisheries, see Gertrud Haidvogl et al., "Historical Ecology of Riverine Fish in Europe," *Aquatic Sciences* 77 (2015): 315–324. https://doi.org/10.1007/s00027-015-0400-0.

52 See e.g., Jens Ivo Engels and Gerrit Jasper Schenk, "Infrastrukturen der Macht – Macht der Infrastrukturen: Überlegungen zu einem Forschungsfeld," in *Wasserinfrastrukturen*, 22–58. https://doi.org/10.1515/9783486781052.22 and Gerrit Jasper Schenk, and Stephanie Eifert, "»Kritische Infrastrukturen« als Ergebnisse individueller und kollektiver Kritikalitätszumessungen – ein Ansatz für die Mediävistik?," in *Was heißt Kritikalität? Zu einem Schlüsselbegriff der Debatte um Kritische Infrastrukturen*, eds. Jens Ivo Engels, and Alfred Nordmann (Bielefeld: transcript, 2018), 47–96. https://doi.org/10.14361/9783839442074.

53 The same holds for the most ambitious multi-series volume on water history: *A History of Water.* Europe is badly underrepresented, especially between ancient Rome and the nineteenth century.

54 András Vadas, "Geography, Natural Resources and Environment," in *The Oxford Handbook of Medieval Central Europe*, eds. Nada Zečević, and Daniel Ziemann (Oxford: Oxford University Press, 2022), 23–38. https://doi.org/10.1093/oxfordhb/9780190920715.013.2.

55 See for the early discussion of these constructions: Sándor Takáts, "Műveltségtörténeti közlemények: a magyar molnár," [Studies in intellectual history: the Hungarian miller] *Századok* 41 (1907), 52–56, Sándor Takáts, "Műveltségtörténeti közlemények: a magyar malom I–II," [Studies in intellectual history: the Hungarian mill] *Századok* 41 (1907): 143–160 and 236–249, and Sándor Takáts, "A magyar faragómolnárok I–II," [The Hungarian millers] in *Rajzok a török világból*, 4 vols [Images from the Ottoman Times], ed. Sándor Takáts (Budapest: Magyar Tudományos Akadémia, 1915–1917), vol. 2. 422–463. See also: Kálmán Lambrecht, *A magyar malmok könyve: történeti anyag* [Book of Hungarian mills: the historical material] (Iparosok olvasótára, 21) (Budapest: Lampel, 1915).

56 See the works by József Papp, Bertalan Andrásfalvy, and Tibor Bellon.

57 For the basically only overview of the question: Zsigmond Károlyi, *A vízhasznosítás, vízépítés és vízgazdálkodás története Magyarországon* [The history of water use, water construction and water management in Hungary] (Műszaki tudománytörténeti kiadványok, 13) (Budapest: Tankönyvkiadó, 1960). See also the chapters written by Károlyi in *A magyar vízszabályozás története* [The history of water regulation in Hungary], ed. Dénes Ihrig (Budapest: [Országos Vízügyi Hivatal], 1973).

58 András Kubinyi, "Städtische Wasserversorgungsprobleme im mittelalterlichen Ungarn," in *Städtische Versorgung und Entsorgung im Wandel der Geschichte* (Stadt in der Geschichte. Veröffentlichungen des südwestdeutschen Arbeitskreises für Stadtgeschichtsforschung, 8), ed. Jürgen Sydow (Sigmaringen: Jan Thorbecke, 1981), 180–190, Sándor Garády, "Buda középkori és újkori vízművei, különös tekintettel a Svábhegy-Városkút kutatására," [The medieval and modern water works of Buda with special regard to Svábhegy-Városkút] *Budapest Régiségei* 32 (1998): 245–265, Imre Holl, "Sopron (Ödenburg) im Mittelalter," *Acta*

Archelogica Academiae Scientiarum Hungaricae 31 (1979): 105–145, Gabriella Fényes, "Középkori vízmű maradványai Budapest Lánchíd utca 19–21. alól: 1/ A lelőhely," [Remains of a medieval water construction under 19–21 Lánchíd Street, Budapest] *Budapest Régiségei* 41 (2008): 193–227, Gyula Siklósi, "Die Wasserversorgung und das Kanalsystem im mittelalterlichen Székesfehérvár (Stuhlweissenburg)," *Antaeus* 26 (2003): 217–244, Ágnes Nagy, "Brunnen und Zisternen im mittelalterlichen Ungarn," *Antaeus* 26 (2003): 343–411 (this volume of the *Antaeus* includes other articles on the topic), and Ágnes Kolláth, "Székesfehérvár vízellátása a középkorban és a kora újkorban," [The water supply of Székesfehérvár in the Middle Ages and the early modern times] In *Határon innen és túl. Gazdaságtörténeti tanulmányok a magyar középkorról* [Within and beyond the borders. Studies in the economic history of medieval Hungary], eds. Boglárka Weisz and István Kádas, István (Budapest: Bölcsészettudományi Kutatóközpont, Történettudományi Intézet, 2021), 351–379.

59 Sándor Garády, "Jelentés az 1936–1942. évben végzett ásatásokról," [Report of the excavations carried out between 1936 and 1942] *Budapest Régiségei* 13 (1943): 418–427.

60 Sándor Takáts, "A komáromi vizahalászat a XVI. században I–II.," [Sturgeon fishing at Komárno in the 16th century] *Magyar Gazdaságtörténelmi Szemle* 4 (1897): 425–445 and 485–509.

61 Degré, *Magyar halászati jog*. On disputes related to fishing rights: István Ecsedi, "Rekesztő halászat a Tiszán és az alföldi kisvizeken," [Kiddle fishing in the Tisza and the smaller waters in the Middle Ages] *Néprajzi Értesítő* 18 (1926): 61–75, József Papp, "Halászati jog a Hortobágy vizein," [Fishing rights on the waters of the Hortobágy] in *Tanulmányok a Hortobágy néprajzához* [Studies in the ethnography of the Hortobágy], ed. Béla Gunda (Debrecen: Kossuth Lajos Tudományegyetem, 1972–1974), 189–205; Bertalan Andrásfalvy, "A paraszti halászati jog a Duna mentén Tolna és Baranya megyében," [Peasant fishing rights on the Danube in Tolna and Baranya counties] *Ethnographia* 81 (1970): 316–326, Bertalan Andrásfalvy, *A Sárköz és a környező Duna menti települések ősi ártéri gazdálkodása és vízhasználatai a szabályozás előtt* [The Sárköz and the ancient flood plain and water management of the settlements along the Danube] (Vízügyi történeti füzetek, 6) (Budapest: Vízdok, 1973), and most importantly, Bertalan Andrásfalvy, *Duna mente népének ártéri gazdálkodása Tolna és Baranya megyében az ármentesítés befejezéséig* [Flood plain economy of the peoples of the Duna valley in Tolna and Baranya Counties until the completion of the water regulation works] (Tanulmányok Tolna megye történetéből, 7) (Szekszárd: Tolna Megyei Tanács, 1975). See also: István Tringli, "Adalék a magyar vadászati jog középkori történetéhez," [To the medieval history of hunting rights] *Herman Ottó Múzeum Évkönyve* 52 (2013): 213–219.

62 Degré, *Magyar halászati jog* and István Sugár, "Az egri püspökség és káptalan középkori vizafogó szegyéi a Tiszán," [Sturgeon fishing traps of the bishopric and chapter of Eger on the River Tisza in the Middle Ages] *Archívum. A Heves Megyei Levéltár Közleményei* 9 (1979): 5–33, Erik Fügedi, "Középkori magyar városprivilégiumok," [Urban privileges in medieval Hungary] *Tanulmányok Budapest Múltjából* 14 (1961): passim. On the fishing rights of the mining town settlers, see Chapter 3.3.

63 László Bartosiewicz et al. "Animal Exploitation in Medieval Hungary," in *Economy of Medieval Hungary* (East Central and Eastern Europe in the Middle

Ages, 450–1450, 49), eds. József Laszlovszky et al. (Leiden and Boston: Brill, 2018), 113–165. https://doi.org/10.1163/9789004363908_007.

64 For examples, see István Viczián, and Csilla Zatykó, "Geomorphology and Environmental History in the Drava Valley, near Berzence," *Hungarian Geographical Bulletin* 60 (2011): 357–377, and Csilla Zatykó, "Aspects of Fishing in Medieval Hungary," in *Processing, Storage, Distribution of Food. Food in the Medieval Rural Environment. Ruralia VIII*, ed. Jan Klápště (Turnhout: Brepols, 2011), 399–408. More recently, see András K. Németh, "Vizek és vízgazdálkodás a középkori Tolna megyében, I–III.," [Waters and water management in medieval Tolna County] *Wosinsky Mór Múzeum Évkönyve* 35 (2013): 121–151, 36 (2014): 223–251, and 37 (2015): 7–39, András K. Németh, "Barátok tava. Tájrégészeti és történeti adatok egy helynév eredetéhez," [The lake of the brothers. Landscape archaeological and historical data on the origin of a placename] *Archaeologiai Értesítő* 138 (2013): 367–374, Zsuzsa Eszter Pető, *Hermits in the Heart of the Hungarian Kingdom: Medieval Monastic Landscape of the Pauline Order in the Pilis* (Archaeolingua. Series Minor, 41) (Budapest: Archaeolingua, 2018), 67–71, and László Ferenczi, "Molendinum ad aquas calidas. A pilisi ciszterciek az állítólagos Fehéregyházán. Történeti, topográfiai és tájrégészeti kutatás a pilisi apátság birtokán," [The Cistercians of Pilis at the alleged 'Fehéregyháza' – Historical-topographical and landscape archaeological study on the estate of the Pilis Abbey] *Studia Comitatensia* [N] 1 (2014): 145–160.

65 Takáts, "Műveltségtörténeti közlemények: a magyar malom."

66 Zsigmond Károlyi, *A vízhasznosítás, vízépítés és vízgazdálkodás története Magyarországon.* (Vázlat) [History of water-use, water-construction and water management in Hungary (A sketch)] (Budapest: Tankönyvkiadó, 1960). See also: *A magyar vízszabályozás története* (the relevant part is the work of Zoltán Károlyi).

67 Károlyi, *A vízhasznosítás*, 59–64.

68 László Makkai, "Östliches Erbe und westliche Leihe in der ungarischen Landwirtschaft der frühfeudalen Zeit," *Agrártörténeti Szemle* 45 [Supplementum] (1974): 45. See also: Walter Endrei, "A magyar malomipar, 1550–1800," [Hungarian milling industry, 1550–1800] in *Műszaki innovációk sorsa Magyarországon: malomipar, vaskohászat, textilipar* [The fate of technological innovations in Hungary: milling industry, metallurgy, textile industry], ed. Walter Endrei (Budapest: Akadémiai, 1995), 48–54. See Chapter 2 for a comprehensive survey of the problem.

69 Andrásfalvy, *A Sárköz és a környező Duna menti települések*, and Andrásfalvy, *Duna mente népének ártéri gazdálkodása Tolna és Baranya megyében.*

70 Károly Takács, "Árpád-kori csatornarendszerek kutatása a Rábaközben és a Kárpát-medence egyéb területein," [Investigation into Árpádian period canals in the Rábaköz and other areas of the Carpathian Basin] *Korall* no. 1 (2000), 27-61 and Károly Takács, "Árpád-kori csatornarendszerek II," *Korall* no. 3–4 (2001): 297–314. For a summary in English, see Károly Takács, "Medieval Hydraulic Systems in Hungary: Written Sources, Archaeology and Interpretation," in *People and Nature in Historical Perspective* (CEU Medievalia, 5), eds. József Laszlovszky, and Péter Szabó (Budapest: Archaeolingua – CEU Press, 2003), 289–312.

71 *Ispánok* (*comites*) are appointees of the kings who governed counties in the kingdom of Hungary. Takács suggested that this service population, under the leadership of the *ispán*, was the social basis for the maintenance of the system.
72 E.g., Moldenhauer, "Mühlen und Mühlenrecht," and numerous studies in *Fleuves et marais*, *L'eaux et conflits*, and *Wasser in der mittelalterlichen Kultur*.
73 Bence Péterfi, "Debates Concerning the Regulation of Border Rivers in the Late Middle Ages: The Case of the Mura River." *Hungarian Historical Review* 8 (2019): 313–335, and Renáta Skorka, "On Two Sides of the Border. The Hungarian–Austrian Border Treaty of 1372," *Hungarian Historical Review* 8 (2019): 290–312. See in detail Chapter 3.4.
74 In detail, see Chapter 1.2.2 on the sources.
75 Vajda, "Okleveles adatok," Vajda, "Hazai vízimalmaink 1301–1325 közötti okleveles adatai," [Charter evidence on Hungarian water mills in the 1301–1325 period] in *Középkortörténeti tanulmányok III.* [Studies in medieval history, III] ed. Boglárka Weisz (Szeged: Szegedi Középkorász Műhely, 2003), 193–213, and Tamás Vajda, "Újabb adatok Árpád-kori vízimalmainkról," [New Charter evidence on Árpádian Period mills of Hungary] in *Medievisztikai tanulmányok V.* [Studies in medieval history, V], eds. Éva Révész, and Miklós Halmágyi (Szeged: Szegedi Középkorász Műhely, 2007), 211–245.
76 Andrea Kiss, *Floods and Long-Term Water-Level Changes in Medieval Hungary* (Springer Water) (Cham: Springer, 2019) https://doi.org/10.1007/978-3-319-38864-9, and Tamás Vajda, "Vízimalmok építésével előidézett, mesterséges áradások a középkori Magyarországon," in *Középkortörténeti tanulmányok 8. A VIII. Medievisztikai PhD-konferencia (Szeged, 2013. június 17–19.) előadásai* [Studies in medieval history, 8. The 8th PhD conference in medieval studies (Szeged, 17–19 June 2013)], eds. Márta Tóber, and Ágnes Maléth (Szeged: Szegedi Középkorász Műhely, 2015), 347–362.
77 István Tringli, "A magyar szokásjog a malomépítésről," [Customary law on construction of mills] in *Tanulmányok a középkorról* [Studies in medieval history] (Analecta Medievalia, 1), ed. Tibor Neumann ([Budapest–Piliscsaba]: Pázmány Péter Katolikus Egyetem, 2001), 251–268.
78 *A Perényi család levéltára 1222–1526* [Cartulary of the Perényi family, 1222 to 1526] (Magyar Országos Levéltár kiadványai, II. Forráskiadványok, 44), ed. István Tringli (Budapest: Magyar Országos Levéltár, 2008).
79 Katalin Szende, "Mills and Towns: Textual Evidence and Cartographic Conjectures from Hungarian Towns in the Pre-Industrial Period," in *Extra muros. Vorstädtische Räume in Spätmittelalter und Früher Neuzeit. Espaces suburbains au bas Moyen Âge et à l'époque moderne* (Städteforschung, A 91), eds. Guy Thewes, and Martin Uhrmacher (Cologne and Vienna: Böhlau, 2019), 485–516. https://doi.org/10.7788/9783412515164.485.
80 Ferenczi, "Molendinum ad aquas calidas."
81 Pető, *Hermits*, 67–70.
82 Ünige Bencze, "Reconstructing a Monastic Landscape: The Example of the Cistercian abbey Cârţa (Kerz, Kerc)," in *Monastic Life, Art, and Technology in 11th–16th Centuries* [Special issue of Annales Universitatis Apulensis. Series Historica], ed. Ileana Burnichioiu (Alba Iulia: Universitatea „1 Decembrie 1918" din Alba Iulia, 2015), 29–44, Ünige Bencze, "A Medieval Pauline Landscape in the Szekler Land," *Transsylvania Nostra* 9, no. 34 (2015): 10–17, Gusztáv Jakab

et al., "The Environmental History of a Former Salt Town in Transylvania (Sic, Northern Romania)," *Acta Archaeologica Academiae Scientiarum Hungaricae* 69 (2018): 185–206. https://doi.org/10.1556/072.2018.69.1.8, and Gusztáv Jakab et al., "Relict Anthropogenic Ecosystem from the Middle Ages: History of a Salt Marsh from Transylvania (Sic, N Romania)," *Environmental Archaeology* 25 (2019): 96–113. https://doi.org/10.1080/14614103.2019.1578547.

83 Viczián, and Zatykó, "Geomorphology and Environmental History," and Csilla Zatykó, "Eltűnt berzencei malmok nyomában," [In pursuit of the lost mills of Berzence] in *Genius loci*, 108–112.

84 Elek Benkő, "The Medieval Water Supply System of Pilis Abbey," *Anteaeus* 37 (2021): 193–224.

85 *Víz és társdalom Magyarországon a középkortól a XX. század végéig* [Water and society in Hungary from the Middle Ages to the end of the 20th century], ed. Gergely Krisztián Horváth (Budapest: Balassi Kiadó, 2014).

86 Tamás Vajda, "Földesúri jogok és kötelezettségek a középkori vízimalmok körül," [Seigniorial rights and dues around medieval water mills] László Gálffy, "Dunai társadalmak: Változások a városi folyamtérben Bécs és Buda között (XIII. század vége–XV. század közepe)," [Danube societies: changes in the urban riverspace between Vienna and Buda (from end of the 12th to the mid-15th century)] and András Vadas, "Vízgazdákodás és háborús védekezés: Csákány és a Vas megyei Rába-mente a kora újkorban," [Water management and war defense: Csákány and the course of the Rába in Vas County in the Early Modern period] in *Víz és társadalom*, 79–110, 111–155, and 207–245 respectively. See also, András Vadas, *The Environmental Legacy of War on the Hungarian-Ottoman Frontier, c. 1540–1690* (Environmental Humanities in Pre-modern Cultures) (Amsterdam: Amsterdam University Press, 2023). https://doi.org/10.1515/9789048552016.

87 Grzegorz Myśliwski, "Utilisation of Water in Central Europe (12th–16th cents.)," in *Economia e energia*, 321–333.

88 Maria Dembińska, *Przetwórstwo zbożowe w Polsce średniowiecznej (X–XIV wiek)* [Grain processing in medieval Poland (10th–14th centuries)] (Wrocław: Zakład Narodowy im. Ossolińskich Wydawnictwo PAN, 1973).

89 Rafał Kubicki, *Młynarstwo w państwie zakonu krzyżackiego w Prusach w XIII-XV w. (do 1454 r.)* [Milling Industry in the State of the Teutonic Order in Prussia in the 13th–15th centuries (to Year 1454)] (Gdańsk: Wydawnictwo Uniwersytetu Gdańskiego 2012), Rafał Kubicki, "Water Mills and Rural Settlement in the Dominion of the Teutonic Order in Prussia," in *Usus aquarum,* 29–50, https://doi.org/10.7788/9783412500894.29, Tomasz Związek, "Kształtowanie sieci młynów wodnych na przykładzie powiatu konińskiego (ok. 1300–1550)," [The formation of a network of water mills on the example of the Konin district (approx. 1300–1550)]," *Studia Geohistorica* no. 2 (2014): 118–142. https://doi.org/10.12775/SG.2014.09.

90 Urszula Sowina, and Wacław Kulczykowski, "The First Urban Water Supply Systems in the State of the Teutonic Knights: The Transfer of Hydrotechnical Knowledge," in *Origines et mutationes. Transfer – Exchange – Power*, eds. Aleksandra Girsztowt, Piotr Kitowski, and Andrzej Gierszewski (Kraków: Libron, 2017), 51–62.

91 Urszula Sowina, *Water, Towns and People. Polish Lands against a European Background until the Mid-16th Century* (Polish Studies – Transdisciplinary Perspectives, 17) (Frankfurt am Main: Peter Lang, 2016).

92 Jerzy Piekalski, "Elementy infrastruktury średniowiecznego Wrocławia," [Elements of the infrastructure of medieval Wrocław] in *Wrocław na przełomie średniowiecza i czasów nowożytnych. Materialne przejawy życia codziennego* (Wratislavia Antiqua, 6), eds. Jerzy Piekalski, and Krzysztof Wachowski (Wrocław: Uniwersytet Wrocławski, 2004), 9–24, Jerzy Piekalski, "The Environment and Living Conditions in Wroclaw (Breslau) in the 16th–18th Centuries," in *Historical Archaeology in Central Europe*, ed. Natascha Mehler (Rockville, MD: Society for Historical Archaeology, 2013), 379–394, and Paweł Cembrzyński, *Zaopatrzenie w wode i usuwanie nieczystosci w miastach stref baltyckiej i sudecko-karpackiej w XIII–XVI wieku* [Water supply and waste disposal in cities of the Baltic and Sudeten-Carpatians zones in the 13th–16th centuries] (Wratislavia Antiqua, 14) (Wrocław: Uniwersytet Wrocławski, 2011).

93 Apart from Sowina's works, see the early social topographies: Mateusz Goliński, *Wokół socjotopografii późnośredniowiecznej Świdnicy*, 2 vols [The sociotopography of late medieval Swidnica] (Wrocław: Wydawn. Uniw. Wrocławskiego, 2000), vol. 1 and Jacek Wiesiołowski, *Socjotopografia późnośredniowiecznego Poznania* [The social topography of late medieval Poznan] (Warsaw: Państw. Wydawn. Nauk, 1982). See more recently: *O rzece i wodzie w życiu codziennym człowieka średniowiecza* [Rivers and waters in everyday life] (Spotkania Bytomskie, 8), eds. Sławomir Moździoch, and Krystian Chrzan (Wrocław: Wydawnictwo Instytutu Archeologii i Etnologii Polskiej Akademii Nauk – Ośrodek Badań Nad Kulturą Późnego Antyku i Wczesnego Średniowiecza, 2015). Cf. Marcin Danielewski, "Korabniki, Niewody i Rybitwy z terenu Kujaw jako osady służebne związane z książęcym regale wodnym," [Korabniki, Niewody and Rybitwy in the Kujawy region as service settlements connected with the ius ducale water system] *Roczniki Dziejów Społecznych i Gospodarczych* 71 (2011): 33–53, and most recently: Tomasz Związek, "The Will, the Mill and the Town. Prolegomena to the Functioning of the Miller Family in the Late Medieval Town," in *Usus aquarum*, 71–86. https://doi.org/10.7788/9783412500894.71. On wastewater management, see Leslie Carr-Riegel, "Waste Management in Medieval Krakow: 1257–1500," (MA thesis, Central European University, 2016) and Rafał Szmytka, "Pollution in Early Modern Krakow," in *Krakow: An Ecobiography*, eds. Adam Izdebski, and Rafał Szmytka (Pittsburgh: University of Pittsburgh Press, 2021), 108–131. https://doi.org/10.2307/j.ctv201xjkz.8.

94 Martina Maříková, "Středověké mlýny v českých zemích (Archeologické a písemné prameny)," [Medieval mills in Czech lands (archaeological data and written sources)] *Mediaevalia Historica Bohemica* 10 (2005): 89–148, Tomáš Petráček, "K otázce datace počátků vodních mlýnů v českých zemích v kontextu změny vnímání dynamiky vývoje českých zemích 10.–12. století," [On the issue of dating the beginning of water mills in the Czech lands in the context of changes in the perception of the dynamics of the Czech lands in the 10th–12th centuries] in *Sborník k poctě Jiřího Kalfersta* [Studies in honor of George Kalferst] (Hradec Králové: Muzeum východních Čeché Hradci Králové, 2014), 249–252, Oldřich Kotyza, "K počátkům vodních mlýnů v českých zemích aneb o

existenci vltavských jezů a hydraulických mlýnů v Praze 10. století. Poznámky k břevnovskému aktu ze 14. ledna 993," [On the origin of water mills in the Czech lands or on the existence of the weirs on the Vltava and hydraulic mills in Prague in the 10th century. Notes on the Břevnov act of 14 January 1993] in *Středověká Evropa v pohybu k poctě Jana Klápště / Medieval Europe in Motion in Honour of Jan Klápště*, eds. Ivana Boháčová and Petr Sommer (Prague: Archeologický ústav AV ČR, Praha, v. v. i., 2014), 461–499.

95 Jiří Andreska, "Development of Fish-Pond Culture in Bohemia," in *The Fishing Culture of the World*, 2 vols, ed. Béla Gunda (Budapest: Akadémiai Kiadó, 1984), vol. 1. 77–83, Frantisek Graus, *Dějiny venkovského lidu v Čechách v době předhusitské*, 2 vols [History of the rural population in Bohemia in the pre-Hussite period] (Prague: Státnínakladatelstvi politické literatury, 1953–1957), vol. 2, 32–35 and 483–486, and Hoffmann, "Economic Development," 660–662.

96 Vlasta Jankovská, and Jan Pokorný, "Palaeoecology of a Medieval Fishpond System (Vajgar, Czech Republic)," *Folia Geobotanica* 37, no. 3 (2002): 253–273, Ivo Přikryl, "Historický vývoj našeho rybníkářství a rybničních ekosystémů: Produkce rybníkářství," [Historical fishponds and fishpond ecosystems] *Veronica* 1 (2004): 7–10, Adéla Pokorná et al., "The Oldest Czech Fishpond Discovered? An Interdisciplinary Approach to Reconstruction of Local Vegetation in Mediaeval Prague Suburbs," *Hydrobiologia* 730, no. 1 (2014): 19–213. https://doi.org/10.1007/s10750-014-1837-1, and most recently, see the overview by Lucie Galusová, "Vodní mlýn jako objekt archeologického výzkumu," [Water mills as objects of archaeological research] *Archaeologia historica* 40 (2015): 267–293. https://doi.org/10.5817/AH2015-1-15.

97 Andreska, "Development," and and Hoffmann, "Economic Development."

98 *Wassermühlen und Wassernutzung.*

99 Lucie Galusová, and Martina Maríková, "Die Baugestalt der Wassermühlen im mittelalterlichen Böhmen und Mähren," in *Wassermühlen und Wassernutzung*, 309–324.

100 Jaroslav Jásek, "An Attempt at an Outline of the Historical Development of Water Supply and Sewerage of Medieval Settlements in the Czech Lands," in *Wassermühlen und Wassernutzung*, 57–76.

101 ÁMTF, vols 1–4. See its recent continutation: Árpád Nógrády, *Sáros vármegye* [Sáros county], 2 vols (Budapest: Bölcsészettudományi Kutatóközpont Történettudományi Intézet, 2022).

102 AOklt and ZsO.

103 MNL OL database. For the most important cartularies, see ÁÚO, DHA, vol. 1, HO, *Erdélyi okmánytár: oklevelek, levelek és más írásos emlékek Erdély történetéhez* [Transylvanian Cartulary. Charter, missals and other written sources to the history of Transylvania], ed. Zsigmond Jakó et al. (Budapest: Akadémiai, 1997–2004), CD, and the cartularies of noble families (not listed here, but in the relevant notes throughout the book).

104 *A középkori Magyarország digitális könyvtára /The Digital Library of Medieval Hungary*, ed. György Rácz. Later integrated into www.hungaricana.hu.

105 For the period until August 29, 1526, see note 109, and for the period between August 27, 1526 and 1570, see the new database of the Hungarian National Archives: https://adatbazisokonline.hu/imgview/reformacio-kora-mnl-ol-1526-1570/.

106 *Lexicon Latinitatis medii aevi Hungariae*, 6 vols, ed. János Harmatta (Budapest: Akadémiai and Argumentum, 1987–2017) and Kornél Szovák, *Supplementum ad Lexicon Latinitatis Medii Aevi Hungariae, A-I* (Budapest: [Argumentum], 2013).
107 This is how far the Sigismundian Cartulary (ZsO) has gotten until now.
108 Béla Némethy, "Vízjogi előzmények a magyar Corpus Jurisban," [Water law antecedents in the Corpus Iuris] *Vízügyi Közlemények* 13 (1931): 126–148 and János Ede Szilágyi, "Vízjog: a vizek tulajdonának és használatának főbb magyar előírásai a nemzetközi tendenciák tükrében," [Water law: the major regulations on the ownership and use of water in Hungary in light of international trends] *Sectio Juridica et Politica Miskolc* 29 (2011): 595–622.
109 On the customary law on mill construction, see: Tringli, "A magyar szokásjog a malomépítésről" and Chapter 3.1.
110 E.g., Laura Lehua Yim, "A Watercourse 'in Variance': Re-situating a Sixteenth-Century Legal Map from Ashbourne, Derbyshire," *Imago Mundi* 68, no. 2 (2016): 147–163. https://doi.org/10.1080/03085694.2016.1171484, William D. Shannon, "Dispute Maps in Tudor Lancashire," *The Local Historian* 42, no. 1 (2012): 2–15, Peter Barber, "Mapmaking in England, ca. 1470–1650," in *The History of Cartography, Vol.3. Cartography in the European Renaissance*, ed. David Woodward (Chicago: Chicago University Press, 2007), 1594–1595, Roger. J. P. Kain, "Maps and Rural Land Management in Early Modern Europe," in *The History of Cartography, Vol. 3*, 706–708. For a collection of similar maps: *Local Maps and Plans from Medieval England*, eds. R. A. Skelton, and P. D. A. Harvey (Oxford and New York: Clarendon and Oxford University Press, 1986).
111 For surviving perambulation sketches, see: Andrea Kiss, "Study on the Historical Geography of the First Extent Perambulation Sketch from the Carpathian Basin," *Zbornik Odsjeka za povijesne znanosti ZPDZ HAZU* 19 (2002): 127–141, Csilla Zatykó, "Medieval Villages and their Landscape: Methods of Reconstruction," in *People and Nature*, 343-375. Cf. Mikhail, "Oriental Democracy."
112 András K. Németh, and Gábor Máté, "Szempontok és példák a középkori eredetű malmok és malomhelyek folytonosságának vizsgálatához," [Notes and examples to the study of the continuity of medieval mills and mill sites] in *Középkori elemek a mai magyar anyagi kultúrában* [Medieval elements in the present-day Hungarian material culture] (Tanulmányok a Kárpát-medence anyagi kultúrája köréből, 1), eds. Anikó Báti, and Zsigmond Csoma (Budapest: Agroinform, 2014), 47–68.
113 For the edition of the first military survey, see *Az Első Katonai Felmérés 1763–1785* [DVD-Rom] [First military survey, 1763–1785, DVD-Rom] (Budapest: Arcanum Adatbázis Kft., 2006). Made available online at: http://mapire.eu.

2

THE SPREAD OF WATER MILLS IN MEDIEVAL HUNGARY WITH A CENTRAL EUROPEAN OVERVIEW

By the late medieval period, many waterways in Western and Central Europe featured water mills, and their associated infrastructure was among the most widespread elements of water management.[1] Water mills were not merely water wheels located along a river; in many cases, they provided an even distribution of the water supply. They included several structural components—mill races, dams, sluices, and mill buildings—all of which affected the water level and flow used by a mill. These buildings often obstructed one another, as explained in the following chapters. Once water mills became standard features along waterways, the question of who owned and controlled the water they used arose. In Chapter 2, the spread of water mills in the Kingdom of Hungary is discussed and contrasted with similar data from other regions of Central Europe.

For over a century, historians have commonly believed that Christianization significantly contributed to the spread of technological innovations during the Middle Ages.[2] The more effective use of power ranks among the most important of these innovations. On the one hand, the spread of the breast harness was essential for maximizing animal power, while on the other, water mills offered a more efficient method of grinding grain than the typical hand mills of the era. Research, drawing on the foundational handbook by Richard Bennett and John Elton cited earlier, attributes the proliferation of mills in the high medieval period to the efforts of religious orders across Europe, primarily the Benedictines.[3] This thesis received partial support from two influential studies published in the mid-1930s, one by Lewis Mumford and the other, noted above, by Marc Bloch.[4] Both historians suggested that, following the Roman period, the reintroduction of water mills in most parts of Europe was primarily associated with religious orders. However,

DOI: 10.4324/9781003596905-2

Adam Lucas recently highlighted the differing interpretations of the two historians: while Mumford contended that the spread of new technologies aimed to enhance time management for the peasantry, Bloch argued that it was primarily intended to boost the income of church estates from the peasantry through compulsory grinding, the aforementioned *ban*, at their mills.[5]

Bloch attempted to provide an overview of the spread of mills across Europe. His limited access to secondary literature allowed him to sketch only a very approximate chronological sequence. Although his chronology ultimately proved inaccurate in many areas, including Ireland and Iberia, his main conclusion—that areas under longer Roman rule used water mills earlier than others—is still widely accepted by historians.[6] Since Bloch's time, however, archaeological research has shown that the use of water mills was much more widespread in the Roman Empire than previously assumed.[7] Despite Bloch's efforts to be comprehensive, some regions of Europe, such as Central Europe, lacked readily accessible secondary literature at the time and were therefore overlooked in the study. Both he and later scholars, when offering an overview of the spread of mills in Europe, often failed to engage with the scholarship from this region, even though understanding the roles of Roman presence and monastic orders in this process would have been an important area to examine.

This chapter of the book does not aim to provide a comprehensive survey of technological advancement in Central Europe. Instead, it focuses on the still-oft-oversimplified association of monastic orders, especially the Benedictines, with the spread of water technologies, including water mills, in Hungary. Here, sources and scholarship about the Kingdom of Hungary lie at the center of the analysis. However, references are made to the strikingly similar questions in Polish, Czech, and, to some extent, Croatian, Serbian, and Romanian scholarship.[8] A rough chronology of the spread of this technical device in Hungary is presented to better understand how the norms for the use of waterways evolved.

2.1 Mills and Benedictines: Some Preliminary Considerations

Before turning to Central Europe and the spread of mills in that region, it is essential to clarify why early monastic communities are often linked to the proliferation of water wheels. Wherever grain production existed, it needed to be ground. For much of the period from the Neolithic era through the Middle Ages, this task was performed with small hand mills, as evidenced by excavated fragments found throughout Europe. However, hand mills were significantly less efficient than water wheels. The amount of grain processed by a hand mill, even a rotary hand mill, barely met family needs, and the quality of the flour produced was not comparable to that of a water mill.

Nonetheless, water mills were not feasible in every location. The placement of early Benedictine and other religious houses on hilltops did not always permit the use of water as an energy source. However, when Saint Benedict of Nursia wrote his *Rule* for the abbey that he established at Monte Cassino, he identified mills as essential components of the monastery.

According to Benedict's *Rule*, mills had to be located within the monastery's walled area, though it is unclear whether he meant human-powered hand mills or water mills.[9] The scarcity of fresh running water may have been a reality for many early monastic foundations. Still, water, used for generating energy and other purposes, was typically considered an essential factor in the location of monasteries during the High and Late Middle Ages. This policy is reflected in the famous St. Gall monastery plan created at Reichenau. Although the somewhat stereotypical representation of the mills on the plan makes it difficult to discern which types were depicted, a detailed study by Walter Horn and Ernest Born demonstrated that they must have been water-fed.[10] This indicates that by the early ninth century, a water mill was seen as an essential component of a Benedictine abbey. However, at the abbey of Reichenau—situated on an island in Lake Constance—the water in proximity to the abbey may have been perceived differently than at many other monastic sites. Still, in this instance, Horn and Born's arguments support the notion that, during the Carolingian period, monastic complexes were envisioned to include water mills. It has recently been shown that water mills were widespread throughout the broader hinterland of Reichenau quite early in the Carolingian period. Moreover, archaeological research has uncovered a water mill complex in Bavaria that dates to the late seventh century. This early example suggests that mills were used in what is now southern Germany from the early medieval period onward.[11]

Christianity, the focus of this chapter, arrived in Central Europe from two directions. Following the Roman influence, Christianity spread from the Byzantine Empire and later from Western Christian regions through missionary activities. Western missionaries in Hungary, Poland, and Bohemia played a crucial role in the spread of Christianity in the tenth and eleventh centuries. However, in Wallachia, Moldavia, and other parts of the Balkans, as well as in East Central Europe before the tenth century, Byzantine missionary activity was certainly more impactful.[12] Despite recent demonstrations of water mill use in the later Byzantine regions of the Roman Empire, there is no evidence that this technology emerged in the Balkans or even in the Carpathian Basin from this direction after the Roman administration withdrew in Late Antiquity.[13] The question remains impossible to answer on the basis of relevant written sources, as there are few from the Inner Balkans before the early thirteenth century. Additionally, there are no related archaeological excavations in Bulgaria, Serbia, or Macedonia either.[14] The approximately 100 water mills known to have operated in Bavaria around the turn of the first millennium, however, leave no doubt that they were widely utilized there. This is the region from which most monasteries in Central Europe recruited members around the year 1000. The earliest monastic foundations in Bohemia, Břevnov and Ostrov, were both linked to Regensburg and Niederaltaich. Although the first monks who arrived from Regensburg indeed originated from Rome, another location where water mills were commonly found in the tenth century.[15] The first monasteries in Poland may also have been established around 1000. Still, their history is somewhat unclear until the mid-eleventh century, and a permanent monastic presence may not predate the 1040s. Nevertheless, in Bohemia and Silesia, Bavarian influence was certainly the most significant.[16]

The situation is similar regarding Hungary. The first missionaries are known to have arrived in the Carpathian Basin in the second half of the tenth century. Saint Wolfgang and Bruno (Prunward) both hailed from southern German regions; the former was raised at the abbey of Reichenau mentioned earlier. The first monasteries, Pannonhalma (founded in 996), Pécsvárad (1015), Bakonybél (1018), and Zalavár (ca. 1019) all emerged during the reign of Prince Géza and his son and successor, King Saint Stephen. The Czech and Bavarian influences can undoubtedly be detected in Hungary, just as in Silesia.[17] Thus, the monks of the first Western monastic foundations in Central Europe were recruited from areas where water mills were, if not widespread, then at least well known.

2.2 The First Water Mills in the Region: Historical Evidence, Linguistics, and (the Lack of) Archaeological Data

As concluded in the previous subchapter, the emergence of the Benedictine order reveals significant variation across the region. Given these temporal disparities, it is increasingly important to demonstrate similar time lags in the emergence of water mills in the area. First, data from Hungary will be examined, followed by a comparison with the available data for the Czech lands, Poland, and the Balkans.

In Hungary, scholars have offered differing interpretations of the spread of water mills in the Carpathian Basin. Analyzing early written evidence—both pragmatic and narrative—based on initial references to mills has sparked debates over whether these mills were powered by water or human muscle. Many prominent historians of the Ápádian period contributed to this question, including Péter Váczy (1904–1994), László Makkai (1914–1989), György Györffy, and Gyula Kristó (1939–2004). In addition to specialists in early statehood, historians of technology also presented arguments supporting various dates.[18] Beyond the written evidence, the issue has also been studied through the integration of other sources, such as historical linguistics and archaeology.

The integration of linguistics into the study of the spread of mills first appeared in scholarly literature almost a century ago. Between the 1950s and the 1970s, linguists debated the origin of specific milling terms, including the Hungarian term *malom* (or *molna*), meaning "mill".[19] Elemér Moór argued that the Slavic origin of the word indicates the early use of water mills among the Slavic population living in the Carpathian Basin when the Hungarians arrived. Moór concluded that the term was adopted by the Hungarians from a Slavic dialect that served as the basis for the Slovak (!) language, and he believed that this adoption may have occurred in the western area of the Slovak–Hungarian language contact zone, specifically in the Little Hungarian Plain. More importantly for this book, based on acoustic changes in the Slovak word for mill (*mlýn*) and the adoption of other water management-related terms, Moór, following in the footsteps of István Kniezsa, dated this process to the ninth and tenth centuries. However, two related issues were never discussed by linguists: first, whether water mills were used in West Slavic areas at that time, and second, why mlýn (and *molna*) could refer only to water mills, not hand-driven ones.

In historical scholarship, it was probably László Makkai who first referred to this linguistic debate in an important German-language article—although he never mentions which works he was referring to—and accepted the ninth-century adoption of the term (and, of course, of water mills).[20] Walter Endrei (1921–2000), a historian of technology, accepted Makkai's view and, building on that dated the appearance of water mills in the region to the eighth [sic!]–ninth-century Carolingian rule in the area of Transdanubia.[21] The dating of the adoption of water mills among West Slavic peoples and Hungarians, well before 1000, is unique in the historiography of the region. Symptomatic of the uncertainty in the dating is that Endrei, in his later works, never again mentions the eighth–ninth-century spread of water mills in the region.[22] Since Endrei, Tamás Vajda (1976–) has challenged the validity of the tenth-century (or, more broadly, the Carolingian) spread of mills, drawing on archaeological evidence in a recent article.[23]

Archaeological data also need to be considered in the Central European context, partly because increasingly precise dating methods often provide insights into the spread of water wheels across various parts of Europe during a period when relevant written sources are lacking.[24] Although no archaeological excavation in the Carpathian Basin has yet uncovered any evidence of water mills predating written records of such constructions, it is worth exploring the research potential of this area as well. In recent years, archaeological dating methods—most notably dendrochronology—have significantly expanded the possibilities for studying the history of various wooden buildings and infrastructure. Many medieval mills have been excavated or observed in Hungary during archaeological research over the last fifty years, but dating them has typically proven a significant challenge.

This is important for this book, as some of the mills were previously dated to post-Carolingian times in earlier scholarship.[25]

Hand-mill fragments may also serve as valuable indicators, alongside the actual archaeological remains of water mills. For many years, most hand mills—specifically the grinding stones—were unearthed at sites on the Great Hungarian Plain. This distribution led to the interpretation that the western parts of the Carpathian Basin—those under Carolingian authority—were technologically more advanced than the eastern plain areas within the Basin, where hand mills—particularly for household flour production—persisted longer. This perspective has been vigorously challenged in recent years as more small grinding stones have surfaced in Transdanubia.[26] One significant site worth mentioning is the village of Kána (on the borders of present-day Budapest), where numerous fragments of hand mills were found. It is unclear exactly when these hand mills were used there. Still, the village likely existed during roughly the same time as the Benedictine abbey of Kána, to which the settlement belonged. The abbey was founded in the mid-twelfth century and operated until the Mongol invasion of 1241–1242. The large number of hand mills found in a village owned by Benedictines during the mid-twelfth to mid-thirteenth centuries clearly indicates that water mills were not the only tools available for grinding grains in the mid-Árpádian period, even within monastic communities in the so-called *medium regni*, the power center of the Kingdom of Hungary, which was then considered the most developed region with an exceptionally dense monastic network.[27]

As mentioned, despite the importance of linguistic and archaeological evidence, written records—despite their issues, which will be discussed later—remain the most reliable for establishing the date of the introduction of mills to Hungary. However, there are some issues when relying on early foundation and endowment charters, where the earliest mentions of water mills might appear. Recently, it has been suggested that the earliest unquestionably authentic document referencing water mills in the Hungarian Kingdom dates to 1009. This document is a donation charter granting land to the newly established bishopric of Veszprém. Although the original has not survived, scholars accept it as genuine early-eleventh-century literacy. Therefore, most experts see it as the first solid evidence of water mills in Hungary.[28] In this charter, King Saint Stephen, besides listing counties under the bishopric's authority, granted land in three counties—Fejér, Veszprém, and Visegrád—as well as in the district of the castle of Kolon, all located in Western Hungary. The charter states that the lands were donated with *omnibus utensibus iugiterque pertinentiis, scilicet famulis familiabusque, pratis, vineis, areis, edificiis, campis, terris, agris cultis et incultis,* piscacionibus, aquis aquarumque decursibus, molendinis, *viis, inviis tam exitibus quam inexitibus.*[29] Judging by this list, there is little doubt that mills, coming as they do immediately after other water-related income, were water-powered.

It is argued here that, despite water mills being mentioned in this context, this does not necessarily mean that mills were operating at the borders of the donated settlements at that time. Three possibilities need to be considered. First, water mills may indeed have been functioning in the mentioned areas. In this case, the Benedictines, along with other religious institutions, might not have had much impact on the introduction of mills, partly because the lands in question were, of course, obviously in the king's hands at the time of the donations, and partly because, until the end of the first decade of the eleventh century, the impact of Benedictine technical know-how must have been relatively limited.[30] Despite royal ownership, research suggests that the church exerted a significant influence on the early organization of royal estates in Hungary.

György Györffy, following the lead of German diplomatic research and Hungarian historian Imre Szentpétery (1878–1950), highlighted that this charter shows apparent similarities to those issued by the Holy Roman emperors, particularly those of Emperor Henry II. This suggests a second possible reason for the appearance of water mills in the 1009 charter. Philological analysis typically considers the endowment charter for Saint Michael's monastery at Bamberg (Michelsberg) as the closest parallel to the donation charter of the bishopric of Veszprém. Although this charter did not list the potential incomes in this way, another charter from the same year, associated with the same notary, listed the donated incomes in the same order as mentioned earlier.[31] Setting aside the debate over the notary associated with the Veszprém charter, it is clear that he originated in the Empire, likely in Northern Italy or Southern Germany. During this period, water mills were standard features on estates in both Northern Italy and areas north of the Alps.[32] The notary, who had drafted several similar donation documents in the Empire under Emperor Henry II for various institutions, may not have fully understood the physical realities of the estates involved. While the pertinence list in the Veszprém charter is very similar, it is not identical to those of imperial charters (see Table 2.1). One could argue that, to some degree, they reflect local conditions, as some documents from this period in the Holy Roman Empire refer to specific infrastructure and legal institutions. For instance, in 1002, in Weilburg, Hesse, a distinction was made between two types of forests—*silva* and *foresta*—as well as a fishing lake (*sagina*), fishing rights (*piscatio*), pasture (*pascuum*), and the common pasture, or *conpascuum*. In 1005, in Gassino near Turin and its surrounding areas, allodial servants and their helpers (*servis et ancillis casalivis*) and willows (*stellareis*) were mentioned. Of these terms, only silva and fishing rights (*piscatio*) appeared in the Veszprém charter, but without the distinctions or other listed terms. This was easily achievable, as notaries often worked with representatives from the endowed institutions, who informed them of what needed to be documented.[33]

TABLE 2.1 The pertinence lists of the Veszprém endowment charter and three imperial and royal donation charters

Veszprém (1009)	*Henry II (1018)*	*Henry II (1002)*	*Arduin (1005)*
omnibus utensibus iugiterque pertinentiis, scilicet famulis familiabusque, pratis, vineis, areis, edificiis, campis, terris, agris cultis et incultis, piscacionibus, aquis aquarumque decursibus, molendinis, viis, inviis tam exitibus quam inexitibus	cum omnibus eius pertinentiis, videlicet servis et ancillis terris cultis et incultis campis pascuis silvis venationibus aquis pratis aquarumque decursibus piscationibus molendinis viis inviis exitibus	cum cunctis eorum usibus et pertinentiis, hoc est servis et ancillis areis edificiis terris cultis et incultis pratis pascuis sive conpascuis silvis forestis venationibus saginis aquis aquarumque decursibus piscationibus molendinis viis et inviis exitibus	omnibus pertinentiis et adiacentiis suis mobilibus et immobilibus rebus quae dici vel nominari possunt, videlicet cum servis et ancillis casalivis ortis vineis campis pratis cultis et incultis silvis stellareis pascuis venationibus aquis aquarumque ductibus molendinis paludibus piscationibus et omnibus rebus

Source: DHA I, 52 and *MGH Diplomata regum et imperatorum Germaniae*, vol. 3 [Henrici II et Arduini diplomata] (Hannover: Hahn, 1900–1903), 488–489 (Heinrich no. 384), 24 (Heinrich no. 21) 713–714 (Arduin no. 10).

However, such donations were intended to give beneficiaries the broadest possible freedom to use the estate they received, including any future income from the land they had acquired.[34] Therefore, a second possibility is that water mills were included because of the income they could generate, a consideration common in various parts of the Holy Roman Empire in the early eleventh century. Partly related to this, a third option could be hypothesized. Carolingian charters often included pertinence lists, quite similar to the one provided for the bishopric of Veszprém. Similar Carolingian documents may also express royal intent to create a more unified settlement network, along with other economic units. In the case of the Carolingians, a network of small agricultural units was organized under villas. The lists of pertinences in such documents illustrate the arrangements the rulers aimed to establish as models of production.[35] The royal endowments to the newly founded bishoprics in Hungary, such as Veszprém, may also follow a similar pattern. However, due to insufficient control sources, it is difficult to confirm this assumption. Nevertheless, in the early endowment charters, unlike some of the Carolingian capitularies and instructive documents, nothing indicates systematic differences or significant variances that would suggest an intentional arrangement in early pertinence lists.

The second and most likely option, which posits that mills were included on the list as a potential future source of income, certainly allows for a partial reevaluation of the scholarly view that sets the latest possible date for the appearance of water mills in Hungary at 1009. However, the arguments above do not provide definitive proof that water mills were not already in operation in the Hungarian Kingdom by 1009. Nonetheless, this point should definitely be considered when attributing the emergence of mills to the Veszprém charter.

As noted above, research usually attributes the spread of mills to the Benedictines, although the earliest mention cannot be directly linked to their activities. Still, several references clearly show that water mills were introduced sometime in the eleventh century. The foundation charter of the Benedictine abbey of Tihany from 1055—the earliest charter from the Hungarian Kingdom preserved in its original form—mentions both mills and millers. King Andrew I (1046–1060) granted several estates from his own holdings to his newly established monastery and later burial site.[36] As in the Veszprém case, the mills were located in royal, not ecclesiastical, lands. However, unlike the Veszprém charter, this document confirms that water mills were already running in Hungary by the mid-eleventh century. When they were first used and how widely they spread remain open questions.

There is one notable narrative source to include in this list of early written references to mills in Hungary. However, in this case, the source indicates the widespread use of hand mills in eleventh-century Hungary. It comes from the *vita* of Gerard, the bishop of Cenad, who died a martyr during the pagan uprising of 1046. According to the so-called *Legenda minor*, written around 1100:

> Once, as he [Gerard] was making haste to the king in order to defend someone, the man of God was put up in a sandy part of the country that was good for keeping cattle. Here, around midnight, one of the women servants, while she was turning the mill by hand to grind some wheat, started to sing a song in order to lighten her hard work; at this, the holy father woke up and, calling his servant, asked him what that was.[37]

The bishopric of Cenad, where the events in the hagiography occurred, had a sparse church network at the time and relatively few waterways suitable for milling. However, this clearly shows that, even at the end of the eleventh century, when the *vita* was authored, hand mills were likely the primary source of flour.

Given the preceding discussion and the sometimes controversial data, the appearance of mills in the Hungarian Kingdom can be traced back to before 1055. However, there are several uncertainties and discrepancies across the sources.

Written evidence from the Czech lands has also been taken into account in contexts similar to those of the early charters that mention mills in Hungary. The first relevant document for the Czech lands is the endowment charter of the Benedictine abbey of Břevnov. The reference to mills in the charter is much more detailed than in the endowment charter of Veszprém. It mentions mills as well as weirs, not just as general pertinences of estates, but in a specific context and at well-defined locations: *cum omni familia et terra sufficienti ac montem ad meridiem, tendentem a bivio quodam, in quo preciduntur molares, Schirnovnice dicto … Duo molendina sub ipso castro Praga et de ipso flumine Wltaue ad tria obstacula molendinorum in eodem loco. … Decimam quoque de omni agricultura in Porecze … et mansum in litore fluminis Wltaue ad horreum construendum et obstaculum in eodem loco ad molendina edificanda.*[38] The document was long viewed as a highly interpolated charter; however, Jiří Pražák recently argued that most of the charter indeed dates to the tenth century.[39] Pražák's work led some scholars to accept that the mills and mill sites mentioned in the charter indicate that, by then, water was regularly used to generate power.[40] Recently, however, a study by Oldřich Kotyza highlighted several significant points, leading him to consider the above part of the charter an interpolation dating to the late twelfth century at the earliest.[41] The most problematic point, according to Kotyza, is the mills' location in the charter. Although the Vltava River is not among the most significant rivers of Central Europe, its discharge was still too large to provide suitable locations for mills to be constructed along. Most medieval mills were built on smaller streams because, technologically, erecting mills along major rivers with their numerous obstacles, channels, races, etc., required greater expertise and, perhaps even more importantly, investment.[42] Despite the many compelling arguments he presents, Kotyza does not deny the potential existence of mills in the vicinity of Prague around the year 1000. According to him, however, if mills were present, they must have been operating along smaller streams near the town and certainly not along the Vltava.

Aside from this controversial charter from 993, the earliest document mentioning water mills dates between 1125 and 1140. Jan Klápšte highlighted this charter, noting that it describes a not-very-wealthy canon's inheritance, which included a mill. The existence of water mills on a smaller estate suggests that, by then, a considerable number of mills had already been built in the Czech lands. Although Únětice—the village where the water mill is mentioned—was not a particularly wealthy landholding, it is located close to Prague, which had already become a hub of ecclesiastical and political power, thereby accelerating regional development.[43] Another source from 1130, similar to the first Hungarian reference dating to 1009, mentions water mills in a pertinence list associated with an estate transaction.[44] However, unlike in Hungary, chronicles also contain essential data. Most notably, the first book of the early twelfth-century chronicle of Cosmas of Prague (completed in 1119) refers to millers (*molendinarius*), who were clearly specialized professionals responsible for operating water mills.[45]

In Poland, the lack of legal evidence from the early statehood period and the absence of references to water mills in the earliest Polish historiography—such as Gallus Anonymus's chronicle—leave insufficient data to determine whether mills existed in the eleventh or early twelfth century.[46] The first written record from roughly the same period as in the Czech lands, 1149, reflects this. Compared to Bohemia and the Hungarian Kingdom, however, systematic analysis of written records in Poland has only been partially conducted.[47] Research tends to focus more on the late medieval period and on the commercialization of the milling industry. Despite the scarcity of data on the eleventh- and twelfth-century spread of water mills in the Polish lands, two important sources provide insights into how grain was usually ground in the late twelfth and the first half of the thirteenth century. The first is the so-called Book of Henryków, authored by its abbot, Peter, which has made the Cistercian abbey of the same name one of the best-known ecclesiastical institutions in medieval Poland. The text includes a detailed history of the monastery's early years, with references to mills and milling. In the first book, Peter recalled that: "Let it be known that in those days here in the circuit, water mills were extremely rare, so the wife of the said Boguchwał the Czech very often stood at the quern to grind."[48] This story, written around 1200, highlights the scarcity of water mills, suggesting that milling in the Henryków region of Silesia (the area closest to the Holy Roman Empire) was predominantly done by hand at that time. The second source is hagiographic and shares notable similarities with the previously discussed legend of Gerard. Like the *Legenda minor* of the martyred bishop, the *vita* of St. Hedwig of Silesia recounts a woman grinding grain with a hand mill—though in this case, the woman was not singing but required a miracle to get the hand mill working again.[49]

The predominance of handmills up to around 1200 may have been even greater further north or east.[50] Archaeological excavations to date have not demonstrated the existence of water mills predating written sources in either Bohemia or Poland.[51] Therefore, there is no direct evidence of water mills being used in these regions before the mid-twelfth century. However, similar to Hungary, several fragments of hand mills were found during excavations in both the Czech lands and various parts of Poland. These findings indeed provide insight into the linguistic debate among Hungarian scholars about whether the Hungarian term for mills (or water mills) was borrowed from a West Slavic language prior to the tenth century.

Research on the spread of water mills in Croatia, Serbia, and Moldavia-Wallachia has been limited; available publications are sparse regarding the technological changes in the wider region. Scholars suggest that widespread water mills existed in present-day Croatia during Roman times. However, most studies cover Dalmatia, which may have had weaker connections to Central Europe in the eleventh century compared to inland areas. Archaeological evidence confirms the presence of ancient water mills near the River Jadro, close to the town of Salona, during the ancient period. In the early eleventh century, this river was again used for milling. Still, there is no information on mill operations between the Roman era and the start of the eleventh century. Though references to mills appear in a tenth-century document, this text has since been identified as a late-medieval forgery. Donations to ecclesiastical institutions indicate that in the eleventh century, water mills operated in Dalmatia's coastal areas, both on the mainland and on the islands; however, no relevant sources from Croatia before the late-eleventh-century personal union with the Hungarians have survived.[52] In the inland Balkans, references to water mills first appear in the late thirteenth century.[53] The influence of Roman technology on medieval milling has been noted in Croatian and Romanian studies. Some scholars argued that water mills were continuously used in the territory of the former Roman province of Dacia. Claims made in the 1970s have been criticized, and most scholars now believe that mills first appeared in this region in the twelfth to thirteenth centuries, when they are first documented in sources related to the Banate and Transylvania.[54]

2.3 Conclusions

Historical scholarship, mainly influenced by Lynn White Jr., has connected the spread of medieval technological innovations—such as water mills—with the rise of monasticism in regions traditionally seen as the "peripheries of Christianity."[55] Central Europe is one area where Christianization and monastic expansion happened over a relatively brief period. Although Christianization started in the Czech lands and Poland several decades before it did in the Principality (later Kingdom) of Hungary, by the mid-eleventh century, Hungary's monastic network was more extensive than in its northern neighbors. The link between church institutions and the spread of water mills has proven to be less clear than historians often suggest.

The earliest references to water mills in the region do not suggest that Benedictines owned these buildings. Instead, they include charters granting church institutions the right to use water for mills or reference existing water mills. The predominance of sources that record royal donations to various church bodies should largely be attributed to the nature of early literacy in Central Europe. Most legal documents from the eleventh and early twelfth centuries were issued in connection with royal resolutions and primarily involved donations. By the middle of the eleventh century in Hungary, not only kings but also other laymen established monasteries and other institutions, enriching the foundations with water mills.[56]

This does not claim that the spread of monasticism and church organization did not influence the emergence of water mills in Central Europe, as it certainly did. The connection between monasteries and mills is probably more complex than previously believed. There is very little written evidence from the eleventh to twelfth centuries indicating widespread use of water mills, even on church lands, and certainly not enough to suggest that grain was generally processed in water mills shortly after Benedictine monasteries appeared in Central Europe. The farming and management of church properties were undoubtedly among the best organized at the time. Therefore, the spread of mills on these estates likely happened more quickly than elsewhere. However, there is no evidence that water mills played a dominant role in the agricultural landscapes of the Kingdom of Hungary from the early eleventh century, or in the Czech lands and Poland from the twelfth century.

Although the link between the rise of monasticism and the spread of water wheels may not be immediately clear, water mills became common in Hungary, Poland, and Bohemia starting around the twelfth and thirteenth centuries. From this time onward, one can gain some insight into how water use was regulated and what norms governed the exploitation of waterways, rivers, and streams. Of course, many of these practices probably originated much earlier; the Magyars used waterways long before Hungary was conquered and a Christian state was formed. However, it is only from the late Árpádian period that we can start to understand some of these norms and practices.

Notes

1 This chapter is a significantly modified version of a paper, published as "Technologies on the Road between West and East: The Spread of Water Mills and the Christianization of East Central Europe," in *The Medieval Networks in East Central Europe: Commerce, Contacts, Communication*, eds. Balázs Nagy, Felicitas Schmieder, and András Vadas (New York: Routledge, 2018), 123–138.

2 For the Christianization of the countries, see: *Christianization and the Rise of Christian Monarchy. Scandinavia, Central Europe and Rus' c. 900–1200*, ed. Nora Berend (Cambridge: Cambridge University Press, 2007). https://doi.org/10.1017/CBO9780511496400, and Nora Berend, Przemysław Urbańczyk, and Przemysław Wiszewski, *Central Europe in the High Middle Ages: Bohemia, Hungary and Poland c.900 – c.1300* (Cambridge: Cambridge University Press, 2013), 110–137. https://doi.org/10.1017/CBO9780511813795, and more recently Florin Curta, *Eastern Europe in the Middle Ages (500–1300)*, 2 vols (Brill's Companions to European History, 19) (Leiden and Boston: Brill, 2019), vol. 1. 470–499. https://doi.org/10.1163/9789004395190.

3 Richard Bennett, and John Elton, *History of Corn Milling*, 4 vols (London: Simpkin, 1898–1904). For the role of Cistercians, with primary focus on their role in spreading industrial mills, see Christie Peters, "Industrial Milling and the prolific Growth of the Cistercian Order in the Twelfth and Thirteenth Centuries," in

Fifty Years of Medieval Technology and Social Change, ed. Steven A. Walton (London: Routledge, 2020), 150–164.

4 Lewis Mumford, *Technics and Civilization* (New York: Harcourt, Brace and Co., 1934), and Marc Bloch, "Avènement et conquêtes du moulin à eau," *Annales d'histoire économique et sociale* 7 (1935): 538–563. https://doi.org/10.1017/S0003441X00017556.

5 Adam Lucas, "The Role of the Monasteries in the Development of Medieval Milling," in *Wind and Water in the Middle Ages: Fluid Technologies from Antiquity to the Renaissance* (Penn State Medieval Studies, 2 = Medieval and Renaissance Texts and Studies, 322), ed. Steven A. Walton (Tempe, AZ: Arizona Center for Medieval and Renaissance Studies, 2006), 89–91.

6 E.g., Finbar McCormick et al., *The Archaeology of Livestock and Cereal Production in Early Medieval Ireland, AD 400–1100* (Reconstructing the Early Medieval Irish Economy EMAP Report 5.1) (Belfast: [no publisher], 2011), 39–42.

7 Örjan Wikander, *Exploitation of Water-Power or Technological Stagnation? A Reappraisal of the Productive Forces in the Roman Empire* (Scripta Minora 1983–84, 3) (Lund: Gleerup, 1984) and *Handbook of Ancient Water Technology* (Technology and Change in History, 2), ed. Örjan Wikander (Leiden: Brill, 2002), esp. his contribution in the volume. See also: Adam Robert Lucas, "Narratives of Technological Revolution in the Middle Ages," in *Handbook of Medieval Studies: Terms – Methods – Trends*, ed. Albrecht Classen (Berlin: de Gruyter, 2010), 979–980. https://doi.org/10.1515/9783110215588.967.

8 For a partial survey of the question with reference to Poland, Bohemia, and Hungary, see: Grzegorz Myśliwski, "Utilisation of Water in Central Europe (12th–16th cents.)," in *Economia e energia, secc. XIII–XVIII: atti della trentaquattresima Settimana di studi, 15–19 aprile 2002* (Istituto Internazionale di Storia Economica F. Datini, Prato. Serie 2. Atti delle settimane di studio e altri convegni, 34), ed. Simonetta Cavaciocchi ([Florence]: Le Monnier, 2003), 321–333.

9 *Monasterium autem, si possit fieri, ita debet constitui ut omnia necessaria, id est aqua, molendinum, hortum, vel artes diversas intra monasterium exerceantur, ut non sit necessitas monachis vagandi foris, quia omnino non expedit animabus eorum.* Sant Benet de Núrsia [Saint Benedict of Nursia], *Regula monachorum* (Subsidia Monastica, 21), ed. Ignasi M. Fossas (Barcelona: Publicacions de l'Abadia de Montserrat, 1997), cap. 66.

10 Stiftsbibliothek St. Gallen, Codex Sangallensis 1092. For a comprehensive analysis of the plan, see the project: Carolingian Culture at Reichenau and St. Gall. Online: www.stgallplan.org/. For the problem of the mill on the plan, see: Walter Horn and Ernest Born, "Water Power and the Plan of St. Gall," *Journal of Medieval History* 1 (1975): 219–258. The main conclusions of Horn and Born, although with some caution, are accepted by Adam Lucas, *Wind, Water, Work: Ancient and Medieval Milling Technology* (Technology and Change in History, 8) (Leiden: Brill, 2005), 208–209. https://doi.org/10.1163/9789047417224. See more recently: Ellen F. Arnold, "The Plan of St Gall and the Power of Water: An Environmental Reconsideration," *Journal of Medieval History* 51 (2025): 515–529.

11 Carl I. Hammer, "'A Suitable Place for Putting up a Mill.' Water Power Landscapes and Structures in Carolingian Bavaria," *Vierteljahrschrift für Sozial- und Wirtschaftsgeschichte* 95 (2008): 319–334. For the earliest excavated mill in the region north of the Danube: Wolfgang Czysz, *Die*

ältesten Wassermühlen. Archäologische Entdeckungen im Paartal bei Dasing (Thierhaupten: Klostermühlenmuseum, 1998) and Wolfgang Czysz, *Römische und frühmittelalterliche Wassermühlen.*

12 See the number of studies by László Koszta on the topic.

13 For mills in the Byzantine Empire, see: K. Th. Raptis, "Water as Power: Early Christian and Byzantine Watermills in Greece: Typology and Distribution," in *1st IWA International Symposium on Water and Wastewater Technologies in Ancient Civilizations: Symposium Preprint Book*, eds. A. N. Angelakis, and D. Koutsoyiannis (Iraklio: National Agricultural Research Foundation, 2006), 109–117, Sophia Germanidou, "Watermills in Byzantine Greece (Fifth–Twelfth Centuries). A Preliminary Approach to the Archaeology of Byzantine Hydraulic Milling Technology," *Byzantion* 84 (2014): 185–201. https://doi.org/10.2143/BYZ.84.0.3049180 and Curta, *Eastern Europe*, vol. 1. 323. For the possible impact on the north of the Balkans, see: e.g., Dumitru Ţeicu, *Watermill in the Banat* (Brăila: Museum of Brăila – "Istros" Publishing House, 2012), 384.

14 Biljana Arandjelović, and Ana Momčilović-Petronijević, "The Water Mills Architecture in the South of Serbia," *AR-Architecture, Research* no. 2 (2010): 59–62.

15 Petr Sommer, "The Monastery of Brevnov," and "The Monastery of Ostrov u Davle," in *Europe's Centre around AD 1000*, eds. Alfried Wieczorek, and Hans-Martin Hinz (Stuttgart: Theiss, 2000), 266, and 267, respectively. See also: David Kalhous, *Anatomy of a Duchy. The Political and Ecclesiastical Structures of Early Přemyslid Bohemia* (East Central and Eastern Europe in the Middle Ages, 450–1450, 19) (Leiden and Boston: Brill, 2012), esp. 143–169. https://doi.org/10.1163/9789004229815. For the mills in Italy at the time, see: Roberta Magnusson, and Paolo Squatriti, "The Technologies of Water in Medieval Italy," in *Working with Water in Medieval Europe* (Technology and Change in History, 3), ed. Paolo Squatriti (Leiden, Boston, and Cologne: Brill, 2000), 258–266. https://doi.org/10.1163/9789047400110_010.

16 Marek Derwich, "The Earliest Monasteries in Poland," in *Europe's Centre*, 334.

17 See Kornél Szovák, "A bencés szerzetesség korai századai Magyarországon," [The early centuries of the Benedictines in Hungary] and Szilveszter Sólymos, "Az első bencés szerzetesek hazánkban," [The first Benedictine monks in Hungary] in *Paradisum plantavit. Bencés monostorok a középkori Magyarországon / Benedictine Monasteries in Medieval Hungary*, ed. Imre Takács (Pannonhalma: Pannonhalmi Bencés Főapátság, 2001), 35–47 and 48–60 respectively.

18 Péter Váczy, "A korai magyar történet néhány kérdéséről," [On some questions of the early Hungarian history] *Századok* 92 (1958): 278, Makkai, "Östliches Erbe und westliche Leihe," 46, Makkai, "A malom mint a középkori Európa erő- és munkagépe," [Mill as the power machine of medieval Europe] in *Műszaki innovációk sorsa Magyarországon: malomipar, vaskohászat, textilipar* [The fate of technological innovations in Hungary: milling industry, metallurgy, textile industry], ed. Makkai (Budapest: Akadémiai, 1995), 33, Gyula Kristó, "A korai feudalizmus (1116–1231)," [Age of early feudalism (1116–1231)] in *Magyarország története I/1. Előzmények és magyar történet 1242-ig* [History of Hungary, vol. 1/1. Prehistory and Hungarian history until 1242] (Magyarország története: tíz kötetben), ed. György Székely (Budapest: Akadémiai, 1978), 1025. For an overview of the historiography of the earliest mention of water

mill in a Hungarian context, see: Tamás Vajda, "Okleveles adatok Árpád-kori vízimalmainkról," [Documentary data on Árpádian-age water mills] in *Medievisztikai tanulmányok. A IV. Medievisztikai PhD-konferencia (Szeged, 2005. június 9–10.) előadásai* [Studies in medieval history. Papers read at the 4th conference of medievalists, Szeged, 9–10 June 2005], eds. Szabolcs Marton, and Éva Teiszler (Szeged: Szegedi Középkorász Műhely, 2005), 193–197 and more recently, see also Tamás Vajda, "Korai bencés apátságaink vízimalmaink: a szerzetesek szerepe a technikai fejlődésben," [Water mills of the early Benedictine abbey of Hungary and their role in technological development] in *Episcopus, Archiabbas Benedictinus, Historicus Ecclesiae. Tanulmányok Várszegi Asztrik 70. születésnapjára* [Studies in honor of Asztrik Várszegi on the occasion of his 70th birthday] (METEM Könyvek, 85), eds. Ádám Somorjai OSB, and István Zombori (Budapest: Magyar Egyháztörténeti Enciklopédia Munkaközösség – Historia Ecclesiastica Hungarica Alapítvány, 2016), 25–45.

19 János Melich, "Malom," [Mill] *Magyar Nyelv* 16 (1920): 61–66 and on his footsteps in the 1950s and 1960s: István Kniezsa, *A magyar nyelv szláv jövevényszavai*, 2 vols [Slavic loanwords in Hungarian] (Budapest: Akadémiai, 1955), vol. 1. 325–327 and 342; Elemér Moór, "Malom és molnár," [Mill and miller] *Nyelvtudományi Közlemények* 67 (1965): 130–138, Dezső Pais, "Pamlény: malom? — és az mn > ml hangváltozás," [Pamlény – mill? – and the mn > ml sonant changes] *Magyar Nyelv* 61 (1965): 455–458, Antal Bartha, "Gazdaságtörténet és szavak," [Economic history and the words] *Magyar Nyelv* 65 (1969): 14–25 and Elemér Moór, "A szóegyüttesek és a csoportos szóhiányok: művelődéstörténeti tények," [Joint words and joint word omissions: intellectual historical facts] *Magyar Nyelv* 70 (1974): 173–181. See also the reflections of Egon Maróti on the debate: *Magyar Nyelv*, "A vízimalom európai elterjedésének történetéhez," [On the history of the spread of water mills in Europe] *Antik Tanulmányok* 22 (1975): 55–74.

20 Makkai, "Östliches Erbe," 44–45.

21 Walter Endrei, "Malom," [Mill] in *Korai magyar történeti lexikon (9–14. század)* [Early Hungarian historical dictionary, 9th to 14th centuries], ed. Gyula Kristó (Budapest: Akadémiai, 1994), 441. For the Carolingian rule in the region, see: Curta, *Eastern Europe*, vol. 1. 101–112 with reference to the basic works on the topic.

22 See for instance his entry "Malom," [Mill] in *Magyar művelődéstörténeti lexikon*, 14 vols [Lexicon of cultural history in Hungary], ed. Péter Kőszeghy (Budapest: Balassi, 2003–2014), vol. 7. 257–258 or Péter Kőszeghy and György Balázs, "Vízimalom," [Water mill] in *Magyar művelődéstörténeti*, vol. 12. 485–488. Also available online in an updated form at: http://mamul.btk.mta.hu/.

23 Tamás Vajda, "Régészeti adatok középkori malmainkról," [Archaeological data on medieval mills in Hungary] in *Középkortörténeti tanulmányok VI.* [Studies in medieval history, 6], eds. Péter G. Tóth, and Pál Szabó (Szeged: Szegedi Középkorász Műhely, 2010), 307–318.

24 See most importantly: *Römische und frühmittelalterliche Wassermühlen im Paartal bei Dasing Studien zur Landwirtschaft des 1. Jahrtausends* (Materialhefte zur bayerischen Vorgeschichte, 103), ed. Wolfgang Czysz (Kallmünz: Verlag Michael Laßleben, 2016).

25 For the critics of some of these archaeological finds: Vajda, "Régészeti adatok," and Vajda, "Árpád- és Anjou-kori vízimalmaink tájalakító hatása," [Water mills

as driving forces of landscape change in the Árpádian and Angevin Periods] in *Micae mediaevales II. Fiatal történészek dolgozatai a középkori Magyarországról és Európáról* [Studies of young historians on medieval Hungary and Europe], eds. Péter Jakab et al. (Budapest: ELTE BTK Történelemtudományok Doktori Iskola, 2012), 60–63.

26 Vajda, "Régészeti adatok."

27 For the hand mills, see: Szilvia Földesi, "Remarks on the Medieval Millstones at Kána," in *Archaeological Investigations in Hungary in 2004*, ed. Júlia Kisfaludi (Budapest: Kulturális Örökségvédelmi Hivatal – Magyar Nemzeti Múzeum, 2005), 54–55. For an overview of the results of the research on the village of Kána, see: György Terei, "Az Árpád-kori Kána falu," [The Árpádian-age village of Kána] in *A középkor és a kora újkor régészete Mgyarországon*, 2 vols [Archaeology of the Middle Ages and the Early Modern Period in Hungary], eds. Elek Benkő, and Gyöngyi Kovács (Budapest: MTA Régészeti Intézet, 2010), vol. 1. 81–111. For the regional differences, see: Beatrix F. Romhányi, "Spatial Transformations and Regional Differences in the Medieval Kingdom of Hungary (1000–1500)," *Hungarian Historical Review* 13 (2024): 339–360.

28 György Györffy, *István király és műve* [King Stephen and his legacy] (Budapest: Akadémiai Kiadó, 1977), 419. (For a shorter German version of the book, see: György Györffy, *König Stephan der Heilige* [Budapest: Corvina, 1988]) Vajda, "Korai bencés apátságaink vízimalmaink," passim.

29 DHA, vol. 1. 52 (emphasis mine – AV).

30 For their role in early farming practices, see: László Erdélyi, "Szent István-kori bencések hatása a föld- és kertmívelésre meg az iparra," [The role of the Benedictines in the age of St Stephen on gardening and industry] in *Emlékkönyv Szent István király halálának kilencszázadik évfordulóján*, 3 vols [Memorial volume on the 900th anniversary of Saint Stephen's death], ed. Jusztinián Serédi (Budapest: Szent István Társulat, 1938), vol. 1. 479–491.

31 *MGH Diplomata regum et imperatorum Germaniae*, vol. 3 [Henrici II et Arduini diplomata] (Hannover: Hahn, 1900–1903), 488–489 (no. 384).

32 For German-speaking areas, see above the works on Bavarian mills, for Italian mills, see: Paolo Squatriti, *Water and Society in Early Medieval Italy, 400–1000* (Cambridge: Cambridge University Press, 1998) https://doi.org/10.1017/CBO9780511583094.

33 Cf. András Kubinyi, "Királyi kancellária és udvari kápolna Magyarországon a XII. század közepén," [Royal chancery and court chapel in Hungary in the middle of the 12th century] *Levéltári Közlemények* 46 (1975): 90. The author is indebted to István Tringli for drawing my attention to this article and the issue of *Empfängerausfertigung*.

34 Berent Schwineköper, " 'Cum aquis aquarumve decursibus' Zu den Pertinenzformeln der Herrscherurkunden bis zur Zeit Ottos I.," in *Festschrift für Helmut Beumann zum 65. Geburtstag*, ed. Kurt-Ulrich Jäschke (Sigmaringen: Thorbecke, 1977), 22–56; Walter Schlesinger, "Die Hufe im Frankenreich," in *Untersuchungen zur eisenzeitlichen und frühmittelalterlichen Flur in Mitteleuropa und ihrer Nutzung*, 2 vols. (Abhandlungen der Akademie der Wissenschaften in Göttingen, Phil.-hist. Klasse 3, 115), eds. Heinrich Beck, Dietrich Denecke, and Herbert Jankuhn, (Göttingen: Vandenhoeck & Ruprecht, 1979), vol. 1. 40–70.

35 See for instance, the infamous *Capitulare de villis*. Published in: *MGH Capitularia regum Francorum* (MGH Legum 2.1), ed. Alfred Boretius (Hanover: Hahn, 1883),

82–91 (no. 32). For the problem, see: Christoph Sonnlechner, "The Establishment of New Units of Production in Carolingian Times: Making Early Medieval Sources Relevant for Environmental History," *Viator. Medieval and Renaissance Studies* 35 (2004): 21–48. https://doi.org/10.1484/J.VIATOR.2.300191, Christoph Sonnlechner, "Raumstrategische, agrarökologische und ökonomische Funktionen von Salzburger Stützpunkten im Frühmittelalter," in *Die Kirche im mittelalterlichen Siedlungsraum. Archäologische Aspekte zu Standort, Architektur und Kirchenorganisation* (Beiträge zur Mittelalterarchäologie in Österreich, 21), ed. Sabine Felgenhauer-Schmiedt (Vienna: Österreichische Gesellschaft für Mittelalterarchäologie, 2005), 119–130, and Christoph Sonnlechner, "Die Etablierung Salzburgs als Netzknoten: karolingische Kirchenstruktur, Raumstrategien und Organisation der Landnutzung um 800," in *Deutsche Königspfalzen* (Beiträge zu ihrer historischen und archäologischen Erforschung, 8: Places of Power – Orte der Herrschaft – Lieux de Pouvoir = Veröffentlichungen des Max-Planck-Instituts für Geschichte, 11/8), ed. Caspar Ehlers (Göttingen: Vandenhoeck & Ruprecht, 2007), 199–226.

36 DHA, vol. 1, 145–152 or the latest edition of the charter: István Hoffmann, *A Tihanyi alapítólevél mint helynévtörténeti forrás* [The foundation charter of Tihany as a source of onomastics] (A Magyar Névarchívum Kiadványai, 16) (Debrecen: Debreceni Egyetemi Kiadó, 2010), 23–29. For the relevant part, see 28: *Sunt igitur aratra XX cum LX mansionibus, vinitores cum vineis XX, equites XX, piscatores X, agasones V, bubulci III, pastores ovium III, subulci II, apinarii II, coquinarii II, sutores II, fabri II, aurifex I, dolatores II,* molendinarii cum molendinis II, *tornator I, vestimentorum ablutor I, cerdo I, ancille X, preter hec sunt emissarii XXXIIII cum subditis equabus, vacce C, oves septingente, porci C, apium vasa L. Preter hec ad necessaria fratrum per singulos annos constituimus de armento regali L poletros. Inter omnes namque sunt servorum ecclesie mansiones CXL.* (emphasis mine – AV)

37 *Tempore quodam dum pro cuiusdam defensione ad eundem regem properaret, in eius regionis sabulosa parte que apta usui pecudum exstat, vir Dei hospitatus est. Ubi circa mediam noctem ancilla quedam dum manibus molam circumferens triticum contereret, duritiam sui laboris cantilena demulcebat, ad quam sanctus pater expergefactus accersito ministro, quid sit, inquirit. – Passio beatissimi Gerardi (Legenda minor) /Passion of the Most Blessed Gerard (Minor legend)* [translated by Cristian Gaşpar] in *Sanctitas principum. Sancti reges, duces, episcopi et abbates Europae Centralis (saec. XI–XIII). The Sanctity of the Leaders. Holy Kings, Princes, Bishops, and Abbots from Central Europe (Eleventh to Thirteenth Centuries)*, eds Gábor Klaniczay, and Ildikó Csepregi (Central European Medieval Texts, 7) (Budapest, New York, and Vienna: CEU Press, 2022), 250–253.

38 For the early history of the monastery, see: Dana Koutná-Karg, "Die Anfange des Klosters Brevnov," in *Tausend Jahre Benediktiner in den Klöstern Brevnov, Braunau und Rohr*, ed. P. Johannes Hofmann OSB (St. Ottilien: EOS Verlag, 1993), 219–230, and also Sommer, "The Monastery of Brevnov."

39 Jiří Pražák, "Privilegium pervetustum Boleslai," in *Milénium břevnovského kláštera (993–1993). Sborník statí o jeho významu a postavení v českých dějinách* [Millennium of the history of the Břevnov monastery (993–1993). Collection of studies on its significance and role in Czech history], eds. Ivan Hlaváček, and Marie Bláhová (Prague: Karolinum, 1993), 13–24, and Kalhous, *Anatomy of*

a Duchy, 140–143. For the edition of the charter, see: CDB, vol. 1. 347–350 (no. 375).

40 Martina Maříková, "Středověké mlýny v českých zemích (Archeologické a písemné prameny)," [Medieval mills in Czech lands (archaeological data and written sources)] *Mediaevalia Historica Bohemica* 10 (2005): 89–148 and Tomáš Petráček, "K otázce datace počátků vodních mlýnů v českých zemích v kontextu změny vnímání dynamiky vývoje českých zemích 10.–12. století," [On the issue of dating the beginning of water mills in the Czech lands in the context of changes in the perception of the dynamics of the Czech lands in the 10th–12th centuries] in *Sborník k poctě Jiřího Kalfersta* [Studies in honor of George Kalferst] (Hradec Králové: Muzeum východních Čeché Hradci Králové, 2014), 249–252.

41 Oldřich Kotyza, "K počátkům vodních mlýnů v českých zemích aneb o existenci vltavských jezů a hydraulických mlýnů v Praze 10. století. Poznámky k břevnovskému aktu ze 14. ledna 993," [On the origin of water mills in the Czech lands or on the existence of the weirs on the Vltava and hydraulic mills in Prague in the 10th century. Notes on the Břevnov act of 14 January 993] in *Středověká Evropa v pohybu k poctě Jana Klápště / Medieval Europe in Motion in Honour of Jan Klápště*, eds. Ivana Boháčová and Petr Sommer (Prague: Archeologický ústav AV ČR, Praha, v. v. i., 2014), 461–499.

42 Hammer, "'A Suitable Place for Putting'," and See Chapter 3.2.2 of the present work.

43 Jan Klápšte, *The Czech Lands in Medieval Transformation* (East Central and Eastern Europe in the Middle Ages, 450–1450, 17) (Leiden and Boston: Brill, 2012), 320–321. https://doi.org/10.1163/9789004226463. For the original donation, see: CDB, vol. 1. 129–131 (no. 124).

44 CDB, vol. 1. 111 (no. 114).

45 Cosmas Pragensis, *Chronicon Boemorum* in *MGH Scriptores rerum Germanicarum* Nova Series, vol. 2, ed. Bertold Bretholz (Berlin: Wedimann, 1923), 14 (cap. 1/5). For the English translation of the chronicle, see: Cosmas of Prague. *The Chronicle of the Czechs*, eds. János M. Bak and Pavlína Rychterová (Central European Medieval Texts, 10) (New York: Central European University Press, 2020). https://doi.org/10.7829/j.ctv1z7kj6j. On the problem of millers, see: Vajda, "Korai bencés," 34.

46 For this, see in the case of Poland: Anna Adamska, "'From Memory to Written Record' in the Periphery of Medieval Latinitas: The Case of Poland in the Eleventh and Twelfth Centuries," and Ivan Hlaváček, "The Use of Charters and Other Documents in Přemyslide Bohemia," in *Charters and the Use of the Written Word in Medieval Society* (Utrecht Studies in Medieval Literacy, 5), ed. Karl Heidecker (Turnhout: Brepols, 2000), 83–100 https://doi.org/10.1484/M.USML-EB.6.090 708020500030500007070102 and 133–144 https://doi.org/10.1484/M.USML-EB.6.09070802050003050007070102 respectively.

47 Myśliwski, "Utilisation of Water," 324. The 1149 charter referred to by Myśliwski was not at my disposal when writing this chapter. Earlier, see: Maria Dembińska, *Przetwórstwo zbożowe w Polsce średniowiecznej (X-XIV wiek)* [Grain processing in medieval Poland (10th–14th centuries)] (Wrocław: Zakład Narodowy im. Ossolńskich, 1973), esp. 63–72.

48 *Sed sciendum, quia in diebus illis erant hic in circuitu aquatica molendina valde rarissima, unde dicti Bogwali Boemi uxor stabat sepissime ad molam molendo.* Edited in: *Liber fundationis claustri Sancte Marie Virginis in Heinrichau (Księga*

henrykowska), ed. Roman Grodecki (Wrocław: Muzeum Archidiecezjalne we Wrocławiu, 1991), 47 (cap. 113). For an English translation of the text, see: Piotr Górecki, *A Local Society in Transition. the Henryków Book and Related Documents* (PIMS Studies and Texts, 155) (Toronto: Pontifical Institute of Mediaeval Studies, 2007), 139. See also on mills here: 40–43.

49 *Vita s. Hedwigis (legenda maior)* in *Monumenta Poloniae Historica*, vol. 4, ed. Aleksander Semkowicz (Lwów: Gubrynowicz i Schmidt, 1884), 556.

50 On the spread of mills in the later medieval period, see: Dembińska, *Przetwórstwo zbożowe w Polsce*. For the Teutonic areas, see: Rafał Kubicki, *Młynarstwo w państwie zakonu krzyżackiego w Prusach w XIII–XV w. (do 1454 r.)* [Milling Industry in the State of the Teutonic Order in Prussia in the 13th–15th centuries (to Year 1454)] (Gdańsk: Wydawnictwo Uniwersytetu Gdańskiego 2012).

51 Lucie Galusová, "Vodní mlýn jako objekt archeologického výzkumu," [Water mills as objects of archaeological research] *Archaeologia historica* 40 (2015): 267–293. https://doi.org/10.5817/AH2015-1-15.

52 *Codex diplomaticus Regni Croatiae, Dalmatiae et Slavoniae. Diplomatički sbornik Kraljevine Hrvatske s Dalmacijom i Slavonijom*, 2 vols (Monumenta historica Slavorum meridionalium, 2–3), ed. Ivan Kukuljević Sakcinski (Zagreb: Ex officina societatis typographicae, 1874–1875), vol. 1, 121–122 (no. 138), 154–155 (no. 189), 170–176 (no. 209) and 181–182 (no. 220). On the mills of Salona, see: Lovre Katić, "Solinski mlinovi u prošlosti," [The ancient mills of Salona] *Starohrvatska prosvjeta* Serija III. 2 (1952): 201–219. On mills in Croatia in the medieval period: Marin Knezović, "Voda u hrvatskim ranosrednjovjekovnim ispravama," [Water in the Croatian early medieval documents] *Ekonomska i ekohistorija* 3, no. 1 (2007): 42–46.

53 Ранко Финдрик [Ranko Findrik], "Увод у проучавање старих воденица [Uvod u proucavanje starih vodenica]," [Introduction to the study of ancient mills] *Саопштења [Saopstenje]* 15 (1983) 95–117 and Biljana Arandjelović and Ana Momčilović-Petronijević, "Arhitektura vodnih mlinov v južni Srbiji," [The architecture of water mills in Southern Serbia] *AR Arhitektura, raziskave* 2 (2010): 59–62.

54 Țeicu, *Watermill*, passim.

55 Lynn White, Jr., *Medieval Technology and Social Change* (Oxford: Oxford University Press, 1962).

56 See for instance from 1061: DHA, vol. 1. 170–174. On the criticism of the heavily interpolated charter: Bernát Kumorovitz L., "A zselicszentjakabi alapítólevél 1061-ből, 'Pest' legkorábbi említése," [The foundation charter of Zelicsszentjakab from 1061 and the earliest mention of 'Pest'] *Tanulmányok Budapest Múltjából* 16 (1964): 43–83.

3

WATERSCAPES AS CONFLICT ENVIRONMENTS IN LATE MEDIEVAL HUNGARY

This central chapter is dedicated to discussing the main research problems of the book, namely the various kinds of conflicts that arose from the use of water, based on medieval legal evidence from the Kingdom of Hungary. The separate subchapters examine different landscapes and water bodies—from small streams to major waterways—to provide a comprehensive understanding of conflicts in rural areas, urban settings, and the vicinity of towns with special economic roles, such as mining towns. It is argued throughout these pages that, from the turn of the thirteenth century, nearly every waterway in the Kingdom of Hungary's inhabited areas was intensively utilized in local economies, resulting in an increasing number of resource conflicts. These conflicts varied significantly across river types. Major rivers, as discussed in the introduction, had to meet needs that differed from those of smaller watercourses. While the former were exploited for transportation, they also supplied industrial and drinking water, served as a source of fish, and, to a lesser extent, provided energy. Smaller rivers and streams played a minimal role in transportation overall, but were critically important in ensuring an even distribution of water-driven grain mills and fishponds.

Different needs necessitated different arrangements along rivers. In the first subchapter (Chapter 3.1), the book examines the establishment of water-use norms for two competing mills along the same waterway, as well as conflicts between landlords whose landholdings were adjacent to the waterway where the mills were constructed. As discussed in Chapter 2, it was likely not until the thirteenth century that lawsuits arising from competing demands for water for powering water mills became more prevalent. Earlier, there had been a necessity for negotiations between landlords who utilized the same waterway. However, it is probable that from the thirteenth and fourteenth

DOI: 10.4324/9781003596905-3

centuries onwards, the different actors felt it essential to secure their rights to use various bodies of water. Based on legal evidence, the first subchapter examines the crystallization of norms governing mill construction in the medieval countryside. It is argued that from the fourteenth century onward, the previously established norms and practices evolved into widely accepted customary laws.

Chapter 3.2 examines the conflicts surrounding water use in urban environments. As major settlements developed in Central Europe, water has been, and remains, one of the most critical resources for local economies.[1] Due to significant population concentrations and the resulting increased demand for resources, more efficient water use became necessary, along with closer cooperation among the various actors using a particular water body. In many ways, towns serve as laboratories for a variety of resource conflicts. This subchapter highlights various cities and waterways in the Kingdom of Hungary to demonstrate the conflicts that arose and how they were resolved locally. Chapter 3.3 discusses water use in a specific town type: mining towns and their surroundings. These towns across Central Europe were established from the thirteenth century onward, primarily by German-speaking individuals (along with Czech, Slovak, and potentially Polish people), many of whom were engaged in prospecting and mining precious metals and other ores. Privileges, some concerning access to water for fishing and milling, played a crucial role in the foundation of these settlements. This subchapter argues that water-related economic privileges were fundamental to the life of such towns and the villages settled under the authority of burghers from mining towns. It aims to illustrate how water use was organized in these towns, where, in addition to processing grain for the rapidly growing populations, waterpower was essential for several other industries, including ore-crushing and timber-sawing for mine shafts. A dense network of waterways existed in the hilly and mountainous areas of Transylvania and Upper Hungary, where most of the mining settlements in the medieval Kingdom of Hungary were located. Despite the dense water network, legal evidence indicates that water was a highly contested resource in these environments.

Chapter 3.4 examines another aspect of water-related conflicts that has been largely overlooked: the role of water in delineating boundaries. Medieval estate boundaries were primarily defined by natural features such as hills, ditches, forests, and meadows. In some cases, trees were marked, or small mounds were constructed to delineate estate boundaries. However, at first glance, none of these methods seems as definitive as boundaries established by rivers and streams. A closer examination of customary law collections and court cases concerning estate boundaries reveals that bodies of water can also be a source of conflict. This subchapter analyzes thirteenth- to sixteenth-century charters from the medieval Kingdom of Hungary, focusing on land ownership issues arising from riverbed changes that transferred property to the opposite shore. It is argued that similar legal disputes were common in medieval Hungary. Although a norm was established in the late medieval period to address such issues, it was often ineffective in resolving problems arising from shifting waterways.

All subchapters are dedicated to specific water-related disputes and provide a general overview of the problem across a wide variety of landscapes and waterscapes in medieval Hungary. Of course, many other conflicts also arose. Still, the forthcoming discussions reveal that water management across various bodies of water involved a complex control system that could ultimately function only by applying certain legal principles and, equally importantly, through continuous negotiations among different actors along the same waterway. The chapter aims to demonstrate that, due to the nature of water as a substrate and the immediate effects of interventions upstream and downstream, conflicts arising from water far surpassed simple property rights issues.

3.1 Legal Regulations and Practice of Mill Construction in Medieval Hungary

As argued in Chapter 2, since the Árpádian period (i.e., from the time of early Hungarian statehood onward), hydraulic works have been undertaken on various scales, similar to those in Poland and Bohemia.[2] These constructions affected both smaller rivers and, over time, more substantial water flows. By the thirteenth century, in the densely populated areas of the Kingdom of Hungary, rights to streams and smaller rivers were increasingly contested, and norms governing their exploitation were well defined. As shown in the Introduction to this book, scholars have devoted some attention to the development of these norms, although many questions remain only partially resolved. What rights did landowners have to rivers flowing through their properties? How could a landowner restrict a water source on their land? How could one manipulate a watercourse to generate income? As discussed in Chapters 1 and 2, one of the primary motivations for manipulating bodies of water in the Middle Ages in the study area was to harness energy for grinding grain and utilizing waterpower for other industrial activities. However, these mills significantly affected river flow upstream and downstream, owing to dams and other structures that impeded water flow. This chapter focuses on the period during which well-defined norms for constructing such obstacles to rivers and streams were established, leading to the development of customary law. It will primarily analyze the term *locus molendini*, which, despite its frequent occurrence in Hungarian legal sources and, to a lesser extent, in other European legal literature, remains underexplored in scholarship.

Searching for a definition of the term *locus molendini* in the scholarly literature is surprisingly unproductive, despite its frequent occurrence in legal evidence from Hungary and elsewhere.[3] In many cases, it appears in lists of pertinences, similar to those discussed in Chapter 2. The principal cartularies of medieval Hungarian legal literacy contain hundreds of additional references in which *loca molendinorum* appear as subjects in legal transactions. Tringli, in his work cited in Chapter 1, provides a rather simple definition for the term, claiming: "mill site *(locus molendini)* denoted a place where a water mill once stood, and by rebuilding or restoring this building, milling could once again take place there."[4] Accepting this definition, most studies in the scholarship build on this understanding of the term.[5] A slightly different perspective emerges in a paper discussing the mills mentioned in the cartulary for a place of authentication (*locus credibilis*).[6] In this work, Tamás Kőfalvi suggested that *locus molendini* referred to a land plot somewhat similar to a modern construction site. He reached this conclusion by studying a charter issued by the convent of Pécsvárad (southern Hungary) in 1469. Based on the charter, he suggested that *locus molendini* was clearly defined in both legal and physical–territorial senses. However, the document does not, in fact, or even self-evidently, support this argument.[7] The situation becomes even more confusing when considering the expression "abandoned mills" (*molendinum desertum*), which also appeared in several late medieval documents, albeit significantly less frequently than *locus molendini*.[8]

In addition to the more recent works cited, earlier legal and technological histories also addressed the terminological problem. Alajos Degré, in his previously mentioned work, already noted the need to analyze the term more closely to determine whether *locus molendini* denoted an actual plot suitable for erecting a mill, or a legal privilege permitting a mill to be built at a specific location.[9] Although he did not provide a definitive answer, he proposed that this term denoted a legal claim or authorization rather than a physical space. In a short monograph on the medieval hydrography of Zala County (western Hungary), József Holub suggested that the term might not be directly connected to the existence of a mill that had previously operated on a site, but rather denoted a certain piece of land, the road leading to the land, and the possible site of a mill building itself.[10] He also made an important point: although the actual site of a former mill was denoted by l*ocus molendini*, the documents, in many cases, specifically refer to the existence of a lost building at the mill site (for instance, *locus molendini antiqui)*.[11] Recently, Tamás Vajda attempted, on the one hand, to follow up on the crystallization of the term as a legal technical term by the thirteenth century; on the other hand, he demonstrated that the medieval documentary evidence allows for various interpretations of the term (which somewhat contradicts the notion of the crystallization of the legal term by the late Árpádian period). He argues that, in some instances, the term refers to the actual locations of earlier mills (*genitivus possessivus*)—as previously suggested by Tringli—or a plot or site suitable for erecting a mill (*genitivus explicativus*).[12] Vajda proposes a third meaning, referring to the actual territory on which the mill building stood, unrelated to its condition as a building and concerned solely with the right of possession. Earlier, he also proposed that mill sites are only referenced when the owner of the broader area around the mill differs from the owner of the mill's specific location.[13] Finally, András K. Németh, an archaeologist and landscape historian, argued that "for some reason, the plots suitable for erecting mills were kept in mind, even if then, no building was there on the plot." He leaves the question open, but this subchapter contends that a precise answer can be given to the "why", which is essential in understanding the formation of customary law on using waterways in medieval Hungary.[14] Beyond Hungary, this issue has arisen in German and French historiography over the past decades, but scholars have struggled to explain its usage.[15] Finally, this question also appears in Romanian historiography, utilizing written evidence from the eastern part of the medieval Kingdom of Hungary, Transylvania. Research has argued that mill sites are well-defined topographic features but do not necessarily include buildings.[16] Given the surprisingly high frequency of locus molendini in Hungarian legal literacy, more attempts have been made there than in other historiographies to understand its precise meaning. However, as evidenced above, historians reached different conclusions, and the rationale for the frequent use of this term remains unclear.

3.1.1 Locus molendini: A Place, a Right, a Claim?

It is impossible to establish the regulation of water use based solely on the surviving royal decrees from medieval Hungary. In the most significant compilations of medieval laws in Hungary, such as Zakariás Mossóczy's *Corpus Juris Hungarici* (first printed in 1584), one finds virtually nothing regarding water-related rights or regulations governing water construction. However, the absence of such rules does not imply that they did not exist in medieval Hungary. By examining an early fifteenth-century lawsuit, István Tringli demonstrated that by that time, an established custom already existed concerning water control in Hungary. He argued that the principle was simple, as expressed in a document dating to 1401: a new mill had to be constructed in a way that its placement did not harm another mill's interests (*molendina taliter debeant edificari, quod preiudicium uni per aliud non inferatur*).[17] In this context, "another" must be understood as the presence of a similar construction, meaning that a new mill should not cause harm to an older, existing one. Tringli further demonstrated that over time, the concept of "loss" (*preiudicium*) was broadened to encompass any economic loss suffered by another landlord's estate, including a reduction in fish catch and damage to ploughlands due to floods. Indeed, in the late medieval period, there were at least as many cases in which the claimed losses were related to ploughlands as to other mills.[18] When dams failed due to flooding or poor construction, they could not only cause significant damage to downstream mills but also inundate the surrounding riverbanks. Consequently, the height of these mill dams became a common source of debate and conflict during the Middle Ages.[19] Of course, this issue was not unique to Hungary. Similar regulations have been preserved, for instance, from the Holy Roman Empire and Lithuania.[20]

Aside from charter evidence, the previously cited summary of medieval customary law in Hungary, authored by Stephen Werbőczy, judge royal and palatine in the early sixteenth century, remains a key source for addressing this question. His work, the so-called Tripartitum, is a significant collection of unwritten yet by the early sixteenth century established common-law elements in Hungary. The fate of this work is undoubtedly interesting. It never became a formally sanctioned legal source; however, until the nineteenth century, it remained the most important reference point in legal disputes.[21]

The primacy of the benefit established earlier—whether material present or not—is also emphasized as a general principle in this work: "First, in respect to time, as has just been discussed: since the other person's privilege was given earlier, it cannot be abolished by a later one, not even through a derogatory clause."[22] This principle was virtually never disregarded. The question is what happened when two landlords simultaneously planned to build mills or other water-related structures near each other. A written agreement from 1461 preserved a similar case. The solution appears to have been logical: the parties agreed to build a mill or a fishery on the same section of the Sárosd River (on the border between Zala and Somogy Counties in western Hungary). They concurred that if either party caused damage to the other's estate (by flooding the other's mill or lands) with their construction, they would not sue each other.[23] This example indicates that where there was no primacy in construction time, the risk was shared by landowners.

To the best of our knowledge, the earliest case in which the role of time in such disputes was clearly articulated is the aforementioned court resolution from 1401. The following pages demonstrate that this principle may have been applied much earlier in medieval Hungarian pragmatic literacy. In doing so, the term *locus molendini* is discussed.

3.1.2 *Locus Molendini in Early Literacy: First References and Terminological Unifications*

It is worth considering when and in what form the notion of "mill site" enters legal literacy. This is important because the earliest references—probably not by chance, as will be noted below—were preserved in highly interpolated or forged charters. These early references to mill sites, with a few exceptions, however, employ terminology other than *locus molendini*. The terminology usually clearly denotes plots on which a mill formerly stood, or rather, a plot of land that would be suitable for erecting a mill.[24] This is similar to tenth- and eleventh-century literacy in the Holy Roman Empire. Although it occurs only once, the term also appears in a charter issued by the Eastern Frankish king, Arnulf.[25]

The first reference to the term in Hungarian legal sources appears in the interpolated foundation charter of the abbey of Zselicszentjakab (part of present-day Kaposvár) from 1061, which is discussed in Chapter 2 on the spread of water mills in the medieval Kingdom of Hungary.[26] There is no need to recapitulate the rather exhaustive historiography of this document here, but it is crucial for the current discussion that the document was transcribed and extracted multiple times from the eleventh to the seventeenth century (1257, 1374, 1377, 1432, seventeenth century).[27] In the document, Otto, *ispán* of Somogy County—and later palatine of Hungary[28]—among his other estates, granted the newly founded abbey "a mill with mill stone and the plot that belongs to the same mill in Kesis village" (*in villa Kesis molam cum molario et terram ad eandem molam pertinentem*).[29] Scholars have suggested that the term *mola* is an early, archaic form of the later *locus molendini*. This earlier term ceased to be used by the thirteenth century.[30] By the late thirteenth century, although the terminology was not fully unified, the most frequently used term for similar arrangements was *locus molendini*. However, right in the middle of the thirteenth century, when research indicated that the document was first transcribed (1257), several other terms for mill sites were still present in legal literacy. For instance, in one of the most significant sources produced by the Benedictines in Hungary at that time, the so-called Albeus-conscription, a complete register of the estates and people who lived within the territory of the Benedictine abbey of Pannonhalma, mill sites are sometimes referred to as *locum aptum ad molendinum*.[31] Documents issued by the cathedral chapter of Győr used the terms *locus molendini* and *terra molendini* interchangeably.[32] The section of the charter of Zselicszentjakab that includes the above reference likely dates to the eleventh century. Nevertheless, the earlier term's relationship to the one(s) used later for mill sites is somewhat questionable, as the transaction solely concerned a piece of land with a mill situated on it.

An estate conscription dated not long after the foundation charter of the abbey of Zselicszentjakab is also significant to the present discussion. The document in question is a forged charter registering the estates of the bishopric of Veszprém (in western Hungary), issued by King Ladislas I on April 29, 1082.[33] It has been established that the document was forged between 1317 and 1327, when the bishopric of Veszprém likely produced the charter to secure its property rights.[34] Although the reference to *loca molendinorum* in the text is not among the earliest mentions of mill sites, it is significant that, in the early fourteenth century, the text's forgers included references to mill sites. They once used a rather unusual form, referring to *vacuum locum molendini*, a term seldom noted in Hungarian legal literature.[35]

The earliest authentic reference to the mill site as a *locus* dates to 1138 and is preserved in the highly detailed estate conscription of the collegiate church of Dömös (in central Hungary, not far from Budapest) ordered by King Béla II. According to the document, there was a "mill and two mill sites" in the yet-to-be-localized village called Uten (*est enim ibi molendinum et duo loca molendini* [sic!]).[36] Although the original charter has not been preserved, the document in which it was transcribed, dated January 20, 1329, is unquestionably authentic.[37] Similarly authentic is a charter of Stephen, son of Adrian, dated to 1158, in which he donated his lands called Barátka (present-day part of Levice) and Szántó,[38] including a *locus molendini* located by the River Kompa in the territory of Szántó, to the nearby Benedictine abbey of Hronský Beňadik after the deaths of his brother Euzidinus and his nephews.[39] There is no basis to question the document's authenticity. It was not until the early thirteenth century that the term appeared again in legal literacy, and even then, it was not used in the same ways as from the fourteenth century onward.

In 1211, mill site was used again in a different form in the conscription of the belongings of the Benedictine abbey of Tihany. The description of the borders of Lusta (Lustah, which later became part of the village of Fok, now called Siófok) estate begins as follows:

> The River called Foc [now called Sió], which flows out from the Bolotyn [Lake Balaton] at its effluence, belongs to the church [the abbey of Tihany]. From this point, it [the river] flows south, reaches the village of Lusta, and encircles the island – thus also fully belonging to the aforementioned church – forming a place suitable for mills.[40]

Therefore, the River Fok encircled an area, facilitating the construction of mills. The Benedictines likely documented this fact for their use. Two things seem clear from the source. First, at that time, no mills stood at this location, and there is no evidence that a mill had ever been present at the site. Second, the term does not appear to hold any legal relevance. The document appears to have done little more than record a specific location, which is typical of similar boundary descriptions.

In 1214, King Andrew II confirmed the donations of Boleslaus, bishop of Vác, to the convent of Leles. Among the numerous donations were estates and their benefits, including two mill sites. Although the document uses the term *locus molendini*, similar to the two twelfth-century instances discussed above, it is again a forgery, a fact that will be shown later to be of major importance.[41] However, references to two mill sites are preserved in the undoubtedly authentic estate conscription of the collegiate chapter of Vasvár (western Hungary) from 1217, where the same Latin term is employed.[42] A mill, a forest called Rögös, and additional mill sites were registered in the confirmation of the donations of the Cistercian abbey of Klostermarienberg from 1225, at Szomód (*Zumuld*) estate, next to the manor house of the abbot of Tata.[43] In 1237 and again in 1239, King Béla IV confirmed that the archbishopric of Esztergom owned the Šahy estate. The estate included a mill site by the River Ipoly.[44] All of these documents used the term that later became more common: *locus molendini*.

However, this does not imply that *locus molendini* had become the exclusive term for referring to mill sites. As part of a much-studied lawsuit between the Holy Virgin Chapter of Fehérvár (today's Székesfehérvár) and the abbey of Pannonhalma, the estates of the former were conscribed. The document consistently uses the term "place for a mill" (*locus ad molendina*), rather than *locus molendini*, when describing arrangements similar to those mentioned in earlier lines of the document.[45] In 1231, Béla, then junior king (the future Béla IV), authorized *ispán* Isaac to build four mills on the River Rábca at Mihályi (in western Hungary) on the four designated places that were "suited for putting up mills" (*locorum factam ad opus molendinorum*). In return, Isaac had to promise to address losses caused by foreign raids, including those from the Holy Roman Empire, as the area marked the frontier between the two polities.[46] In 1239, a testimonial issued by the palatine of Hungary, Dionysius of the Tomaj clan, resolved a dispute over land transactions in Szatmár County. Among the lands in question was "a mill site with a house plot that was suitable to put a mill" (*locum molendini cum loco curie apto ad statum molendini*).[47]

Based on this analysis of documents referring to mill sites up to the mid-thirteenth century, it can be concluded that, even when the term *locus molendini* appears in the sources, it neither prevails in the surviving evidence nor denotes a specific legal arrangement. The reference to a mill site in the 1061 endowment charter of the abbey of Zselicszentjakab, as well as the reference in the estate conscription of the abbey of Tihany, almost certainly signifies something other than a place where once a mill stood, but rather indicates plots suitable for mill construction, given the hydrography. The meaning is neither clear in the 1138 reference in the charter of Dömös nor in the donation of Stephen, son of Adrian, from 1156. The most relevant question at this point is whether thirteenth-century references to *locus molendini* suggest the existence of earlier standing mills. The question can only be answered affirmatively if a reference to a mill predating can be found in another document referring to a mill site on the same estate. (Of course, even in that case, the two are not necessarily the same.) Additionally, it may also be indicative if the document mentions a mill site that refers to a previously standing mill or elements of a mill's infrastructure, such as a mill race, a dam, or a mill road.

The author of this book was unable to identify any earlier document that described a functioning mill at any of the settlements that later referenced mill sites during the Árpádian period. Of course, there are contrasting examples in which mill sites are later described as functioning mills. For instance, by 1237, two mills were operating in the village of Megyehíd, which had two mill sites (*duobus locis molendinorum*) in 1217.[48] There are also instances in which no mill was built on a mill site for an extended period, and documents separated by considerable time intervals refer to a *locus molendini* at the exact location. The village of Rum in Vas County provides such an example.[49] A document from 1250 noted a mill site in Rum, although the charter was proven to be a forgery. Nonetheless, an authentic document from 1263 that introduces the new landlord of Rum to his property also mentions a mill site, as does a document from 1274, when the then depopulated Rum was donated by King Ladislas IV to a certain Doroszló, son of Doroszló, along with its mill site by the River Rába.[50] There are two additional Árpádian-period charters worth noting that reference earlier mill infrastructure documented at mill sites. The first is one of the earliest surviving documents issued by the place of authentication of the abbey of Pécsvárad in 1256. During the perambulation of Janus estate, one of the boundary markers was an "old mill site" (*vetus locus molendini*).[51] In light of this practice, references to "old" or "ancient" (vetus, antiquus) mill sites usually denote the location of a once-operating mill. The second significant document is also a perambulation charter. In registering the borders of Koksó (*Koxo*) in 1262, one of the noted boundary markers is a broken millstone lying next to a mill site.[52] It was common, until modern times, to discard the millstone when it was no longer in use, and large stones were frequently used as boundary markers, suggesting that, in this case, the *locus* may indeed have marked a site where an earlier mill stood.[53] However, this document is also likely a late medieval forgery.[54]

The early references to mill sites raise several points worth considering. It is clear that, despite the varying forms of the term *locus molendini* in Árpádian-period sources, there are no grounds to suggest that up to the mid-thirteenth century, references indicated previously existing, lost, or abandoned mill buildings. Up to the mid-twelfth century, as demonstrated in Chapter 2, there were few references to water mills altogether. Their numbers likely remained low until the late Árpádian period, not only in Hungary but also elsewhere in Central Europe. When documents issued before the mid-thirteenth century mention mill sites, they probably do not refer to previously existing mills but rather to places suitable for erecting a mill. Their appearance in documents may have served to identify these sites and mark the value of a particular estate or plot.

It is clear, even from the partial survey above, that many of the early references to mill sites where the term *locus molendini* is used are found in forged charters. In most of these charters, various reasons suggest they were forged; however, the use of this term may also indicate a later origin, as it rarely appears in charters in this form before the end of the first third of the thirteenth century.[55] The most relevant question for the present investigation however is why numerous forged charters mention mill sites in different locations. In the next subchapter, this issue is revisited by examining mill sites during the period when the term *locus molendini* became prevalent and carried a specific meaning.

3.1.3 Terminological Uniformity: Uniformity in Meaning?

By the end of the Árpádian period, the term for mill sites had become almost exclusively *locus molendini*, a situation that remained unchanged until modern times. The question arises as to how much homogeneity in terminology signifies unity of meaning. The following lines discuss the different meanings that emerge in the legal evidence from fourteenth- to sixteenth-century Hungarian sources.

One of the meaning groups revealed by examining evidence from the Angevin period onward comprises documents in which the term refers to a place where a mill once stood, or at least to a location where earlier construction—either mill race digging or mill-house construction—had been carried out.[56] As noted above, this is how the term *locus molendini* is typically identified in scholarly research. Some documents, such as land divisions (*litterae divisionales*) or donations (*litterae donationales*), refer to all the elements comprising the infrastructure of a mill located on an estate. However, since the mill was not in operation at the time the document was issued, the building complex was referred to as a mill, with a mill site.[57] In several cases, though, only a few prior works testify to the former landowner's plans to build a mill at the site, such as a mill race to divert water toward the mill wheel(s).[58] Sometimes, entire rivers were diverted into a new channel to supply water for milling, but the work stopped after the new channel was created. For instance, in 1467, when the Pauline monastery of Gombaszög (present-day Slavec in Slovakia) exchanged properties, the property they gave away included several mill sites along the "new bed" of the River Hernád (for this bed and the later-built mills, *see* Figure 3.1). However, no mill operated along this artificial riverbed before it was transferred to a new owner in 1467.[59]

Documents that reference abandoned (*locus molendini deserti*) or old (*locus molendini antiqui*) mill sites should also be interpreted in accordance with the most logical and widely accepted meaning of the expression *locus molendini*. One case is particularly noteworthy because it shows that for contemporaries, a lost mill held a different meaning than a mill site. In 1358, King Louis I (the Great) granted lands to the Pauline abbey of Kékes (present-day Pilisszentlászló in central Hungary). The monks received

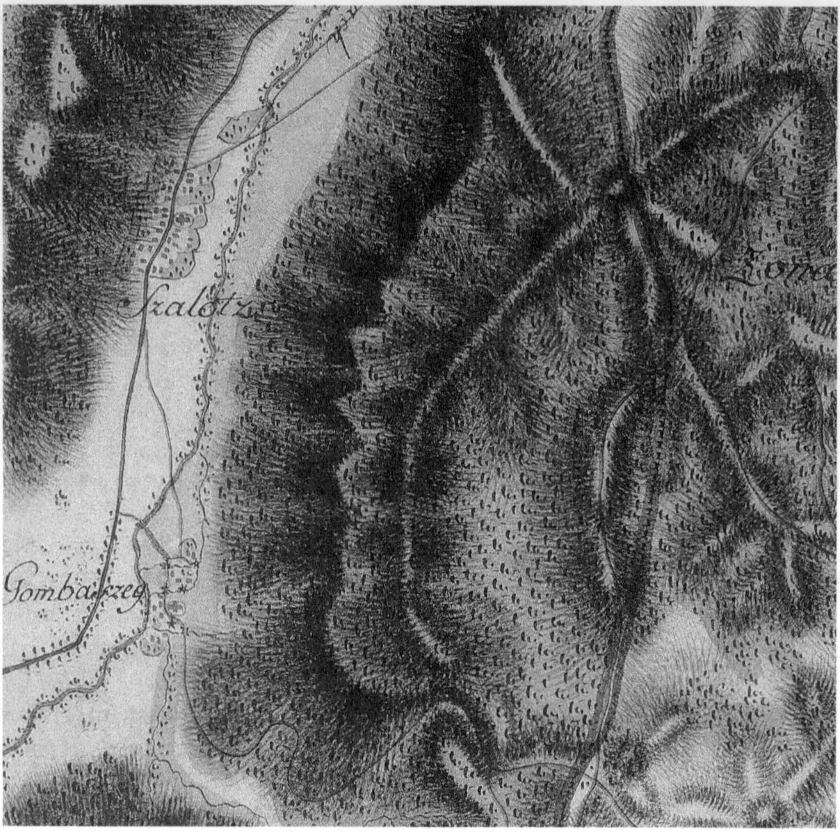

FIGURE 3.1 Gombaszög with the location of the Pauline abbey marked with the red cross and the water mills along the artificial riverbed of the Hernád (First Military Survey).

Source: Vienna, Österreichisches Staatsarchiv, Kriegsarchiv KPS Kartensammlung B IXa 527. XX.8.

lands surrounding their newly founded monastery, including the stream bearing the same name as the site of the abbey—Kékes—a lost and ruined mill (*molendinum patiens ruinam et desolatum*) as well as another mill site (*locum alterius molendini*). There may have been a reason the two pieces of land were described differently.[60]

In other cases, toponyms for mill sites suggest that functioning mill buildings once stood on those sites. Like many other instances, the locations were referred to as someone's molna (the old Hungarian term for mills), such as *Conthmolna* ("the mill of Conth") or *Myskeemolna* ("the mill of Myske").[61] To some extent, references to mill sites with a specified number of wheels are similar. The first such reference dates to 1216, from a supposedly forged confirmation charter. This document mentions a mill site with a single wheel.[62] A confirmation charter issued by the convent of Pécsvárad from 1469 refers to a mill site with three or four wheels,[63] while another from 1511 lists an abandoned mill site called *Zekemolna* ("the mill of Zeke") with two wheels.[64] These precise wheel counts likely refer to the number of wheels used by the previous mill at the same site (and make the 1216 document mentioned above even more likely to be forged).

The last group of documents that refer to previous mills or earthworks, in some form, consists of those that, despite not mentioning any currently standing infrastructural elements, refer to rebuilding a mill at a mill site.[65] One additional group of charters is worth considering. Several documents reference *locus molendini* that were either suitable for constructing certain types of mills or where specific types of mills had previously stood on the site. Without delving into the details of the technological history of mills, the Hungarian legal material and the customary law collection of Werbőczy, discussed above, refer to two fundamentally different types of mills: undershot (*molendinum subtus currens*) and overshot (*molendinum desuper volvens*).[66] The types of mills are associated with waterways of distinct characteristics. Some of these documents, although they do not mention pre-existing mills, describe *locus molendini* for either an overshot or an undershot mill. This reference may imply that either local geographic conditions only allow for a specific type of mill—correlated with the value of the piece of land[67]—or that the previous building on that site was either an overshot or an undershot mill.[68]

Cases that lack reference to the physical reality of any prior construction work will be briefly discussed below. The most evident examples are donations and other legal documents that reference empty mill sites. Such examples include the forged estate-conscription of the bishopric of Veszprém from 1082, mentioned above, which is relevant for terminologies from the 1310s and 1320s. However, the document does not contain the only early reference to empty mill sites. For instance, a document from 1279 issued by the chapter of Esztergom uses the same term (*locis molendinorum adhuc vacuis*).[69]

Another document group, where the term has a different meaning, refers to plots suitable for erecting mills. As previously discussed, in the twelfth to thirteenth centuries, although different expressions were used, this meaning of the term prevailed. In these instances, the hydrographic and/or legal conditions provided a basis for building a mill. Such references can be found in lease agreements, as it was in the interest of both the landlord and tenant for the contracts to describe the detailed rights associated with the land.[70] A plot where a mill could be built for personal use was, of course, much more valuable than a comparable piece of ploughland. Thus, it was essential for the tenant to have the right to use the water, and naturally, the landlord could demand higher rent if there was a possibility of constructing a mill on that property.

Some cases in this group certainly deserve further attention. The first to be discussed is a lease agreement between a burgher from the medieval mining town of Pukanec and the Benedictine abbey of Hronský Beňadik. In this contract, along with several similar ones examined in more detail in Chapter 3.3.5, dated 1345 (or 1346), a burgher leased a mill site by the Büksavnica Stream, a relatively insignificant body of water flowing near Nová Baňa. According to the agreement, the mill site the burgher acquired was the nineteenth (!) such location found upstream from where the Büksavnica flows into the River Hron. While the contract does not provide information on the precise location of this mill site within the less than 20-kilometer stretch of the stream, it is improbable that nineteen (or more) mills could function without disrupting each other's access to water.[71] In another contract issued by the same institution, the abbey leased a mill site adjacent to the even smaller nearby Verence Stream, the sixth such site from its confluence.[72] The folios of the First Military Survey discussed in Chapter 1.2.2 systematically depicted mills and mill sites alike.[73] Even though, as has recently been demonstrated, mill sites typically remained visible for centuries, only five mills (either abandoned or operational) existed along these two streams in the late eighteenth century (*see* Figure 3.2).[74] It is more plausible that the Benedictines designated locations they considered suitable for constructing a mill, some of which they subsequently leased. The Benedictines were involved in similar leases.[75] References to plots suitable for mills are also found in documents beyond those from Hronský Beňadik and Pécsvárad. In one example from 1402, the nobles of the market town of Szakácsi (today's Nagyszakácsi) granted a mill site and one and a half acres of land to the local Pauline abbey for building a mill. The one and a half acres of land were necessary because the waters of the small Horohalya Stream had to be diverted towards the mill site, as the channel from the stream would run through these lands. The context suggests that no prior work had been done to provide water for a mill at that location.[76]

For this book, the most intriguing interpretation of *locus molendini* may be documents referring to mill sites of free choice. The first similar example may have been preserved in a land division charter from 1345. In this document, two brothers from the village of Berencs (present-day Rétközberencs in northeastern Hungary), the sons of Stephen, son of Keled—Dominic and Ladislas—divided the estate they had inherited. Under their agreement, in addition to certain plots, Dominic had the right to select a site on the river Sebesér (Sebusher, a vernacular Hungarian placename meaning "fast current") for a mill.[77] In 1445, John of Zredna (Vitéz), the famous humanist bishop of Oradea, granted one of his men, Nicholas, a piece of land called Kochuba to settle peasants there. Along with these lands, Nicholas received a

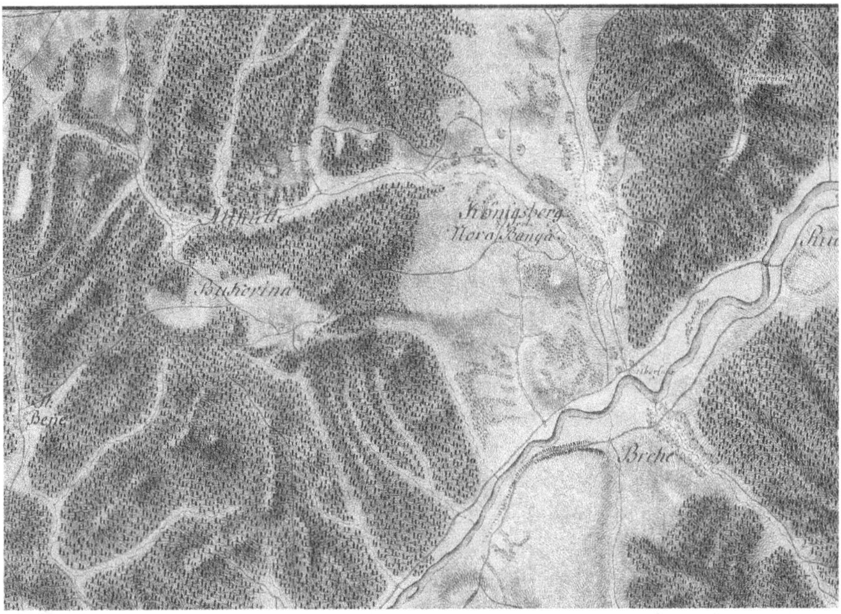

FIGURE 3.2 The surroundings of Nová Baňa with the Büksavnica (the one running west–east) and Verence Streams (running north–south) on the First Military Survey, with the two mill sites on the latter.

Source: Vienna, Österreichisches Staatsarchiv, Kriegsarchiv KPS Kartensammlung B IXa 527. XII.11.

mill site by the Fekete-Körös River. However, the exact location of this future mill was not specified; the beneficiary had the right to choose it.[78] Finally, in 1478, a charter concluding a lawsuit over some estates in Heves County left a certain Ladislas of Szőlős (*Zewles*) in possession of a mill by the River Tarna. In exchange for that mill, however, the sons of Michael of Besnyő were granted the right to select a mill site anywhere along the same river to build one for themselves.[79]

3.1.4 Mills and Mill Sites: How Do They Relate to Each Other?

Why do charters refer to mill sites at all? As discussed above, the most important principle of customary law regarding water mills and dam construction was that existing structures had to be protected from the effects of new ones. The key factor appears to be that legal customs from the fourteenth to the sixteenth centuries did not distinguish between a mill and a place for a mill (a mill site). A new mill had to be compatible with the existing mills and align with all the *loca molendinorum* surrounding it. An example from 1315 supports this assumption. When discussing this case, István Tringli demonstrated that by the early fourteenth century, an established legal custom prohibited a landowner from endangering others by constructing multiple mills along the same stretch of river.[80] From the perspective of this subchapter, however, this case offers another, equally important lesson. In a court case before the bishop of Veszprém, the cathedral chapter of Veszprém and the nobles of the village of Jutas were preparing to divide the lands and a mill site belonging to the sons of a certain Bagar. During the division process, a third party, Master Mikse (*Mykse*), objected, claiming that constructing a new mill at this specific site would harm his interests because the land in question lay between two of his mills.[81] It turned out, however, that a mill had indeed previously stood at the *locus molendini* in question. The Mikse family could not prevent the construction of that mill at that moment, although in 1315, Mikse saw an opportunity to question the legitimacy of the mill site. Instead of challenging the right to build a mill at that location, he chose to question the legitimacy of the original mill construction from decades earlier. The bishop accepted Mikse's objection and prohibited the construction of a mill at that site. It appears this prohibition was not enforced, as documents from the second half of the fourteenth century confirm the existence of a mill called *Bogarmolna* (meaning "mill of Bogar" or "Bagar") at the same estate.[82] For the present investigation, however, it is the reasoning within the lawsuit that warrants attention. Mikse attempted to obstruct the reconstruction of the mill not by insisting there was no mill at that time, but by arguing that it had initially been constructed unlawfully in that location. The reasoning of the charter makes clear that by the early fourteenth century, the legal standing of that piece of land had not changed, regardless of whether it concerned a physical mill or a place for a mill or not.

If this assumption is true, it becomes understandable why, from the early fourteenth century onwards, mill sites appear in legal evidence—especially in charters related to property disputes—in increasing numbers, as well as in forged Árpádian-period charters produced during the Angevin period and afterwards. Proof of a pre-existing mill site on a particular plot of land would have required that later-built mills align with that former building plot. Although the parallel is somewhat distant, Mathieu Arnoux has identified similar trends in the mills constructed in the western Carolingian Empire and in Francia during the ninth to twelfth centuries, in which mill sites likewise served as points of alignment for later mills.[83]

With the gradual expansion of water-related infrastructure, including mills, dams, and fishponds, in the Kingdom of Hungary, the protection of water rights likely became crucial. One way to secure such rights was to have a mill built by a landlord on a river before anyone else did along the same stretch of water, ensuring that any future mills, ponds, or dams upstream and downstream would have to align with it. However, this work involved certain expenses. It was certainly cheaper to—after some groundwork or even without any actual intervention—have a charter issued by a place of authentication or another authorized institution confirming the existence of a mill site at the estate in question, as it could potentially secure the landowner's future rights to the water body.

By analyzing the meaning of the legal technical term *locus molendini*, we actually examine the formation of customary law governing the use of waterways in medieval Hungary. This term is prevalent in medieval legal literature throughout Central Europe and signifies a crucial step in securing rights to water bodies. From the fourteenth century onward, there was a marked increase in the use of written documentation in medieval Hungary, mirroring trends in other parts of Central Europe.[84] This era also witnessed the growth of water-related infrastructure, underscoring the finite nature of water resources; consequently, the importance of safeguarding property rights in water use became increasingly apparent. Disputes over access to water resources, as well as concerns about the jeopardization of each other's water-related livelihoods, were widespread across the landscape. The examples provided above primarily focus on rural areas within the Kingdom of Hungary; however, as will be discussed in the following subchapters, urban environments are certainly not exceptions in the context of legal disputes over water rights.

3.2 Rivers in Urban Landscapes: Rivers, Streams, Moats, and Disputes

Not only did towns in medieval Europe represent major concentrations of population, but they also served as high-density hubs of craftsmanship, resulting in greater energy consumption.[85] Pre-industrial societies relied on various power sources, yet mills—whether powered by water or wind—remained the most effective until the eighteenth century, when coal gradually emerged as the dominant source of industrial energy. In medieval Central Europe, both water, as discussed in Chapters 2 and 3.1, and wind power were utilized. In the Kingdom of Hungary, windmills never gained significance, primarily because the Carpathian Basin is relatively well protected from winds. According to medieval sources, only a few windmills operated in Hungary, and even in areas devoid of rivers suitable for water mills, such as the Danube-Tisza Interfluve, they appeared relatively late compared with other regions of Central Europe. One of the few medieval references comes from a town discussed below: Sopron. Sources mention windmills in greater numbers only from the post-Ottoman period (specifically, the turn of the seventeenth century) onwards. Thus, the most important and effective source of industrial power in towns during the Middle Ages was undoubtedly running water. Although the Carpathian Basin generally has a relatively dense network of waterways, the demands of towns can lead to water-resource scarcity. The next two subchapters of the book argue that access to water for various purposes has caused numerous conflicts within urban landscapes and that competition for water resources dates back to the Middle Ages. In this first subchapter, the book examines the types of water sources used for milling in medieval urban environments within the medieval Kingdom of Hungary. Following this overview, a special kind of town, mining settlements, will be analyzed, where, as will be argued, access to milling water was a critical issue.

Water is one of the most important factors in deciding whether to settle in a particular area. Major waterways in Europe and elsewhere provided the most effective and cheapest forms of transport in premodern times and continued to do so until the emergence of railway networks.[86] Consequently, many of the significant population concentrations in medieval Central Europe are situated near navigable rivers; in other words, few major population centers lack a nearby waterway.[87] Of course, in areas with unique "local energy", like mineral resources, this could vary. The first short subchapter briefly examines legal issues in the use of major waterways as sources of waterpower in medieval Hungary, illustrated by the towns of Buda, Óbuda, and Pest.[88]

3.2.1 Major Waterways and Mills: The Example of Buda, Óbuda, and Pest

Harnessing the energy of waterways was anything but simple. These challenges are certainly true for major rivers, where the inherent potential energy poses a constant threat to water mill operations. As noted in Chapter 2 on the spread of water mills in Central Europe, the earliest water mills were built on small rivers. However, the location of these mills does not mean that medieval societies neglected the energy of major rivers. Although they likely emerged somewhat later than fixed mills in Western Europe, ship mills also began to appear on rivers.

The earliest European reference to a ship mill dates to the sixth century. Procopius, in his account of the Gothic siege of Rome in 537, tells of how floating mills were invented:

> But after the aqueducts had been cut, as I have stated, the water no longer worked the mills and the Romans were quite unable to operate them with any animals owing to the scarcity of food in a time of siege; indeed, they were barely able to provide for the horses, which were indispensable to them. So Belisarios hit upon the following device. Before the bridge that I just mentioned as being connected with the circuit-wall, he fastened ropes from the two banks of the river and stretched them as tight as he could, and then attached to them two boats side-by-side and two feet apart,

> precisely where the flow of the water comes down with the greatest force because of the arch of the bridge, and placing two mills on either boat, he hung between them the mechanism by which the mills are turned. Beyond these, he fastened other boats, each attached to the one next in order, and he set the waterwheels between them in the same way for a great distance. So by the force of the flowing water, all the wheels in turn were made to revolve independently, and thus activated mills with which they were connected and ground sufficient flour for the city.[89]

Although the chronicler likely used the invention of ship mills to highlight the skill of Belisarius and his people, it nonetheless indicates that the technology was already known in the sixth century. This unique evidence, however, does not imply that similar constructions were widespread in the Mediterranean or Western Europe. They were probably quite rare in the region until the High Middle Ages, akin to traditional water mills. Ship mills nevertheless appeared north of the Alps as well—Carolingian sources refer to several on the Rhine.[90] There may have been areas where they were even more common than fixed mills.[91] Their presence in Central Europe and Hungary, similar to the introduction of water mills in general, has been attributed to differing dates by various scholars.[92] Both Polish and Hungarian sources attest to their existence by the end of the thirteenth century at the latest.[93] The first mention in Hungary dates to 1292 and the second to 1305. Both references concern ship mills operating on the section of the Danube by Óbuda, Buda, and Pest.[94] By the late thirteenth century, this area had the highest population concentration in Hungary, necessitating the largest number of water wheels within this relatively limited region. The use of ship mills is logical for these three settlements, as the area was not particularly rich in small rivers and streams suitable for fixed-site mills (*see* Figure 3.3).

In the fourteenth and fifteenth centuries, the number of ship mills along the rivers of the Kingdom of Hungary likely grew significantly, as sources increasingly recorded their presence on several rivers and at smaller settlements than Pest or Buda. However, focusing on Buda, Óbuda, and Pest, the growing number of ship mills, as noted in sources, posed a potential threat to two other important forms of water exploitation: navigation and fishing. Ship mills were not confined to a short section of the Danube near Buda, Óbuda, and Pest, but occupied increasingly longer stretches along the shorelines, thereby harming the interests of both fishermen and ferrymen. The owners of these mills did not select these sites to harm others; instead, these locations were the best spots on the river for milling. First, near the Buda and Óbuda shore, ship mills could operate year-round due to the proximity of hot springs along the shores. The icing on the river began significantly later than on the Pest side. Even in modern times, most ship mills were near Buda and Óbuda for this very reason. The ship mills rarely stood next to

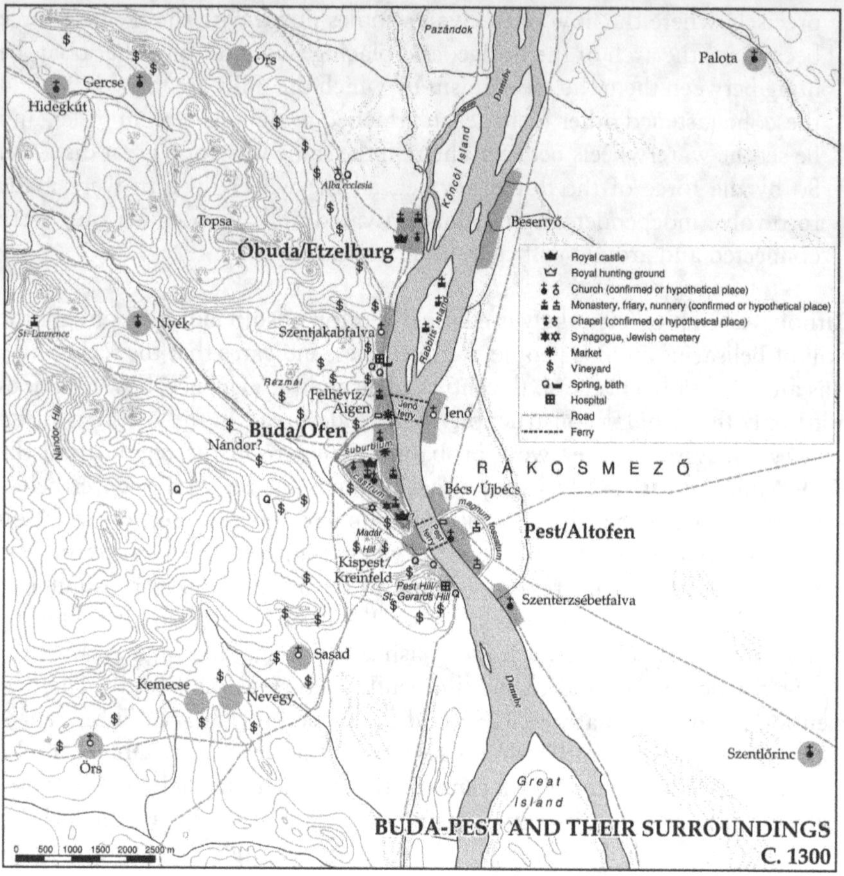

FIGURE 3.3 Pest, Buda, and Óbuda and their surroundings in ca. 1300 (cartography by Béla Nagy).

Source: András Végh, "Buda-Pest 1300 – Buda-Pest 1400. Two Topographical Snapshots," in *Medieval Buda in Context*, 178. https://doi.org/10.1163/9789004307674_009.

one another simply because they could easily damage each other. They were typically fixed to a large stake driven into the riverbed. These stakes, dating back to medieval legal literacy, were referred to in vernacular Hungarian as *szelep*. Originally, each ship mill probably had its own *szelep* (a stake), but at some point, likely in response to growing demand for milling, more and more ships were joined, with only the first, most upstream vessel connected to the stake itself. This crowding resulted in a chain of floating mills that obstructed river crossings and fishing along long stretches of the river.[95]

Already, the second reference to ship mills mentions conflicts regarding their locations. In a document dated 1305, the Dominican nuns of Margaret Island (*Insula leporum*, or "Island of the Hares" in its medieval name) presented a complaint against the Knights Hospitallers of Felhévíz regarding the chapter's placement of their ship mills above the ferry at Jenő.[96] In addition to their dissatisfaction with the mills' location, the nuns also claimed that millers harassed people crossing the river at the Jenő ferry. Due to the growing number of ship mills, the owners of the ferries and harbors complained about having further new stakes (*szelepek*) installed along the Danube by the millers of Buda's suburb. In 1462, the Buda chapter and the Dominican nuns of Margaret Island again filed a complaint against the burghers of Buda, including a certain John Ács (*Alch*, meaning carpenter in Hungarian), regarding the presence of numerous ship mills operating between Felhévíz and Jenő, the latter being one of the two most important crossing points over the Danube in present-day Budapest, as well as a major source of income for the nuns.[97] The mills and the stakes that held them along the riverbanks made crossing and using the harbors difficult.[98] Not only was crossing the Danube increasingly challenging due to the rising number of ship mills, but fishing was also affected. The noisy machinery discouraged fish from inhabiting areas near these constructions, thereby affecting longer stretches of the Danube. Fishing in present-day Budapest—that is, from the southern end of the Szentendrei (or by its medieval name Rosd) Island to the Csepel (or by its medieval name Great) Island along the Danube—was a privilege of the chapter of Buda.[99] However, according to a complaint filed by the chapter, three men—George Révész (meaning ferryman in Hungarian), Paul Aurifaber (meaning goldsmith), and the master of the Saint Nicholas Hospital of Pest—installed ship mills next to the wall of Újbécs (*Beechwara*), where there had never been mills before.[100] The new mills caused losses for fishermen, as their catch was halved. The document even specifies the types of fish typically caught in this part of the river, namely pike and sturgeon.[101]

Despite complaints from institutions with interests in ferries and fishing along the Danube near Buda, Óbuda, and Pest, ship mills steadily increased in number. Most of these, owing to the hot springs, were located on the Buda side, although several were also situated near Pest. Notably, there is a charter concerning a ship mill in the southern suburb of Pest, Szentfalva (or Szenterzsébetfalva).[102] In a donation dated 1490, Stephen Szapolyai (then eternal chief *ispán* of Szepes and father of the later King John Szapolyai) granted a site for a ship mill on the Danube to acknowledge the services of a nobleman, Michael Temesvári Bodó. The designated location for the mill amounted to no more than a stake, "known as a *szelep*."[103] The charter of Szapolyai, likely not by accident, stipulated that no tax could be assessed on the building, and most importantly, that no one—presumably the people of Szentfalva—could disturb the mill's functioning. The residents of Szentfalva held a significant share in Danube fishing, suggesting that this clause likely referred to potential conflicts of interest between fishermen and the ship mill owner.

The last, in the context of ship mills on the Danube near Buda, Óbuda, and Pest, is likely the clearest indication of the challenges these constructions posed for river crossing. By the Jagiello period, Hungarian diets, with very few exceptions, took place in the "capital" of the Kingdom, Buda (and Pest).[104] While the lesser nobility gathered in the fields of Rákos, outside the walls of Pest, the aristocracy assembled at the Buda Castle, although representatives of the "chambers" met regularly. During these times, they had to cross the Danube to discuss matters. On one such occasion, a group of noblemen traveling from Buda to Pest across the Danube unexpectedly crashed into a ship mill, causing the noblemen to fall into the water and drown.[105] This tragic event was remembered even in later decades, as the humanist chronicler Antun Vrančić referenced it, although he dated it to the aftermath of the 1514 peasant uprising.[106]

Most documentary evidence on ship mills refers to their presence along the Danube near medieval Buda, Óbuda, and Pest. This is not because this was the only area in which numerous ship mills existed during the medieval period, but rather because it is here that their presence gave rise to conflicts of interest that led to court cases. If not in the late medieval period, then certainly by the sixteenth century, ship mills—sometimes several ship mills—operated by every major settlement along the Danube.[107] Even if ship mills were not uncommon, most milling in the Kingdom of Hungary, even in urban environments, occurred in fixed mills built along smaller streams.

3.2.2 Streams and Mills in Urban Environments: The Cases of Zagreb and Felhévíz

As discussed above, many major population concentrations in Central Europe, such as Buda and Pest, Kraków, Wrocław, and Prague, lie along major rivers that are difficult to harness; smaller streams also played a crucial role in local economies. These smaller bodies of water provided essential resources for local populations, including milling and the construction of industrial complexes, bathing houses, tanneries, and other facilities. Due to their limited runoff and relatively large number of users, streams and minor rivers in urban areas often become sources of conflict.[108] Two examples will be examined here. The first concerns Zagreb and the small stream that delineated the boundary between the two legally separate settlements that formed the medieval town. The second concerns Felhévíz, an aforementioned northern suburb of Buda, and the exploitation of its hot springs to feed mills and bathhouses.

Zagreb serves as a notable example. It was the smaller water bodies flowing through the heart of the settlement, not the Sava, the main river near the town,[109] that were crucial to the town's economy. Zagreb also illustrates waterways as settlement boundaries, a theme explored further in Chapter 3.4.

During the Middle Ages, the city comprised two distinct parts, Kaptol and Gradec, established at significantly different times. The two settlements were situated very close together and were separated by a small stream called Medveščak. Each settlement possessed different privileges, legal status, and autonomy, yet they also shared certain functions.[110] This situation would have created challenges for the operation of the town(s) regardless, but the ambiguous rights of the institutions and residents of Kaptol and Gradec regarding the stream that divided them occasionally led to violent conflicts.

The Medveščak, or by its medieval name Cirkvenik (*Cyrkvenich*) Stream, collected the waters of the Medvednica Hills, directly north of Zagreb.[111] The small stream valley formed the border between two parts of the settlement, which developed on two small but well-defined elevations. Kaptol, founded at the turn of the eleventh century and owned by its bishops and the cathedral chapter, occupied the hill to the east of the Medveščak. At the same time, Gradec, established by settlers after the Mongol invasion in the 1240s, lay on a plateau at the top of the Grič elevation west of the stream.[112] The inhabitants of both settlements sought to exploit the stream. However, the limited flow along its bed permitted only a small number of mills and bathing houses to operate. At least two obstacles made conflicts inevitable. First, the cathedral chapter seems to have owned land on the right bank of the Medveščak; among other things, they possessed at least three mills on the side that otherwise lay in the hands of the Gradec citizens.[113] This situation is clarified by a donation charter from 1257, in which a canon of the chapter, a certain Peter, donated three mills to the nearby Cistercian abbey. The recently settled citizens of Gradec, however, were highly dissatisfied with the donation and filed a complaint before the court of the ban of Slavonia, Stephen Gutkeled, alleging that it was unlawful. Stephen decided that two mills—the best and the worst of the three—would remain with the Cistercian abbey, while the third would be given to the townspeople of Gradec.[114] This last mill remained in Gradec's possession until it was confiscated by the ban of Slavonia, Nicholas Hahót, due to the treason of its owner in the 1340s.[115] The mill was then awarded to Stephen and Ákos, sons of the previous ban of Slavonia, Mikcs (*Mykch*, or Mikac—as he is referred to in Croatian historiography).[116]

Another problem that hindered the two communities' sharing of the water resource was a mill race dug at the initiative of the Gradec community. This construction diverted much of the water from the original Medveščak riverbed and created a second, artificial stream branch. The latter was referred to as Pretoka or Prekopa, while the main bed of the Medveščak was referred to as Matična. In 1346, the chapter filed a complaint against the citizens of Gradec, in which the canons claimed that the channel destroyed lands owned by the chapter.[117] The representatives of the citizens of Gradec, however, testified that the race was dug within the limits of their land. The issue was complicated by a fire that destroyed all documents concerning the chapter of

Zagreb's privileges and possessions.[118] However, there is an important point that the chapter never raised during the court case against the citizens of Gradec. Even if the channel that Gradec had dug lay within the border of the settlement, the citizens did not have the right to endanger farming on the chapter's lands.[119] This mill race proved durable, remaining in use into modern times. It can be identified on dozens of maps, including one of the earliest representations of the town drawn by Nicholas Angielini, a well-known architect and mapmaker of the sixteenth century (*see* Figure 3.4).[120]

The frequent disputes between the two institutions are readily understood, given the intensive exploitation of the small stream that flows between them. A perambulation from 1328 provides a detailed account of the mills that stood on the two branches (one natural and one artificial) of the stream. While surveying the boundary, the surveyors proceeded from south to north along the Medveščak (*see* Figures 3.4 and 3.5).[121] The first relevant point for the present chapter was the mill of the parish priest of the Saint Emeric Church. Moving upstream, the following two points of relevance were the two mills of the "abbot of Zagreb." These mills were certainly identical to those of the Cistercians, granted to them in 1257 and retained following the lawsuit against the citizens of Gradec. Above these two mills, the next point in the perambulation was the mill of the bishop by the same stream, followed by three more mills owned by the chapter, and then again, a mill of the bishop, followed by three additional chapter-owned mills. Above this, the two branches mentioned earlier—the artificially dug channel and the old river stream bed—converged again. On the artificial mill race excavated by the citizens of Gradec, which the document notably calls the "real bed of the river" (*verum meatum ipsius rivuli*), the citizens owned two further mills. In contrast, on the natural and likely occasionally dry bed of the stream, there was a "mill of the hermits" (*molendinum Heremitarum*).[122] The hermits can be identified as Paulines, whose hermitage lay next to the town. Finally, mills (or a mill with multiple wheels) owned by the citizens of Gradec stood on lands (*predia*) that Nicholas, son of Syrgarinus, received from the chapter—this one also operated by the natural riverbed. Thus, there were at least fourteen mills and a mill site established along a few-kilometer stretch of the not particularly bountiful waters of the divided Medveščak stream. These waters must have been among the most intensively utilized water flows in the territories studied in this book.

The use of these waters remained a hotbed of intense (even bloody) disputes between the citizens of Gradec and the chapter of Zagreb throughout the late medieval period. However, in 1392, they attempted to reach an agreement intended to resolve these disputes, which, as demonstrated later, was ultimately unsuccessful. In that same year, they nonetheless agreed that neither the chapter nor the citizens of Gradec could establish mills in the lower section of the Medveščak stream—from the Dominican friary to the Pisani

FIGURE 3.4 Gradec, Kaptol, and the Medveščak Stream with the mill race on the manuscript map by Nicholas Angielini.

Source: Nicholas Angielini, *Mappae geographicae regni Hungariae et terrarum adiacentium*. For the manuscript, see Vienna, Österreichische Nationalbibliothek, Handschriftensammlung, Cod. 8609. 17v.

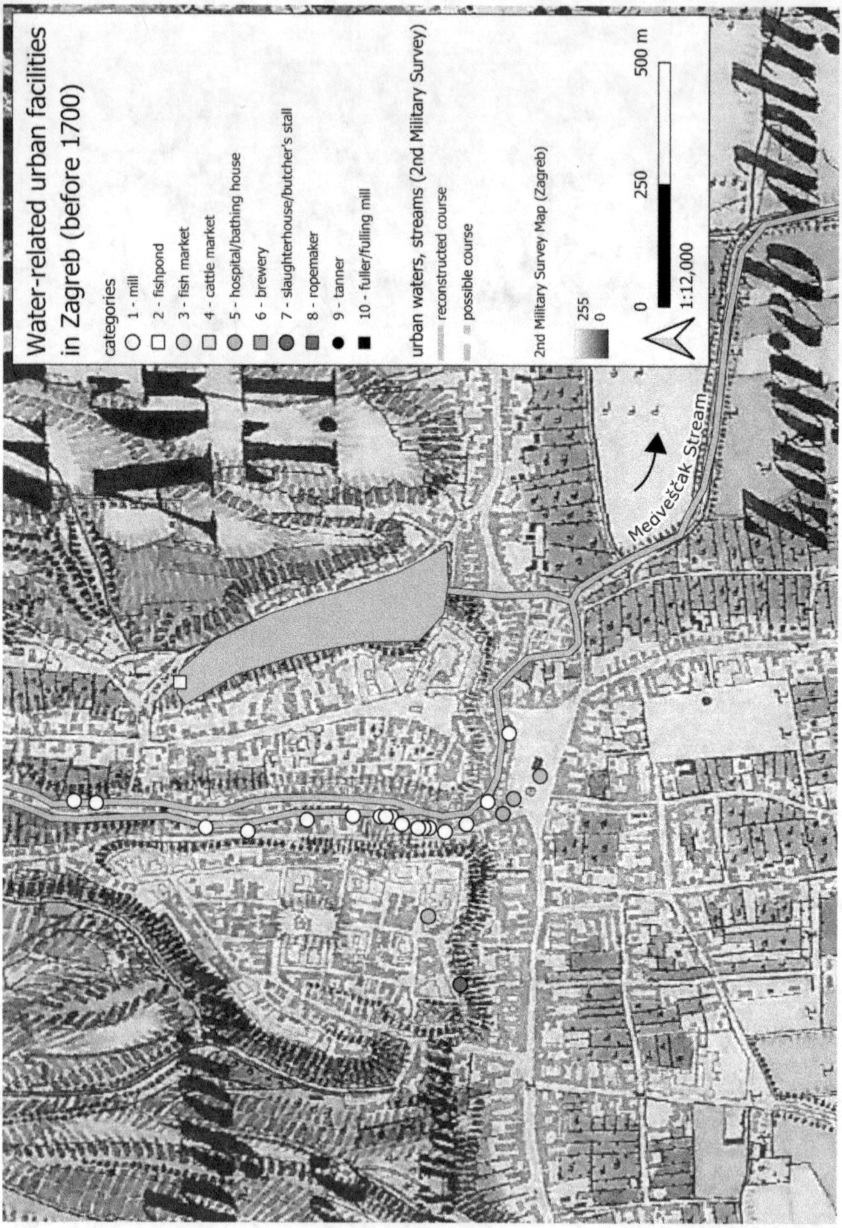

FIGURE 3.5 The mills of Zagreb in the Middle Ages (drawn by András Vadas and László Ferenczi).

or Krvavi (meaning "bloody") Bridge.[123] This area represented the southern (lower) section of the stream. The bridge was located directly downstream of the confluence of the artificial mill race and the former natural stream of the Medveščak. Upstream from this point, since the chapter had operated its mills in this area since the thirteenth century, only they had the right to construct mills up to the mill of the Chupor family, which was situated in the southern part of Nova Villa or Nova Ves (meaning new town in Croatian). Above this point, up to the village of Gračani, a tenant peasant village founded by the citizens of Gradec, the latter town retained the right to operate mills. Two institutions, the Cistercians and the parish priest of Nova Villa, were allowed to maintain their mills, as they had existed there since "old times."[124] The mill of the latter settlement is the same as the above-mentioned mill of the Paulines that the parish priests of Nova Villa acquired in 1376 in exchange for other properties.[125]

In the 1420s, the conflicts between Kaptol and Gradec appeared to have reignited and taken on more violent forms than before. In 1422, documents confirm bloody clashes between the residents of the two settlements.[126] The following year, the citizens of Gradec were accused by a certain George, canon of the chapter of Zagreb, of, among other things, digging a ditch to redirect the flow of the stream.[127] Similar accusations arose in 1475, but this time in reverse. The inhabitants of the chapter town seemed to have diverted the original river's flow toward their mills and fields, leaving the artificial channel used by the citizens of Gradec dry.[128]

The example of the mills and the series of related disputes in Zagreb highlights the significance of small urban rivers and streams for local economies. The River Sava, flowing near Zagreb, was utilized for both fishing and mill construction. It also played a key role in the town's southern defenses. However, the Sava mills were never the main drivers of the local economy, as they were located too far from the densely populated areas.[129] Therefore, despite considerable runoff and the potential for constructing ship mills on the Sava, the Medveščak stream remained central to the local economy and to local conflicts.

3.2.2.1 Felhévíz and Its Hot Springs

The case of Felhévíz is similar to Zagreb's in that the settlement primarily benefited not from the major river flowing nearby but rather from a smaller stream. Felhévíz and its waters, as noted above, represent a special yet hardly unique example, not only because it is located next to the major town (in this instance, Buda and Óbuda) but also because the water flow at the center of its economy was supplied by hot springs. A visit to a thermal bath on the Buda side is a must for anyone visiting modern Budapest, as many of these baths were established during the Ottoman period.[130] The thermal springs that feed these spas were known and utilized long before the Ottomans arrived in the region in the mid-sixteenth century. The most significant evidence of this earlier knowledge is the settlement of Felhévíz, whose Latin name was Aquae calidae superiores ("upper hot springs"). The area between medieval Óbuda and Buda was rich in thermal springs, the use of which dates back to Roman times.[131] The waters of some of these springs were likely dammed to create a small pond as early as the thirteenth century. This pond remains and is known as Malomtó (meaning "Mill Lake"). It created a stable water level on the small stream that flowed from the pond toward the Danube (*see* Figure 3.6).[132] This stream, called Malom ("Mill") Stream, was approximately 100 meters long and ran at a relatively favorable location. Felhévíz was a suburb of both Buda and Óbuda, with Buda's primary wheat

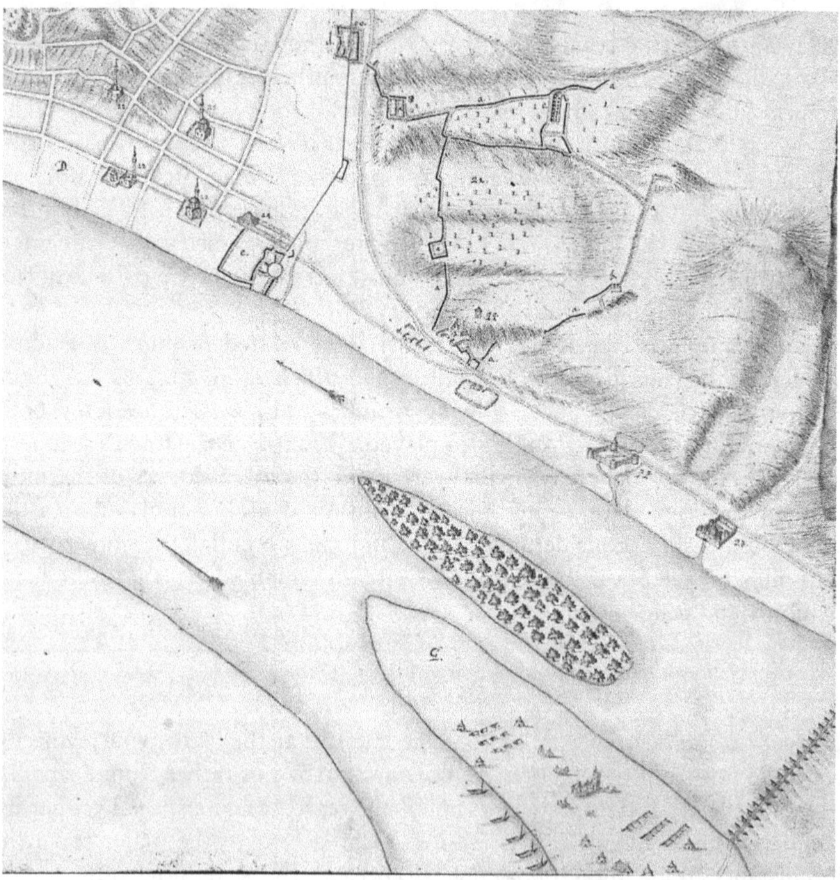

FIGURE 3.6 Felhévíz and the Malomtó (Millpond) as represented in a late seventeenth-century illustration (by H. Bredekow).

Source: Marburg, Hessisches Staatsarchiv, Wilhelmshöher Kriegskarten, vol. 6. no. 20.

market located directly next to Felhévíz. Due to the stream's potential to provide year-round waterpower and its proximity to the market, it became a highly contested resource. The contentious nature of the stream is well-documented, given its specific circumstances. The sites suitable for milling, as subjects of royal donations, were gradually transferred to ecclesiastical institutions. These institutions, as seen in the prior case of Zagreb and its chapter, were keen to secure all possible related income. This led to a conflict of interest among landowners along the short Malom Stream. The combination of friaries and chapters serving as landowners in this area has produced extensive documentation.

While the archives of medieval Buda are largely lost, the records of mendicant and other ecclesiastical bodies that operated in the area of present-day Budapest are much better preserved. This is especially true for the documents of the Dominican nuns of Margaret Island, who had numerous interests in the water-related income around their nunnery.[133] In the case of Felhévíz, not only are the sources exceptionally well-preserved, but one of the leading experts on medieval Buda and Pest, András Kubinyi, dedicated a monographic study to the medieval development of this settlement.[134] Following in his footsteps, other studies have explored the mills established here.[135]

The earliest institutions interested in utilizing the waters at this location were the Dominican nuns of Margaret Island and the Holy Spirit Hospital. The latter institution was situated close to the pond at the source and was most likely founded on a royal initiative. The hospital not only had mills on the Malom Stream but also constructed a bathing house, a common addition to hospitals in medieval Europe. It is reasonable to assume that the presence of the hot springs was the primary factor in their decision to settle at Felhévíz.[136] The earliest reference to a mill in the area predates the first records on the Holy Spirit Hospital, which can only be confirmed from the end of the first third of the fourteenth century onward.[137] The first mill established by the pond belonged to a certain Peter, son of Uza of the Kurszán clan, who donated his mill at Felhévíz to the Dominican nuns sometime before 1276.[138] The next mill mentioned in the sources, situated along this short stream, is particularly interesting from the perspective of the discussion in Chapter 3.1. The mill, called Hevezfeö ("hévízfő", meaning the "head of hot spring"), probably stood by the outflow of the pond. However, at the time of the donation (or leasehold) by King Charles I, there was no mill in existence. The document detailing the transaction did not refer to the site as a *locus molendini*, which could have implied the previous existence of a mill, but rather as an "empty mill plot" (*...nos quemdam vacuum fundum molendini Hevezfeö vulgariter nuncupatum*).[139]

The mill was donated to the citizens of Buda, who built not one but two wheels there, according to the relevant documents, which stood in separate buildings.[140] Two wheels in two buildings can hardly be anything other than two mills. Nonetheless, the source speaks of one complex, most likely to avoid accusations, that had received a plot for only one mill. None of these mills remained in the hands of the citizens of Buda for long; in the 1360s, after a prolonged lawsuit, they came under the ownership of the Poor Clares convent of Óbuda, an institution that enjoyed strong support from the mother of King Louis I, Elisabeth of Poland.[141] At least one more mill was erected at this location. This was in the possession of the royal castle of Óbuda from the mid-fourteenth century onwards. The exact location of the mill, however, remains unclear. The earliest references to the mill mention it standing between two other mills, which the document likely identifies as two of the aforementioned mills on the Malom Stream.[142]

Later documents, however, are less clear about the mill's location. In 1389, it is described as being near the Holy Spirit Hospital,[143] while in 1420, it is mentioned as being close to the Danube and the lands of the Poor Clares of Óbuda.[144] The mill's localization is further complicated by two documents dated 1430, which refer to the mills as being by the bathhouse of the Dominican nuns of Margaret Island.[145] It seems that with the increasing use of the stream, not all mills could operate profitably, leading some to be abandoned or fall into disrepair. This situation is suggested by a donation from King Sigismund of Luxembourg, in which three mills or mill sites were granted to the Poor Clares of Óbuda. The nature of Sigismund's donation is somewhat unclear, as the documents sometimes refer to three mills and at other times to mill sites.[146] It is likely that the nuns then built new mills or rebuilt existing ones, because when the mills were again illegally confiscated on the order of the treasurer, Michael Gúti Ország, through a member of his noble retinue, Nicholas Szász (*Zaaz*), they were referred to as standing mills.[147] One of these mills was subsequently given to Anthony, a glassmaker from Buda.[148] Despite the Poor Clares' attempts to regain possession of their mill, they ultimately agreed to lease it (again referred to as a mill site) to Anthony.[149] András Kubinyi has shown that the Poor Clares similarly made great efforts to recover the remainder of their former mills (then referred to as mill sites).

The fifteenth century saw several additional lawsuits at Felhévíz involving landowners. These lawsuits, however, concerned only property rights and did not directly address access to water. By the early sixteenth century, due to the abundance of mills, numerous lawsuits were filed in the royal courts regarding rights to utilize the waters of the Mill Stream. The Poor Clares rebuilt at least one mill, and their third mill site was also leased and used. Another ecclesiastical institution, the Buda chapter, is documented as a mill owner in a document dated to 1481. According to this record, their mill was seized by the court judge of Buda, Benedict Pider, under the orders of King Matthias.[150] It is likely not a mere coincidence that the central heraldic symbol on Pider's coat of arms, acquired shortly before, was a mill wheel—presumably a nod to his previous occupation. Pider's mill was situated between the mill of the Óbuda castle and a mill owned by the Poor Clares, implying the aforementioned, non-localized mill of the Óbuda royal castle, which also stood somewhere along this short stream, likely upstream from the Poor Clares' mills.[151]

The first of the three lawsuits brought about by pressure on the stream's usage began in 1502. The Poor Clares filed a complaint against a man named Caspar Zsemlyesütő (meaning "bun-baker") for moving the sluice of their mill. As evident from this document, Zsemlyesütő— whose relative, Andrew, also owned a mill at Felhévíz—had raised the height of the sluice, causing the stream water to back up so much that it flowed back to the upstream mill of the nuns, disrupting its successful operation.[152] The nuns argued that the height of their mill sluice was the same as it had been in "old days" (*veteri et antiquo*), which, in light of what has been discussed in Chapter 3.1, is not surprising. The efforts of the Poor Clares proved successful in the long run, as in 1509 Zsemlyesütő was prohibited from causing further damage to the nuns' mill property. The rebuilding, or rather, the expansion of the nuns' mill into a two-wheeled mill that followed this decision, was an opening move in a new lawsuit.

The other two ecclesiastical institutions, the Dominican nuns and the chapter of Buda, sued the Poor Clares for lowering the water levels in the Malomtó (Mill Lake), which they claimed belonged to all the mills. The nearly fifty (!) witnesses—including several bun-bakers and millers—testified that both the Dominican nuns' mill and the chapter of Buda's mill had lost a significant portion of their water supply, making it nearly twice as long to grind the same amount of grain as before.[153] The document reveals the existence of an elaborate water system, including underground water cisterns (*cisternas subterraneas*) and channels (*canales*) that supplied water to the Poor Clares' mill.[154] The palatine, Emeric Perényi, whose court heard the case, ordered that the Poor Clares restore the hydrological conditions that had existed prior to their intervention. Shortly afterward, a group that included several millers investigated whether the Poor Clares had fulfilled their obligation, but, unsurprisingly, found that the work had stalled.[155] Likely linked to the completion of the restoration work on the Poor Clares' mill, the lawsuits continued in Felhévíz. In 1511, Caspar Zsemlysütő sued the nuns again, stating that the mill he leased, the royal mill (the previously mentioned mill of the castle of Óbuda), experienced losses due to the restoration work done by the nuns. Even though the investigation showed no decrease in the output of Zsemlyesütő's mill, he received 20 gold florins from the Poor Clares as compensation. In contrast, another institution, the chapter of Buda, probably due to their earlier losses, received 50 gold florins from the nuns.[156]

The numerous charters documenting the enduring struggles to control the waters of the short Malom Stream illustrate the importance of small water bodies, even near major rivers such as the Danube. Hot springs, of course, were not the only valuable resources at Felhévíz; across Europe, wherever water froze in winter,[157] these waterways were among the most contested resources because of their value for constructing water mills and bathhouses.[158]

3.2.3 *Urban Moats and Water Use*

As demonstrated in the previous two subchapters, even in the Kingdom of Hungary, where urbanization—and the number of people living in major settlements—lagged significantly behind those of Western Europe and Byzantium, water use was a critical issue in urban spaces, and local economies tapped into every available water source.[159] Research has explored a variety of uses for rivers and streams with different characteristics. Still, another type of water body, prevalent in many towns across Europe, has been discussed in the literature on water use only partially: urban moats.

Moats surrounding medieval towns (and castles) were crucial elements of their defenses. However, by the late medieval period, most cities in Western and Central Europe had expanded significantly beyond their walls. With the ever-growing population in the suburbs and increasing economic activity in towns, moats gradually blended into the urban landscape. Not everywhere, and not always to the same extent, but in many cases, moats became integrated into urban economies. They were used in at least three areas of economic life: for milling, less frequently for fishponds, and as ports. These functions of moats have received less attention in scholarship compared to their military and architectural aspects, as available sources seldom mention their non-defensive uses. This subchapter, building extensively on cartographic and visual sources, will provide an overview of the water management of moats and will highlight how the secondary uses of these primarily defensive structures brought conflicts to the fore. After presenting an overview of the role of moats in local economies, this subchapter examines two towns in medieval Hungary: Sopron and Prešov. It argues that moats were also systematically exploited in medieval urban economies and should not be viewed solely as defensive elements.

Few monographs and edited volumes discuss urban moats, but there is a considerable wealth of literature on the urban fortifications of individual settlements.[160] These works were published either as standalone studies or as chapters in "town biographies." Urban archaeology has focused on the study of the fortifications that protected medieval towns. Urban moats and their surroundings were among the few areas accessible for archaeological excavations in many towns, as they were not necessarily developed for real estate purposes. Consequently, numerous studies from Germany, the Czech Republic, Poland, and, to a lesser extent, Hungary address the construction histories of the urban fortifications of specific settlements, based on archaeological data, sometimes supplemented by analyses of iconographic and/or written evidence. However, there is a gap in the study of the economic aspects of medieval (urban) moats. Richard Holt, in a previously quoted overview of the aquatic economies of medieval England, briefly refers to the construction of moats around castles, manor houses, and towns. He highlights two essential aspects of moats. First, in England, moats served as prestige investments as well as defensive structures, and second, they were utilized by local economies.[161] He shows that in England, most moats relied on groundwater, with only the larger ones being supplied by flowing water through ditches. He cites the example of the moat of medieval Birmingham and illustrates its use as a mill pond. However, he suggests that Birmingham's moat was more of an exception than the norm, a claim that may not be entirely accurate. Scholars in England have also emphasized the role of moats as ports.[162]

In France, the historian of technology, André Guillerme, studied moats in a comparative perspective. Selecting settlements in present-day France, he analyzed the relationship between towns and water. The monograph employs methods from two fields: the history of technology and environmental history. Guillerme is among the few authors to address the varying needs of local economies and defensive requirements concerning moats. He demonstrates that the period starting with the Hundred Years' War in France introduced fundamental changes in the use of urban waters. From that point on, military considerations overshadowed economic interests for an extended period.[163] This is pertinent in the context of the Kingdom of Hungary, where a somewhat similar process is observable during the Ottoman-Hungarian wars of the sixteenth and seventeenth centuries. A widely debated monograph by Charles Coulson is also noteworthy, not primarily in relation to towns, but rather castles and their moats. He views defenses constructed around castles in late medieval France, England, and Ireland as mainly prestige investments.[164] In doing so, he departs from traditional approaches that study these structures from an architectural history perspective.[165] Seeing castle fortifications as symbols of prestige, he also explores the potential economic benefits these investments may have provided. Coulson cites several examples of moats that were incorporated into local economies. However, other works rarely include specific references to how the moats were used or how water was regulated to maintain a consistent water supply for them. Before providing a brief overview of research on urban fortifications and moats in Hungary, it is essential to mention a valuable research resource that is extensively used in this subchapter: the volumes of the European Historic Town Atlas Series. The over 550 atlases produced for individual settlements are among the best tools for studying the economic exploitation of moats in premodern Europe. Moats are consistently represented in the topographic reconstructions, and where sources permit insights, mills and occasionally even fishponds and other economic features of moats are noted in the volumes.[166]

There is a significant gap in the scholarly literature on the defensive structures of medieval Hungarian towns, particularly regarding moats.[167] In recent years, fortifications in Transylvania, including urban defenses, have been discussed in some publications.[168] There is also a less comprehensive catalogue of the castles and urban fortifications in sixteenth-century Hungary; however, these works are far from exhaustive in documenting urban defense structures.[169] A few regional castle cadastres have been compiled over the past decades, but they rarely mention moats. Most of the relevant literature focuses on the fortification systems of specific towns, such as Pest, Trnava, Košice, Sopron, and Kőszeg, where archaeological excavations have been conducted.[170] Despite recent excavations on urban fortifications, aside from the urban moat of Sopron, little information about moats has been uncovered.[171] The dissertation of an architect, Péter Rabb, offers an overview of urban fortifications in medieval Hungary. While he briefly mentions moats, his work has limited applicability since the author completely disregarded primary sources and largely overlooked archaeological findings.[172] Nevertheless, as argued in the following subchapter, moats were common in premodern urban settings, and, likely, few served solely military defensive purposes.

3.2.3.1 Building Moats, Using Their Waters: A Short Overview

By the late medieval period, most major towns in Western Europe were not only surrounded by walls but also had a protective moat—an indispensable element in defending settlements amid the widespread use of firearms.[173] This is nicely reflected in the vernacular architectural treatise by the Italian humanist, Filarete. According to his work, the building of the ideal town begins as follows:

> I have ordered that, the first thing tomorrow morning, everyone is to come with mattocks, shovels, and picks. I want you to be the first to begin the work of digging the moats. These will be ten braccia away from the wall. I think they ought to be 30 braccia wide if the dimensions are satisfactory to you. As soon as this is done, the moat should be walled up on both sides. The wall toward the city will be as high from the ground as the battlements of the wall, that is, three braccia high from the ground level, and then the battlements. On the opposite side, it will only be as high as ground level, no more, no less.[174]

Although the Italian treatise was likely never a basis for the construction of complete towns in medieval Hungary, the text was not altogether unknown here either. Filarete's work was translated into Latin and preserved in one of the codices of King Matthias's renowned library, the Bibliotheca Corviniana.[175] For the present enquiry, the most important lesson of the text is that, in the fifteenth century, it was evident that a town required a more complex fortification system than merely a wall. As with many fortification elements, the role of moats in a community's self-representation as a prestige investment should also be considered.[176] Moats, however, represented significant investments that yielded little short-term economic benefit. This is likely the reason the author of a chronicle detailing the lives of bishops of Auxerre chose a rather unconventional approach to explain why Hugo of Noyers, an influential bishop at the turn of the twelfth and thirteenth centuries, ordered one of his smaller residences, Varzy, to be encircled by a moat:

> Because it was unprotected and vulnerable to the attacks of plunderers, [Hugo of Noyers] rebuilt the old walls, and constructed moats and high towers to fortify Varzy, his famous town. Next to Saint Eugene's Church, he erected new buildings for the bishopric with strong walls, great houses, towers, and bastions. To protect the bishops, he built even stronger walls, with more solid walls, towers, and outworks. The inner keep boasted towers, fortifications, and advanced defenses of impregnable strength, surrounded by wide, spring-filled ditches. Thanks to the abundance of [water] sources that feed the moat, it is not only a small augmentation of the fortifications, but because of the multitude of fish they provided, by the mills the bishop built there, and other profitable assets, they greatly improve the entire establishment.[177]

There is limited information on the costs of fortification moats. Still, based on an account from Sopron in 1540 (discussed in Chapter 4) and similar sources from Austria (Freistadt), their construction likely represented a significant financial burden. Unlike town walls, moats had the potential to generate income since their waters could be exploited in the ways noted above. This was not merely a potential source of exploitation; in dozens of towns across Europe, moats were considered a well-defined, predictable source of income. To assess the prevalence of moats in local economies, volumes from the European Historic Town Atlas series have been analyzed. A total of 337 atlases were examined in preparing this subchapter of the book. This sample provides adequate geographic coverage to determine whether the use of moats as milling waters and fishponds was the norm or the exception. Of the 337 studied settlements, 207 possessed at least partially artificial, that is, dug, protective ditches, and of these, 136 had moats filled with water for at least part of the year. Among the roughly 50% of towns with moats, there were attempts to integrate these features into local economies. However, in the atlases, due to a lack of sources, the use of moats as fishponds was only seldom mentioned, while their role in discussing and representing mills was more systematic (as indicated by cadastral maps). In the 76 towns where moats had associated mills, there were two different placements for mills on moats (as in the case of Sopron, which will be discussed hereafter). Most mills were constructed on a somewhat arbitrarily chosen section of the moat. However, in roughly one-third of the cases, mills and their sluices served as the primary means of regulating water levels in specific stretches of moats. In some instances, mills were found in both positions along urban moats. Ensuring a constant flow in moats was crucial because, if water levels remained low for extended periods, they could easily silt up, making their clearance a costly endeavor (as demonstrated by the example of Sopron). Most towns attempted to maintain moats with constant water circulation.[178] The permanent flow of water was even more essential in moats used for fishkeeping. While town atlases seldom reference this, there are indications that fishkeeping in moats occurred at individual settlements from relatively early periods. To mention two well-known examples, sections of the moats at both Quedlinburg and Vienna were rented out to different fisheries, suggesting that the practice was not unique.[179]

Urban history in Hungary over the last fifty years has focused on the definition of a town. If one considers the presence of stone walls as one of the most essential features of a *civitas*, there were only a few dozen towns in the region before the fifteenth century. Even by the early sixteenth century, there may not have been more than about fifty such settlements.[180] Information about the moats surrounding these towns is less available, as most were filled in by modern times while these towns expanded well beyond their medieval limits, similar to other parts of Europe. In most cases, one can assume whether these moats were filled with water or remained dry based on hydrogeographic conditions. Although there were only about fifty towns surrounded by stone walls in medieval Hungary, several towns and villages featured earthen and wooden fortifications, known as palisades, often combined with moats. If one includes these settlements, there were likely more than 100 protective moats surrounding settlements and castles in the Kingdom of Hungary by the end of the Middle Ages. The gradual Ottoman expansion beginning in the sixteenth century initiated a new wave of fortification construction that also affected moats in Hungary. To offset the costs of maintaining moats, towns in Hungary exploited these water bodies, as demonstrated by two examples discussed below: Sopron and Prešov.

3.2.3.2 The Case of Sopron

The first example is the town of Sopron, located in the western part of the Kingdom of Hungary.[181] This town has a distinctive history of settlement, tracing its origins to the Roman fortification of Scarbantia. A wall enclosed the Roman town during the Late Roman period. After the Romans withdrew from Pannonia in the early fifth century, the fortification remained abandoned for centuries; however, in the tenth century, when Hungarian settlers began to inhabit the area, it was still visible. Consequently, a royal stronghold was established there in the eleventh century, incorporating a timber-and-earth rampart into the existing Roman fortification. Several villages supporting the military and administrative populations were established in the vicinity of the fortified area. However, in the thirteenth century, the settlement's role changed. The fortified area was converted into a royal town, with plots of regular sizes being demarcated both within the walls and in the suburbs.[182] The walled section, with its substantial civic population, became one of the best-protected towns in Hungary, featuring a triple wall system and a moat. The town archive contains the richest holdings of materials related to the Middle Ages and the Early Modern period in present-day Hungary. Archaeological research has also been conducted within the fortification area.[183]

Although Sopron had more than 4000 inhabitants in the late Middle Ages, it was not rich in rivers. Two streams flowed through the town: the northwest-southeast-running Ikva stream, which carried slightly more water and crossed the town in its northern section, and the smaller Bánfalvi or Rák stream, which ran west to east. Both the moat and the streams were used to build water mills from the Middle Ages onward.[184] Due to the area's relief, the Bánfalvi stream supplied water to the town's moat. The water from the stream circulated through a system of sluices around the moat.[185] Although it's not usually documented, in the case of Sopron, a document provides a clear indication of when the moat was built. In 1340, King Charles I supported its construction by allocating the town half of the customs revenue collected at Lake Fertő.[186] The grant only mentions *fossatum* but does not clarify whether it refers to a moat or just a ditch. One of the most important features was a sluice at the northernmost point that regulated water levels on the western side. It was part of the Foregate building complex (*see* Chapter 4 as well). According to a fragmented account book from 1437, a mill was also connected to the building by then.[187] This was likely the largest mill on the moat, although it was certainly not the only one. In 1438 and the following year, the so-called Hospital Mill (*Spitalmühl*) also appears in the sources. It was located on the western side of the moat in a relatively favorable position, right next to the grain market (*Kornmarkt*).[188] From the sixteenth century onward, at least a further mill was built on the moat; sources recurrently refer to a fuller's mill. According to an entry in the town's account book from 1589, the fullers had a new mill built in the southwestern part of the moat, which operated with three wheels.[189] The number of mills increased further in the seventeenth century as well (*see* Figure 3.7).[190]

The moat in Sopron was used not only for milling but also for keeping fish, as sources attest. The first reference to fish dates to the fifteenth century

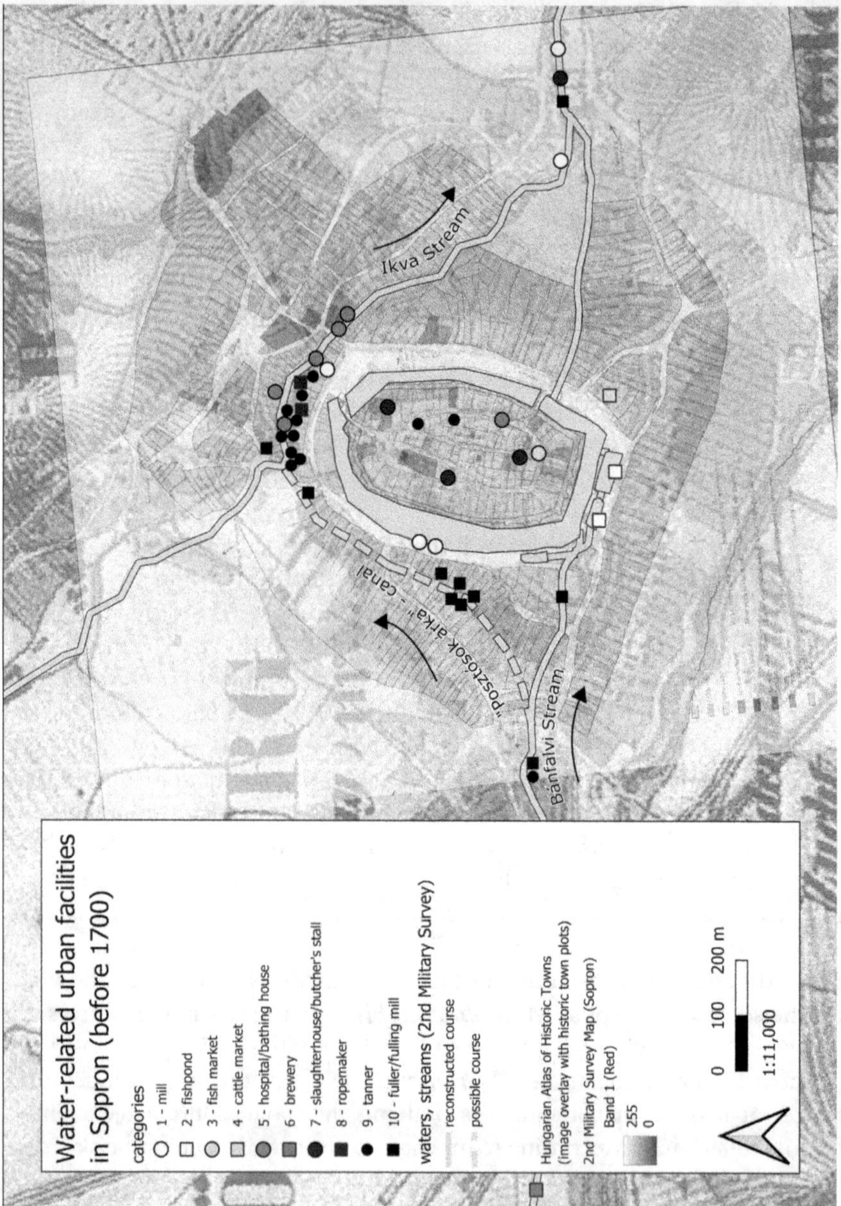

FIGURE 3.7 The water-related urban infrastructure of Sopron before 1700.

and appears in the so-called Notebook (*Gedenkbuch*), a town book of mixed content. According to an entry from the year 1496, the town council decided to relieve Jacob Pilsner of his duties overseeing the fishponds (or rather, small inlets).[191] Based on this entry, it is clear that the three basins in which fish were kept had already been created. The basins were located between the row of houses along present-day Ógabona Square and Újteleki Street and were connected to the moat and Bánfalvi stream water systems. Fish keepers were in use during the late medieval period and were likely still there in the later centuries. This is supported by a complaint from 1570, which stated that the mill located at the houses by Lange Zeile had caused significant losses to the fisheries and should therefore not be reinstated. Lange Zeile, connected to the southern part of the Inner Town, was likely associated with the previously mentioned ponds.[192] The exact cause of the conflict remains unclear, but it may have been related to the amount of water the mill owners permitted to enter the canal or to pollution from the mill. Fish keepers are not only mentioned in these sources but also identifiable in early modern town views (*see* Figure 3.8). The final reference to the fisheries in the moat dates back to

FIGURE 3.8 Zacharias Michel's Bird's eye view of Sopron in 1700 (after Jankó, Kücsán, and Szende, *Sopron*, Plate C.4).

Source: Sopron, Soproni Múzeum (Museum of Sopron) KP 54.509.1.

the early seventeenth century. The last entry of the unique (auto)biography of the town's mayor, Christoph Lackner, pertains to the fisheries in the city:

> in the same year [in 1618] to the great benefit and in a good manner he had the fishpond cleaned which had been filled with mud from the side of the new tower, and had it strengthened all around with pales and fences.[193]

The lake referenced here is located in the southern part of the town and is easily identified in some town views, remaining there well into the eighteenth century.[194]

Although written sources rarely mention conflicts over water use in the moat and its feeding stream, the Bánfalvi, the location of various industries and crafts within the town indicates significant pressure on the limited water source they provided (*see* Figure 3.7). Efforts have been made to separate polluted waters from cleaner ones. Evidence of the management of contaminated waters dates back to 1625, when thirteen Lutheran textile makers from the town of Jihlava (Iglau) in Bohemia settled in Sopron.[195] They were granted both the freedom to practice their religion and the opportunity to form a guild. A small canal, known as the *Posztósok árka* (meaning Ditch of the Clothmakers), was dug from the Bánfalvi Stream to supply water for their workshops. This water, likely containing a significant concentration of pollutants, was discharged into the upper reaches of the Ikva, just upstream from where the town's baths and numerous industrial workshops were situated.[196] Consequently, the Posztósok árka served as a water source for all three of the city's baths, which were located among the workshops of various artisans. However, the Posztósok árka helped protect the downstream sections of the Bánfalvi Stream and the western and southern parts of the moat from sewage influx, where the associated fishponds certainly needed clean water.

It was also important to have clean water for the fulling process at the mills in the southwestern and western parts of the moat. However, in the area downstream from the mills, the water from the moat could certainly not be used for industrial or domestic purposes. Evidence indicates that the moat was predominantly used for sewage disposal along its northern and eastern sections. Several sewage outlets were constructed in the Middle Ages, which drained wastewater from the walled town only into the moat.[197] The sources refer to these drains as *mörin*, which made it impossible to use the moat water in most sections around the town. *Mörins* are mentioned in the provisions of the so-called First Land Register (*Erstes Grundbuch*) every few years.[198] They facilitated the drainage of sewage and rainwater from the inner-city streets and helped keep the fish market clean in the southeastern part of the walled town.[199] Some of them were covered up very early, in the sixteenth century, likely due to their odor, as alluded to by Lacker's later autobiography quoted above.[200] Although their primary function was drainage, a recurring issue at town council meetings was the dumping of rubbish into the moats.[201]

As shown above, only certain areas of the town and the upper reaches of the watercourses were safeguarded from sewage. Along the Bánfalvi Stream, in the western part of the city, fish farms needed clean, slow-flowing water. Under normal runoff conditions, providing clean water for the industrial workshops would have been challenging, except for the aforementioned Posztósok árka. In the urban section of the Ikva, water quality may be poorer due to the large number of workshops and wastewater discharges from the Posztósok árka.

It is evident that the moat in Sopron was closely integrated into the local economy and represented a significant part of the town's water management. Despite the relatively rich source material, the only known conflict over water use was between the mill owners along the *Lange Zeile* and the fishery owners. In Sopron, different industries had to coordinate their water use along the moat and in other areas. In the following subchapter, however, it will be demonstrated that the same methods of utilization also created challenges when it came to managing water in the moat.

3.2.3.3 The Case of Prešov

This subchapter will focus on the town of Prešov, located in the northeastern part of medieval Hungary. Similar to Sopron, the use of defensive moats in the local economy is well documented. In this case, the analysis of the mills and the moat also enhances our understanding of the urban topography and the town's development. The town began to grow somewhat later than Sopron, with the first reference to the settlement dating back to the thirteenth century. At that time, it was already under royal authority, although there is no data on the settlement's broader freedoms, which were probably not significant at the time. Nonetheless, it is important to note that the settlement core had already existed at the same site where the town later emerged.[202] It likely grew into a significant settlement in the late thirteenth and early fourteenth centuries due to its location along an important trade route connecting Lesser Poland with the Kingdom of Hungary. By the late medieval period, the town of Prešov was part of a group of Hungarian towns that enjoyed the widest possible liberties (that is, free royal towns), although it was never among the most populous settlements in the country.

The history of the urban fortifications of Prešov is arguably among the best known in medieval Hungary. This does not primarily stem from their conservation, as virtually nothing remains of the medieval fortifications, but rather from the highly detailed written evidence preserved in the local archives. In addition to medieval documentary evidence, early modern visual sources significantly enhance our understanding of the fortification's structure. Both Hungarian and Slovak researchers have shown interest in the town's medieval history; however, owing to the survival of extensive medieval sources, only a few questions have been addressed by existing scholarship. Since the archivist Béla Iványi's foundational two-volume publication of the medieval records from Prešov's archive, research has drawn on the substantial evidence it presents.[203] Drawing on this rich source material from the early twentieth century onward, several studies have examined various aspects of the town's history, from its ethnic composition and legal history to elements of urban topography. Until recently, the most important works were based on the testaments that survived in relatively high numbers from the late medieval period. However, these documents do not contribute to our understanding of urban fortifications or their economic exploitation.[204] Recently, data were gathered on the history of the town's fortifications, based on the cartulary published by Iványi.[205] However, he did not mention the moats. Since he focused on the Middle Ages, he also overlooked town views, a rich seventeenth- and eighteenth-century source that is essential for identifying some of the medieval topographical elements associated with moats.[206]

Several questions remain unclear regarding Prešov's fortification moat, which may be crucial for the town's history as a whole. Exactly when was the town encircled by a moat? Did it surround the town, and was it filled with water? Following this line of inquiry, how was the moat utilized in the local economy? As is demonstrated in the upcoming subchapter, sources on the construction and expansion of the moat in Prešov enhance our understanding of the growth of certain Central European towns associated with long-distance trade routes in the fifteenth century, and more importantly for this book, with the conflicts that the use of the moat caused within the urban community.

In the thirteenth century, Prešov may not have been much more significant than a village. Yet, its favorable location along the eastern route connecting the emerging centers of Kraków and Buda likely contributed to its growth as a trading post, alongside Košice.[207] Košice likely became important somewhat earlier than Prešov, emerging at the end of the thirteenth century, and continued to grow in privileges and significance throughout the late medieval period.[208] With Košice's rise, Prešov may have gradually gained importance as well. However, the available sources contain no explicit references to the town's early history. The fact that the first known initiative to surround the settlement with a wall dates to 1370 indicates that Prešov also grew relatively quickly. This initiative is documented in a mandate issued by King Louis I the Great to the nobility of Sáros County, instructing them to provide lime, stone, and wood for the construction of Prešov's walls.[209] While some scholarship suggests otherwise, the document does not mention either a ditch or a moat, which were likely dug only somewhat later.[210] By the early fifteenth century, however, Prešov was indeed surrounded by a moat. This is supported by a tax list from 1428, which refers to taxpayers of the houses "by the ditch" (*in fossato*).[211] Additionally, the town's expenses for the construction of the ditch in 1429 further confirm the presence of a more elaborate fortification system. As will be highlighted later, the mills described as standing by the moat leave no doubt that by this time, at least on its western side, the town was protected by a moat.[212]

Another source from King Sigismund's reign provides important details on the fortifications, particularly the moat. In a charter from 1435, the king authorized the rebuilding of the town's fortifications in response to the Hussite threat.[213] The walls of the town, as well as the moat, were subsequently extended. However, the extension of the moat created a problem. At that time, at least two water mills relied on the moat's water. When the moat was extended, the mills had to be relocated. Sigismund permitted the town to dismantle the two mills and reconstruct them elsewhere in Prešov. Because most of the mill parts were made of wood, moving them was not considered impossible in the Middle Ages.[214] The water mills were not originally built by the town but were owned by two burghers, Peter Langwart and Peter Blume, as indicated by the tax lists; both were affluent citizens.[215] According to Sigismund's authorization, these men had the right to rebuild their mills at another location. The mills were indeed relocated, but the town itself undertook to rebuild the water mills. Seeing an opportunity in the relocation, the town decided to compensate the burghers with a share of the mills' capacity.[216] Only a limited share of the mills' income went to their former owners, and the town did not hold a majority stake in the mills. Most of the income from the two mills was allocated to various ecclesiastical institutions in Prešov from that time onward. The town gained full possession of the mills in 1494, half a century later. These conditions persisted until modern times.[217]

From the perspective of urban topography, the localization of the original sites of the mills and their new locations from the 1430s onward is of major importance. Understanding the mills' original site can help identify other topographical features and contribute to our view of how the town evolved.[218] According to documents regarding the relocation of the water mills, one was initially located right next to the Upper Gate (*molendinum ante superiorem portam*). Some sources indicate that the mill operated throughout the fifteenth century, but information becomes more plentiful from 1494 onward, when the building was finally acquired by the town. The exceptionally detailed accounts of Prešov from the late fifteenth century refer to this mill as the Upper Mill (*Obermoell, molendinum superior*; see Figures 3.9 and 3.10).[219] Besides this mill, which can be precisely located next to the Upper Gate, the location of the other mill, which was affected by the extension and referred to as the Middle Mill (mo*lendinum medium*),

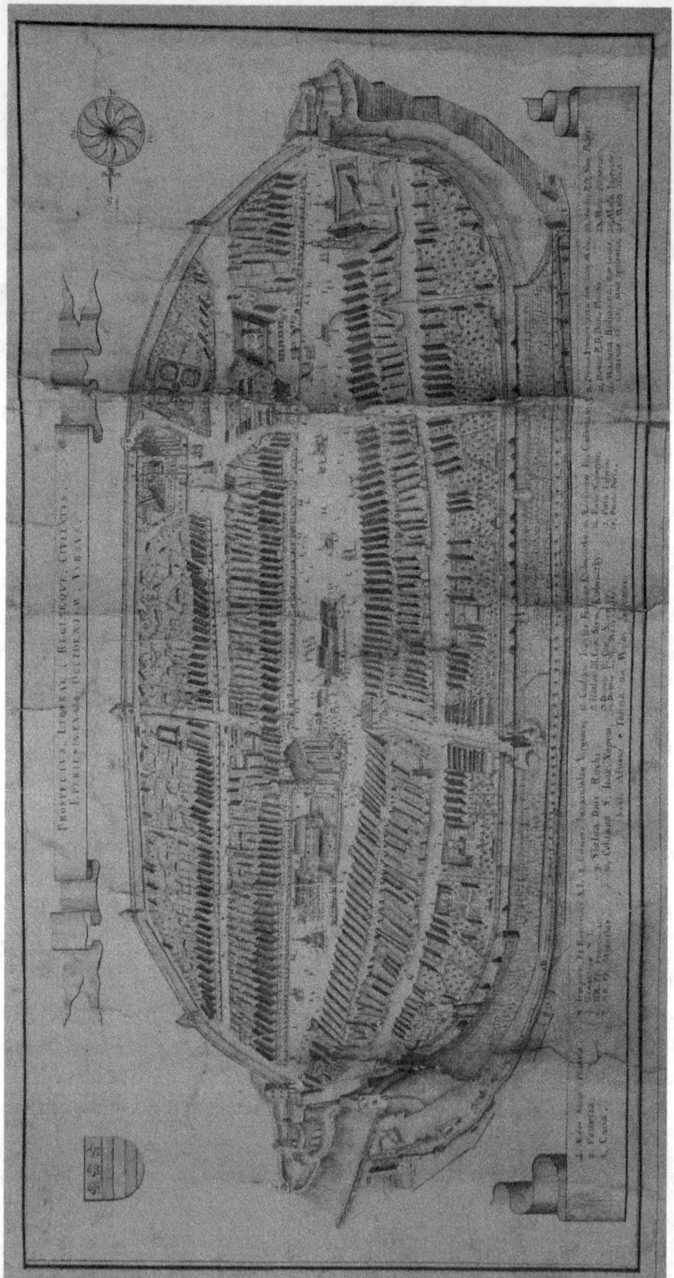

FIGURE 3.9 View of the town of Prešov from 1768 by Gaspar Caspar.

FIGURE 3.10 Detail of the view of the town of Prešov with the Upper Mill (no. 23) from 1768 by Gaspar Caspar.

can also be determined with reasonable certainty. A fragment of an urban account from 1429 mentions expenses related to the bridge near the Middle Mill. This bridge likely stood next to the so-called Middle (or Small) Gate of the town, close to which the Upper Mill stood. From the same year, the urban protocol mentions a street by the Middle Mill adjacent to the town wall.[220] However, the relocation of this mill poses some challenges. Unlike the Upper Mill, it was likely moved somewhat farther from its original location. Instead of being in front of the Middle Gate, it was relocated to the lower part of the town after the expansion of the settlement, thus adopting the new name of the mill: Lower Mill (*Niedermoell, molendinum inferior, mola vero civitatis inferioris*).[221] This suggestion regarding the location of the mills at different periods can be corroborated by one of the earliest detailed views of the town, dated 1768. This town view marks the locations of the Upper and Lower Mill, both likely operating at those sites from the mid-fifteenth century onward (*see* Figures 3.9 and 3.10).[222]

Besides these two mills that operated for centuries in the western section of the town's moat, additional mills were also built in the town during the Middle Ages. The town's records recurrently record a fulling mill (*molendinum pannificum*). While the exact location of this mill remains unknown, due to the lack of other water sources within the town, it likely utilized the moat's waters. The accounts from the early sixteenth century also reference a smith's stone (*de lapide vero quo utuntur fabri or cotes fabrorum*).[223] This mention, based on its modest income, implies a possible connection between this mill and the fulling mill, or perhaps to the Lower Mill.[224] Starting in the fifteenth century, alongside the two flour mills and the fulling mill, there are recurring mentions of a gunpowder mill in Prešov. It remains unclear whether this mill used water as an energy source for producing saltpeter. Additionally, it is unclear whether this was a standalone mill that produced gunpowder exclusively, a wheel within another mill, or a flour mill temporarily repurposed for that use. Several instances from medieval and early modern Hungary suggest that the latter scenario may be more plausible. Based on accounts from 1533/1534, Iványi proposed that the building was situated next to a friary. The only mendicant friary operating in the town during the late medieval period was a Carmelite institution. This building was located by the town gate along the road leading to Solivar, a settlement situated somewhat southeast of Prešov.[225]

Medieval sources do not clearly specify whether a moat also existed on this side of the town. The water in the section of the moat by the mills on the west side of the city originated from the nearby River Torysa (records do not note mills by this larger river). This channel can be easily identified on early modern maps, including relevant sections of the First Military Survey from the late eighteenth century. The eastern section of the moat was likely not connected to the River Torysa or the western section of the moat. However, according to an eighteenth-century description of the town by Matthias Bel, it appears that the entire town was then encircled by a moat.[226] If this were accurate, given the local geography, the water in the eastern section of the moat likely originated from a source other than the River Torysa, as this area is at a slightly higher elevation.[227] The water may have originated from the hills east of the town. It is also possible that the gunpowder mill near the town wall was a dry mill—a possibility that might also apply to the fulling mill. There is no definitive evidence regarding the energy source employed by these industrial complexes. However, a dry mill definitely existed in the town, as evidenced by an entry in the town accounts from 1521.[228]

The account books refer to two additional mills, but neither their locations nor their power sources is clear. The first mill to appear chronologically in the sources was one that ground the malt to provide the burghers with beer. As sources mention this mill as being a "malt mill and a wine house" (*malzmühle und weinhaus*), the building must have housed one of the town's pubs.[229] Sources from the 1510s also refer to another mill (*de molendino bursatili*), which operated for only a couple of years; its location and whether it was a water mill are unknown (*see* Figure 3.11).[230]

The moat in Prešov, owing to the lack of other water sources within the town's walled area, became a crucial part of the local economy. It appears

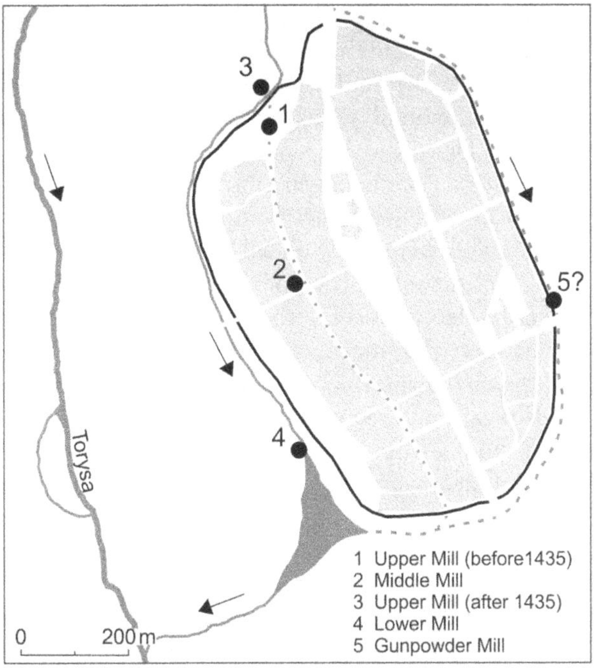

FIGURE 3.11 Topography of Prešov in the late medieval period, with the relocation of the mills and the probable direction of the western section of the urban moat before and after 1435.

that the citizens who owned plots along the moat sought to capitalize on their easier access to water. However, this access became complicated when the town walls were extended. Citizens could partially extend their rights to have mills by the moat, invoking the principle of previously acquired rights, but the town recognized an opportunity in the relocation process to claim at least part of the milling income.

3.2.3.4 Moats, Urban Topographies, and Mills: Some Conclusions

This subchapter was meant to demonstrate that urban economies in premodern Europe utilized not only major waterways and minor streams but also urban moats to generate income. Moats were part of the defenses of towns that, despite representing a significant investment and often symbolizing the status of settlements, produced no income. As their military purposes were only temporary, urban economies gradually began to exploit them in various ways. The military's needs differed from those of individuals who used moats for economic gain.

Of the two cases examined in detail, Sopron provides rather general lessons, primarily that many local industry stakeholders sought access to the moat water. Fishermen and the nearby fish market, often located close to the fisheries, as well as the grain market and the cloth makers and fullers, all aimed to take advantage of the waters surrounding the town.[231] The insights from the Prešov example are at least twofold. One concerns the town's topography and overall urban water use. First, King Sigismund's permission and the two agreements with the mill owners, Peter Langwart and Peter Blume, help to date the extension of the town walls and the streets themselves. During this period, a new street and a line of houses were likely added to the existing settlement, necessitating the relocation of the mills. The documents concerning the mills highlight the fifteenth-century development of the town. While the Upper Mill likely required only a few meters of movement, the Middle Mill had to be relocated farther south, thereby facilitating a significant westward expansion of the town. The records regarding the extension of the fortifications, along with the detailed account books from the late fifteenth and early sixteenth centuries, enabled the collection of data on the town's mills. They also revealed that, despite its relatively small population compared with other free royal towns in Hungary, Prešov housed many such complexes. Whether some of these mills used water or another energy source remains unclear.

This survey of the moats and the exploitation of the waters surrounding Sopron and Prešov also highlighted some general features of urban water management. Population concentrations, such as walled towns in Hungary, had to use the water within their borders efficiently to provide energy for various industries, and, of course, water was necessary for several crafts. This demand for waterpower led to an increasing number of mills being constructed near moats during the Middle Ages. Researchers have identified several similar buildings in towns across Western and Central Europe. The moats of several further cities in the Kingdom of Hungary were also utilized. The examples discussed above are among the best-studied, but the waters surrounding the castles of Moson and Óvár also supplied major mills with waterpower.[232] A section of the moat in Cluj, Transylvania, was also used as a mill race in the late medieval period. In this case, however, this semi-artificial riverbed likely predated the establishment of the moat around the town.[233] From the sixteenth century onwards, with the appearance of the Ottomans and new fortification techniques, military needs became more significant than those of local economies. Nevertheless, partly because of the strategic importance of locating mills near castles and towns, and the prolonged, intermittent wars, mills and fisheries continued to operate in moats into modern times in the Kingdom of Hungary.

3.3 Rivers in Mining Areas: Abundance of Water versus Lack of Access?

Mining towns form a particular settlement type in the medieval Kingdom of Hungary.[234] These settlements were granted privileges related to prospecting for and exploiting metallic ore within their territories. This aspect was a common element among such towns, in contrast to the many and varied privileges a settlement could acquire from kings or other landlords. The geographical distribution of these settlements is closely associated with the mountainous regions of the medieval Kingdom of Hungary, most notably the territory known in the premodern period as Upper Hungary (largely coinciding with present-day Slovakia) and parts of Transylvania.

In the Middle Ages, the distinction between mining towns and other privileged settlements was not as clear-cut as it is often portrayed in modern scholarship. These towns were never officially associated with each other. However, two groups of mining settlements in Upper Hungary frequently formed alliances from the late medieval period onward to achieve their political and, in particular, economic goals. Modern scholars refer to these two town groupings as Lower Hungarian mining towns (*Alsó-magyarországi bányavárosok* in Hungarian), namely: Kremnica, Banská Štiavnica, Banská Bystrica, Nová Baňa, Pukanec, and Ľubietová, and Upper Hungarian mining towns (*Felső-magyarországi bányavárosok* in Hungarian), namely Gelnica, Smolník, Rudabánya, Jasov, Telkibánya, Rožňava, and Spišská Nová Ves (*see* Figure 3.12). The formation of these alliances at the very end of the Middle Ages does not mean that these towns constitute a clear group of settlements characterized by common features in their privileges, economies, political life, or appearance. There are several fundamental differences in how each town developed, while there are also many towns involved in mining that never became part of these alliances. This is particularly true for towns in Transylvania, where the various mining centers never formed alliances with one another.[235]

The settlements discussed in this upcoming subchapter of the book share at least one common feature: they were founded and/or developed to exploit the ores discovered in their territories, primarily gold and silver. Most of these settlements were established at the initiative of kings in the thirteenth and fourteenth centuries and received privileges formalized in charters. These privileges related to ore exploitation often included additional economic rights beyond prospecting rights. Some of the privileges also include forest and water resources. The importance of forests for timber is almost self-evident for a newly established settlement; however, for these settlements, timber was also essential for lining mine shafts, constructing crushing mills, sawmills, and numerous other buildings associated with mining.[236] This chapter, however, focuses on another natural resource mentioned in several foundation charters: water. This resource, as argued, was crucial to mining towns.

In the first part of the subchapter, the focus will be on charters of privilege granted to mining towns—and to some other royal towns—to determine whether access to water is at issue. The aim is to understand the reasons

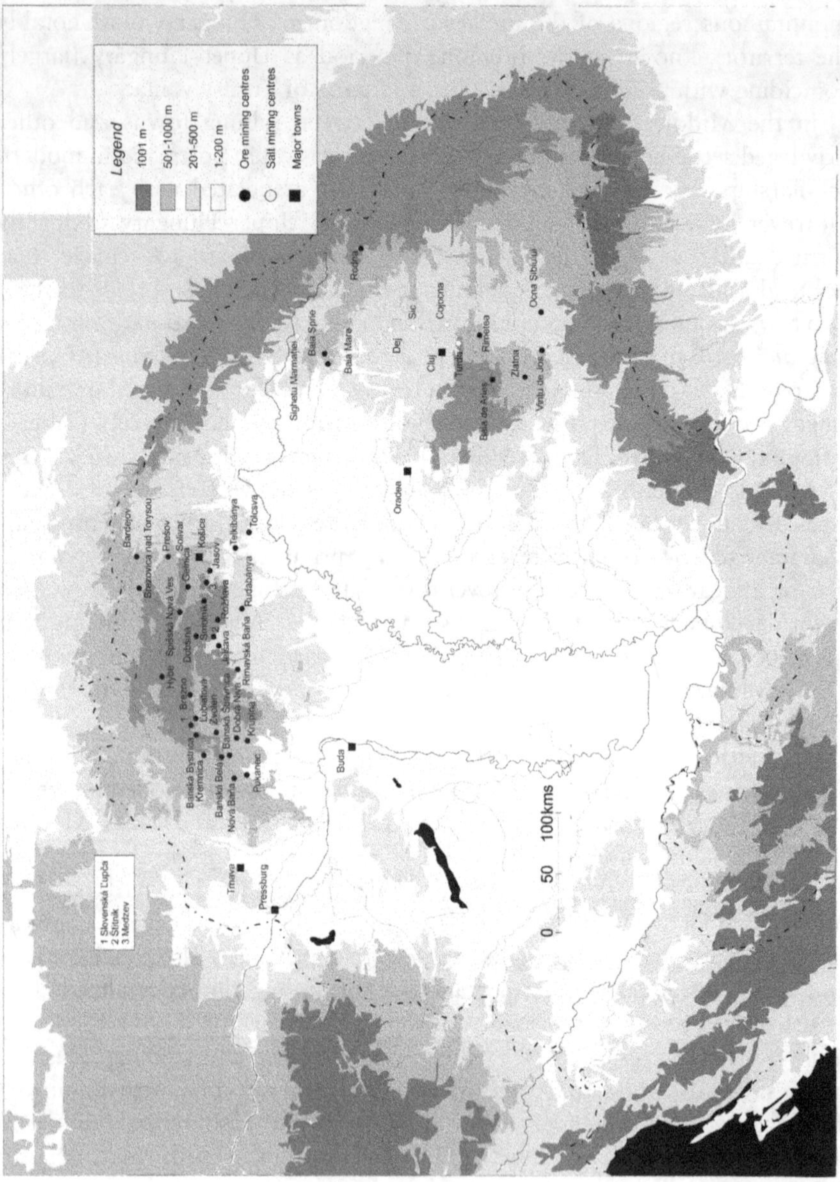

FIGURE 3.12 Mining towns in medieval Hungary.

for the absence of references in most of these privileges to the right to build water mills or, more generally, to exploit water. This silence is all the more surprising because water rights and access are particularly pronounced in the pragmatic textual sources related to these mining towns, especially in documents issued by their authorities.

Following this discussion, the subchapter uses the examples of the water mills from two mining towns, Kremnica and partly Nová Baňa, to highlight the significance of water power exploitation in the lives of the settlements and to examine the associated problems. By studying documents that provide information on the early water mills in these towns, it is possible to understand who had the right to build and own mills in these two towns and in the villages founded by Kremnica in its vicinity. By analyzing the privileges and early documents of mining towns, this subchapter argues that although privilege charters do not mention the right to build mills, settlers in these towns did exercise this freedom.

3.3.1 Mining Town Privileges and Water Use

There is significant literature on the privileging of settlements and the attraction of foreign settlers in the thirteenth and fourteenth centuries. Recently, several studies have specifically addressed the privileges of mining towns in Hungary. Thus, scholars find themselves in a relatively good position to examine the medieval privileges issued to mining towns. The overwhelming majority of the initial charters and later privileges were published in cartularies over the last century and a half, and most were included in high-quality cartularies.[237] Therefore, it is not difficult to gather the most important mining town privileges from the thirteenth and fourteenth centuries. The task becomes somewhat more complicated during the reign of King Matthias and the Jagiellonians; nevertheless, most privileges for these towns date to before 1400. From the perspective of this subchapter, the early settlement phase is particularly important. The collection (included in Appendix 2) not only features settlements later described as the Lower and Upper Hungarian mining towns, but also every mining settlement that acquired a relatively broad range of privileges. The foundation of most of these settlements dates to the century between 1250 and 1350. Since many of the towns receiving privileges during this period inherited the freedoms of their "older brothers", they share commonalities.

As noted earlier, these charters of privileges often grant the freedom to exploit natural resources within the limits of the settlements. The necessity for timber to initiate mining activities, along with the requirement for timber to construct dwellings, makes the rationale behind obtaining permission to exploit timber resources quite clear. Therefore, acknowledging these privileges was almost as crucial as recognizing the freedom to mine. A prominent

Hungarian economic historian, Boglárka Weisz, recently demonstrated that mining towns, in addition to what transpired within their borders, typically received authorization to cut timber within a specified radius from the forests for lining mine shafts or other mining-related purposes (but usually not for constructing dwellings for the inhabitants). However, the extent of this area may vary significantly depending on location, needs, the presence of forests within settlement boundaries, and other factors. Instances exist where zones of one, one-and-a-half, two, and three miles were granted in the privileges. The two-mile zone was the most commonly found in the charters. Despite this trend, the privilege charter of Baia de Arieş from 1325 refers to the one-and-a-half-mile zone that the town received as the "customary zone" for mining towns.[238] It is likely more significant that the mining folk in these towns could utilize the forest resources in these areas, provided they do not infringe on the interests of other parties—that is, other landowners whose property fell within the specified radius. This stipulation aligns with the principle in the Hungarian legal tradition mentioned earlier, where previously acquired rights take precedence over subsequent claims.

Gusztáv Wenzel, a pioneer in the study of mining history in Hungary and a noted expert in source editing, demonstrated that the freedoms granted were a frequent source of conflict, as illustrated by the Dóczy family and their ongoing struggles with the neighboring mining towns.[239] However, without the presence of forests and timber near the towns, opening a mine would have been significantly more expensive. By the Early Modern period, the depletion of forest resources had become a problem in the immediate vicinity of mining towns. This is well reflected in the altarpiece of the Assumption of the Virgin Mary Cathedral at Rožňava (completed in 1513). As the hinterland for the nativity scene, the painter chose to depict the Rožňava landscape itself. Because of timber harvesting for the mine shafts excavated in the surrounding hills, the painter portrayed a barren hill stripped of trees. The image may be somewhat exaggerated, but it clearly indicates that cheap, locally accessible forests were heavily exploited by the mining settlements (*see* Figure 3.13).[240]

In the life of these towns, however, water was equally essential. In addition to fishing, which was significant for some mining towns, especially along the River Hron, water power was also crucial. In addition to grinding grain to produce flour, water power was utilized in at least three other areas of economic life in these settlements. Sawmills were necessary to supply the mines with timber, and the processing of the major stone blocks brought to the surface began with crushing and grading the resulting rubble, with ore-crushing mills holding primary importance. Furthermore, water power was used to lift materials from the mine shafts to the surface and to pump water from the shafts.[241]

The importance of ore-crushing mills is best highlighted by the fact that their number was comparable to that of grain mills. Sawmills also played a

FIGURE 3.13 The so-called Saint Anne altar at the Virgin Mary Cathedral at Rožňava (detail, photo by author).

significant role, as they provided the most economical means of preparing the timber required for mine shaft construction. The presence of groundwater in deeper mine shafts caused serious problems in mines across Central Europe from the fourteenth to the fifteenth centuries.[242] Water power, or, in its absence, animal power—particularly horsepower—could also be harnessed for these machines. Finally, like horse mills, water power was used to bring the ore to the surface.[243] Among the various ways to harness water power for mining, all were likely present in the medieval Kingdom of Hungary. However, some applications emerged much later than the establishment of mining towns at the turn of the thirteenth and fourteenth centuries.

In addition to the uses of water power in the mining industry, the local population also needed access to flour. Supplying some of the mining towns with flour—especially the most populous ones, like late medieval Banská Bystrica, Banská Štiavnica, and Kremnica in Upper Hungary, as well as Baia Mare in Transylvania—required several wheels that processed not ores but grain. Without comprehensive tax lists, estimating the population of these towns is challenging. However, assuming that one mill wheel could provide flour for approximately 200 to 300 people, some mining towns had at least five wheels, and in some cases up to ten, used to grind grain; the populations of some of these settlements could have exceeded 2000 to 3000. Furthermore, other uses of water power, such as the textile and paper industries, gunpowder manufacturing, and, last but not least, brewing, should not be overlooked.[244]

Although, as argued above, water power played a crucial role in the effective functioning of mining and in supplying the related population, the privileges of Hungarian mining towns, with two exceptions, fail to mention settlers' right to build mills on rivers within their town limits. This gap in the texts does not only pertain to water rights. Several charters of privileges listed in Appendix 2 fail to specify the rights of the founded settlements. They remain silent not only on economic matters but also on legal and ecclesiastical issues. However, they grant the same freedoms as earlier-inhabited mining settlements. Water-related freedoms are similarly unspecified in the privileges granted to mining towns, which reference the liberties of other royal towns.[245] The adoption of the freedoms of other mining settlements or royal towns is particularly interesting, as the various gold and silver mining towns that served as models were sometimes given highly differing privileges. As mentioned above, the freedom to exploit either forest or water resources was anything but consistent.

Fishing and the rights to mill construction, two distinct forms of water use, need to be discussed in detail, even though there are only two privileges related to these activities. The extent to which newly established mining settlements had the right to use the waters within their boundaries is particularly intriguing, as numerous charters of privilege granted to hospites (foreign settlers) during the Árpádian period and under Angevin rule granted them the right to use water for fishing and other purposes.[246] Additionally, several privileged settlements, apart from mining towns, only received partial rights to access the waters within their limits.

The earliest mining town privilege relevant to this subchapter was granted to the settlers of Banská Bystrica in 1255 by King Béla IV. The privileges granted by the king included rights to forest resources and water. However, with respect to water-use rights, fishing in the River Hron, known for its abundant fish, was prohibited. Nonetheless, as part of the rights to prospect for precious metals, the king also listed, although only at the end of the list of potential prospecting activities, gold-panning in the rivers, along with other water-related rights for those moving into the area. The specific details of these later rights were not outlined in the privilege charter. It is argued in the following pages that these rights may have encompassed the right to construct mills along the rivers.[247]

In other mining towns, privileges—those freedoms pertaining to water-use rights—primarily focus on fishing. In the extensively debated charter of privilege granted to Partizánska Ľupča in 1263, King Béla IV granted the settlers the right to fish freely on the River Revúca and, as in Banská Bystrica, conferred the right to prospect for gold, silver, and other metals in its waters.[248] It remains unclear how to interpret the free fishing rights in the Revúca, given that the River Váh flowed closer to the settlement. It is worth noting that the similarly ambiguous charter granted to the settlers of Hybe by King Béla in 1265 contains provisions akin to those of Banská Bystrica regarding prospecting. Yet, it explicitly forbids fishing in the Váh.[249] The charter for the settlers of Gelnica, reissued in 1287 by King Ladislas IV, also addresses fishing rights. In this document, following the privileges granted by Ladislas himself, as well as by Béla IV and Stephen V, unlimited fishing rights were granted to the town's fishermen within the settlement's boundaries.[250] In 1318, the settlers of Ružomberok, while receiving the general privileges of Partizánska Ľupča and being given fishing rights, were nonetheless prohibited from fishing in the River Váh.[251] The lack of full access to all waterways for fishing may be related to the fact that fishing was a significant source of income for royal counties up until the mid-fourteenth century. Charters issued to *hospes* settlements often included fishing bans, whether general or specific to particular equipment. It is crucial to note that the 1318 privilege charter of Ružomberok, which banned fishing on the River Váh, permitted the town's inhabitants to fish in the River Revúca, similar to what was granted to Partizánska Ľupča half a century earlier (in 1263).[252] It is not only in the Kingdom of Hungary that settlers received rights to only some of the rivers within their borders. In medieval Polish urban privileges, this limitation appears more consistently, exemplified by a charter granted to the town of Dobczyce in Lesser Poland, which allowed fishing only in the ruler's absence.[253]

The situation is more ambiguous in the 1340 charter granted to the inhabitants of Ružomberok, which granted them full fishing rights. However, a few lines later, the document mentions the right to fish on the Revúca, but the Váh is not mentioned. The 1340 charter of Ružomberok is significant not only for its fishing rights but also as one of the two privilege charters granted to mining towns that recognize the right to build mills. When water use rights were mentioned, they were most often in the context of fishing or prospecting rights. However, the 1340 privilege expanded the rights of Ružomberok's inhabitants, granting dwellers the right to construct mills.[254] Understanding this grant is crucial for this subchapter. Did this right to utilize water for milling signify the acquisition of a new right that the settlers did not have before, or was it an existing right that was clarified? This question is equally intriguing for understanding what the remaining water rights, not specified in the Banská Bystrica charter, may have been.

One possible interpretation is that the burghers of Ružomberok lacked a mill, or that the mill they used had not been built by them. The former scenario is improbable, as a span of twenty-two years passed between the initial liberties and their extension, during which the town could not have functioned without a working water mill. Furthermore, it is certain that by 1325 the settlement was inhabited, as a burgher of Ružomberok appears as a witness in a privilege charter issued to another *hospes* community.[255] The other interpretation is that mills were built between 1318 and 1340. Still, before 1340, their construction was a seigniorial right, and they would have been constructed by Doncs, the *comes* of Zólyom County (until 1338), who initiated the settlement process.[256] This latter possibility, despite the absence of surviving written evidence for a mill, seems unlikely, as there are no parallels to it in other mining towns. There is also no evidence of the king, nor their *comites*, constructing mills in any of the towns founded by them to exploit the ores discovered in their territories. The 1340 charter should instead be understood as indicating a claim by the burghers to secure a right they had already been exercising: the free construction of mills to grind grain and process ore. This idea is supported by the fact that in 1340, after Doncs's death, the settlers of Ružomberok successfully persuaded King Charles I to issue a privilege that included the right to prospect without any reference to the *ispán*'s 1318 privilege.[257] (This question will be revisited when discussing the lawsuits connected to the mills owned by the settlers of Nová Baňa.)

Importantly, none of the early privileges specifically forbade settlers from building mills, as this was among the water-related freedoms mentioned in the context of the Banská Bystrica charter. Alajos Degré, in his monograph cited above on fishing in medieval Hungary, suggested that these privileges were limited to drawing water from rivers and watering animals. However, nothing indicates that water rights were restricted to these two elements alone.[258] It is quite possible that, when the charter issued to the burghers of Ružomberok in 1340 referred to their right to build mills, it was not an extension of rights but rather a clarification of an already existing, but poorly defined, privilege.

The fact that, apart from the water privileges granted to the burghers of Ružomberok, only a handful of settlements—not just mining towns, but privileged settlements in Hungary as a whole—acquired the right to build mills does not imply that others were denied this right. Instead, it suggests that, until the mid-fourteenth century, it may not have been considered essential to specify this right in charters.[259] The inclusion of this right within the privilege in 1340 can likely be linked to the increasing number of mills established along the Revúca and other rivers in the region, indicating competition for water resources.[260] Furthermore, following the clarification of customary law on mill construction discussed in Chapter 3.1, it became increasingly vital for burghers to construct mills without significant disputes over water-use rights.

This tendency appears to be corroborated by another privilege, similar to that granted in 1340 to the burghers of Ružomberok. The document in question is a charter issued in 1376 by King Louis I to provide privileges to the Transylvanian mining towns of Baia Mare and Baia Sprie. The antecedent of the charter is an earlier privilege issued by the same king in 1347, which refers back to an even older privilege that was supposedly later consumed by fire. The 1347 document, contrary to earlier literature, has survived.[261] It contains detailed information on rights to forest use but remains silent on the benefits associated with water.[262] While in Upper Hungary, fish caught from rivers, especially the River Váh, may have supplied the royal court or the populations of the county castles; these aspects may not have played a role in Baia Mare.[263] It is therefore unlikely that local fishing regulations would have been stricter. The absence of a reference to the right to use the water does not imply a ban on fishing or the construction of mills.

The 1376 privilege, however, brings the question of water rights to the fore. The charter secures the rights of Baia Mare and Baia Sprie to build water mills on the rivers within their city limits. In addition to the water mills themselves, the document lists the benefits the king granted to the burghers. These, apart from the mills, included "forges, furnaces, laundries, allodia and other buildings." The meaning of two of the Latin words in the list of buildings, *casa* and *balneum*, is anything but self-evident.[264] The latter is likely easier to interpret due to the context. Nevertheless, from the late medieval period onwards, bathhouses have been found in numerous mining towns. It is unlikely that the privilege concerned these institutions; rather, it was a location for panning.[265] The exact meaning of the other word, *casa*, however, is not evident at all. The word *chaza* or *casa* appears in other contemporary documents pertaining to mining towns.[266] On the one hand, a document from Kremnica, issued in 1331 and discussed in detail below, suggests that the word denoted a building directly associated with water use.[267] On the other hand, late medieval accounts of Banská Bystrica repeatedly refer to *casae* in the form of *locus casae*. The wording in the 1376 document, *locum casae*, suggests a more specific building than a storehouse for tools, as has been suggested earlier, especially since the building needed a special *locus*, probably directly by running water.[268] Last but not least, the 1376 charter of privilege listed this benefit among various water-related rights, which also suggests that the building in question was directly water-dependent. The best guess is that it was a forging mill.[269]

Accordingly, the charter permits the burghers of Baia Mare and Baia Sprie to utilize the waters of the towns for milling. This right was likely significant for the mining and processing of ores. King Louis I granted this right to the two mining towns, not as a special liberty, but in accordance with the "customs of other mining communities" (*ritu aliarum nostrarum montanarum*). These words are crucial to this subchapter because, to date, the only known privilege that provides this right is the 1340 charter of Ružomberok. However, even that charter did not specify the water-related benefits with the same level of detail as the 1376 document. Thus, the overall picture remains unclear. Based on the analysis of documents predating the 1376 privilege, there is no basis to suggest that the right to erect mills was commonly held. Nevertheless, considering what has been stated regarding the possible interpretations of the rather general privileges to use water and the note from the charter of 1376, it can be argued that the settlers of the mining towns exercised the right to use water for their benefit or at least to construct mills during both the late Árpádian and Angevin periods. In the following subchapter, privileges granted by kings to those outside mining communities are analyzed to assess how common it was in the thirteenth and early fourteenth centuries to grant settlers the right to build mills, thereby supporting either the assumption or its opposite.

3.3.2 *Early Urban Privileges and Water Use*

As mentioned, it was not only mining town privileges that rarely addressed the issue of mill construction.[270] Aside from the charter of Vinohradiv from 1262, Árpádian period charters of privilege granted to *hospes* communities remain silent on the settlers' rights to build their mills. Nor do numerous extant documents mention fishing rights.[271] Except for the mining town charters discussed, only six—specifically, the 1261 charter of Újhely,[272] the one of Vinohradiv (1262), the charter granted to the *hospites* of Győr (1271),[273] those given to the settlers of Vasvár (1279, 1283),[274] and the privilege of Bratislava (1291)[275]—refer to such rights. The privilege of Újhely permits the town's settlers to fish in the River Bodrog; however, the text does not mention other bodies of water or related rights. This omission is particularly significant, as the river adjacent to the newly settled town was not the Bodrog but the smaller Ronyva River. The waters of the Ronyva had been used for milling since at least the founding of Újhely, if not earlier. The first two mills along the river were royal gifts to the Paulines and the Augustinian friars, respectively. The failure to specify the River Ronyva in the privilege may reflect the early establishment of the Paulines in the region and the donation of the mill to the friars. The town quarter inhabited by the Paulines enjoyed immunity until their dissolution in the late eighteenth century, which may indicate their presence predates the settlement of the *hospites* of Újhely.[276] Given these distinct aspects of the area where people settled, it seems pretty clear why kings did not grant unrestricted access to water resources within the borders of Újhely.

The right to use water is pronounced in the privilege charter of Vinohradiv, mentioned above, in the context of the freedom to build mills. The otherwise brief charter also permits the community to fish in their rivers and construct a port. The fact that this privilege provides such detail on water-related matters, akin to the Győr charter, may reflect the significance of water as a vital resource for the settlements. The 1271 privilege granted to the settlers of Győr is crucial because it defines access to various resources in perhaps the most precise terms of any contemporary charters from Hungary. This precision may stem from the settlement's status as the county seat and the bishopric by that time, necessitating that the various social groups living within the town's borders be reflected in the document.[277] According to the Győr charter, settlers were allocated two plots of land surrounding the castle for their use, along with all associated benefits. These included parts of Malomsok[278] bordered by the Danube and an uninhabited parcel belonging to the Seraphin monastery of Hronovce,[279] located by the River Rába. While the charter certainly included the right to fish in the river, the proximity to the River Rába suggests that the construction of mills might also have served as an important source of income for the settlers. The settlers of Vasvár received their privilege from King Ladislas IV in 1279. In it, the king granted the people living in the chapter's territory the rights to fish and utilize the forests along the Rába—an area known for its rich forests even into modern times[280]—as well as any other benefits connected to the latter.[281] However, other benefits mentioned in the charter pertained only to forests, not water. The same holds for the partially recapitulated privilege issued to the Vasvár settlers in 1283.[282] Lastly, the charter provided to the settlers of Bratislava in 1291 granted them the right to fish—specifically in the Danube—of which they had to deliver one-third as tithes to the *ispán* of Pozsony County.

In summary, while early urban and *hospes* privileges sometimes referred to the right to use water within a specific area in a relatively detailed manner, most regulations pertain to fishing rights, and these regulations only indirectly reference the construction of water mills. Why might this be so? Erik Fügedi, in his still fundamental study on the thirteenth-century charters of privilege, argued that despite their donations, rulers typically remained the owners of the land itself. According to Fügedi, kings granted only specific parcels of land (i.e., plots) to tenants, along with certain other rights, such as access to forests or water. This pattern of donation is evident in the privilege from Újhely, in which the issuer, at that time a junior king, retained some waters in his possession and endowed church institutions with related rights and benefits as a landlord. The same applies to the *hospes* community of Győr. In some areas of the town, settlers were not entitled to use water for their own benefit, whereas in other lands granted to them, the king provided all associated privileges, such as the benefits of forests and waters. The latter pattern appears to have become increasingly common in such royal donations, as reflected in the royal charter privilege granted to Ružomberok in 1340.[283]

However, the picture remains unclear based on what has been discussed in this subchapter so far. Two almost entirely different conclusions can be drawn. The first possibility is that in most cases, the king retained his rights as a landlord and provided the settlers, or some of them, with various benefits, either as donations or leases. These may have included rights to fish in rivers, which in some cases were granted only to fishermen, or to build mills along certain rivers, benefiting either the entire community or privileged institutions. The other possibility is that, although most of the charters fail to address the issue in detail, settlers did receive all rights and privileges related to a particular piece of land unless otherwise specified.

As noted in earlier research, early Hungarian urban privileges rarely reference the adoption of foreign legal traditions, such as the Magdeburg Law or other urban privileges.[284] Despite the absence of direct references, many of the urban privileges in the Kingdom of Hungary, populated by German settlers, incorporated elements rooted in these legal traditions.[285]

The contracts and foundation charters that have survived in the Polish territories and Bohemia, however, are often much more detailed regarding issues of water and forest resources than their Hungarian counterparts. In most cases, the *sculteti*, *advocati*, or the community received rights to water, including the construction of water mills. One notable example is the well-known and much-studied foundation charter of Kraków from 1257, which explains in great detail the rights of different groups to water, whether for fishing or for building mills. The document is exceptionally detailed regarding the right to use the waters of the River Vistula, the principal river surrounding the early residence of the Piasts on Wawel Hill. According to the charter, the settling agent (*advocatus*) was granted significant rights regarding water use, particularly for constructing mills. The issuer of the charter, Bolesław V the Chaste, donated four mills to the *advocatus*. These mills were built before the foundation of Kraków: two were held by the king himself; one was owned by the Knights Templar of Miechów; and the other had previously belonged to the Cistercian abbey of Jedrzejów.[286]

Aside from that, which may be even more critical for this subchapter, the duke granted the *advocatus* the right to construct additional mills within the boundaries specified in the charter, after which the owner was required to pay an annual tax of one *ferto* silver to the duke. Under the privilege, the mills need not be built or maintained by the *advocatus* personally, but the rights to them could be transferred to others, along with the obligation to pay one *ferto* silver in taxes. As a result, the rights to build and maintain these mills shifted to the burghers. The case of Kraków is not unique; similar phrasing appears in numerous privilege charters from the thirteenth and fourteenth centuries throughout Central Europe, including agreements between the abbey of Hronský Beňadik and the settlers of the mining areas surrounding the Benedictine lands, which will be discussed later.

The Kraków charter, however, contains one of the most detailed descriptions, and not by chance. The area to which Bolesław V invited settlers was anything but a no-man's land. Several institutions had previously operated in the region, as had settlements that predated the Mongol invasion of 1241.[287] Therefore, it may have been crucial to clearly outline the rights of the various parties to the different sections and banks of the river. The situation was somewhat different regarding most of the settling processes in Hungary discussed above. Here, with a few notable exceptions, the foundation charters did not mention pre-existing settlements or institutions, suggesting that there was no significant competition for resources in these areas at the time of settlement.

Accordingly, it is proposed here that although the privileges in the Kingdom of Hungary seldom mention the right for the settlers or settling agents of newly founded settlements to build water mills, it was common to exercise this freedom. To support this assumption, the following two subchapters examine early water mills in the mining towns of Kremnica and Nová Baňa. The charters of privilege for these settlements are either unknown or do not grant the settlers the right to build their mills.

3.3.3 Kremnica and Its Water Mills

The case of the water mills in Kremnica seems particularly telling. On one hand, changes in the number of mills over time within the town's borders are exceptionally well-documented; on the other hand, there is data regarding the existence of such buildings shortly after the town's foundation. In the following pages, surviving documentary evidence on water mills within the town limits is examined to demonstrate how, when, and by whom these structures were built, as well as their importance to the town's development.

Before delving into the details of the mills' development in Kremnica, a brief overview of the town's foundation is presented, as it relates to the water-related infrastructure. Kremnica's documented history begins in 1328, when, on November 1, King Charles I granted privileges to the settlers. The Hungarian monarch enfranchised settlers from Kutná Hora in Bohemia. Their migration to Hungary was likely initiated during diplomatic negotiations between King Charles I and John of Luxembourg, the king of Bohemia, in 1327. The miners and minters—the former primarily Germans and the latter Italians—along with any future settlers, were granted rights similar to those described earlier, including access to the surrounding forests.[288] However, the document contains a provision that was quite unusual in the medieval Hungarian legal context. Specifically, in matters not addressed by the privilege, the rights of Kutná Hora are deemed authoritative.[289] Consequently, the charter does not cover several aspects of the town's foundation, such as the right to utilize the waters within the settlement's limits. Generally, the burghers of Kutná Hora were entitled to the privileges outlined in the summary of the medieval Bohemian royal mining law (*Bergrecht*), known as *Ius regale montanorum*.[290] This legal text, however, also remains silent on water rights and water mills within the boundaries of mining towns. Despite this lack of information, as in Kremnica, many burghers in Kutná Hora constructed and owned mills within the town's territory.[291]

The first document referring to water mills in Kremnica was issued in 1331, shortly after the town's foundation. By this time, a significant number of mills were operating within the boundaries of the newly established settlement.[292] The document is a pledge contract concluded between Stephen Marsilii and Ieclinus of Olf, both likely burghers of Kremnica, in the presence of Lupoldus (*Hyppolit*), the master of the mint and chamber count of Kremnica, as well as the town's judges and jurors.[293] While the name Marsilii suggests Italian origin—an uncommon group discussed in relation to mining towns in Hungary—Ieclinus may be one of the settlers who arrived from the town of Olpe in the Rhineland.[294] According to the contract, Ieclinus of Olf agreed to repay his debt, comprising a substantial amount of cash and other goods, as specified in the document, while Marsilii accepted several immovable goods from Ieclinus as collateral. As is customary for similar pledge contracts, the document, in addition to the debt schedule, lists the immovable estates pledged. Among these properties, and to identify the buildings, several mills are mentioned in the contract in two areas within the town, known as the Soler[295] and Collner[296] (*see* Figure 3.14), both formed by branches of the Kremnický Stream, the only significant waterway within the borders of Kremnica.[297]

The document lists the following water mills:

- A mill in the Soler valley owned by Ieclinus[298] with one-sixth of the river[299];
- The mill of Kadold[300] in the Soler valley[301];
- Two mills and a furnace owned by Ieclinus in the Soler valley, around the mill of Kadold[302];
- A mill next to the estate of Mazaldrus, owned by Ieclinus, in the Collner valley;
- The mill of the judge of Pukano (Pukanec)[303] in the Collner valley;
- The mill of Ieclinus next to that of the judge of Pukano in the Collner valley;

FIGURE 3.14 The mills and the urban topography of Kreminca in the mid-fourteenth century (after György Györffy).

- Ore stamp mill called Kolbe in the Collner valley above the previous one;
- Ore stamp mill owned by Nicholas Putner, probably also in the Collner valley.

The list above provides important lessons for this subchapter. First, water mills were certainly constructed within town limits from the time of settlement onward. Based on the number of mills, and especially their ownership by burghers in other towns, it is quite possible that some of the mills even predate the issuance of the foundation charter. This situation is not exceptional, as will be demonstrated by the example of the mills within the boundaries of Nová Baňa. The half-dozen grain mills, along with additional ore stamp mills and furnaces at Kremnica, suggest that the settlement became significant within a very short period. Considering that, by its nature, the contract could not have recorded every mill, the town's population may have grown rapidly, indicating one of the most successful town foundations in the region at the turn of the thirteenth century. Second, the ownership of these mills is important. Typically, the situation is unclear, as there is little information about the individuals mentioned in the documents cited above and similar ones. However, the mills were either in the hands of the burghers of Kremnica or owned by burghers in other nearby mining towns.[304] Third, although this is only partially relevant to the main argument, the document is one of the most critical sources on the early topography of the mining town. Based on the contract, it seems clear that the town grew rapidly from the start, and that the settlement spread over a relatively extended area along two small valleys, one running north–south and the other east–west. In the town's topography, water was the key factor, providing the foundation for the town's economy.

The number of mills and their relative locations could be identified based on the document; however, the source does not specify their exact positions along the two streams. The aforementioned historian, György Györffy, along with Slovak researchers, primarily use this contract to visualize the town's extent at the time of its founding. Györffy suggested that there were five water mills along each stream. In the case of the Collner valley, this number is likely accurate. In the Soler valley, however, there were likely only three mills, two of which had a single wheel, while the third, in addition to two wheels that ground grain and a third that crushed ore, was in the same building.[305] Having wheels for different functions housed under one roof was not unusual; in fact, it was pretty standard in the late medieval period, as mentioned in the context of the mills of Prešov as well. Most importantly, all these mills were already operational within three years after the foundation charter of the town was issued.

Aside from the mills constructed within the town limits of Kremnica itself, from the mid-fourteenth century onward, mills were also established in villages founded by or with the involvement of the Kremnica burghers near the town.[306] One of the earliest may have been in the village of Kunešov (known as Szentmihály in medieval times), settled by a man named Vernher of Potska in 1342. In addition to initiating the settlement, he and his heirs were granted jurisdiction over the villagers.[307] The settlers did not receive the rights of Kremnica, but rather those of Žilina, which dated back to 1321.[308] However, the settlers of the newly established Kunešov enjoyed a freedom not specified by King Charles I for Žilina but additionally granted to the locator and the settlers of Kunešov—the right to build along the water.[309] This likely indicates that, shortly after Kremnica's foundation, its burghers recognized the importance of securing this right, which, as previously discussed, was not specified in the town's charter of privilege but was nonetheless treated as an existing right early on.

In the coming decades, numerous new villages were established in the vicinity of Kremnica, and the founding contracts indicate the liberties typically enjoyed by the settlers. Like Kunešov, the village of Sklené was established with the support of Kremnica by the nobles of the later abandoned Felsőmutna under the leadership of *locator* Peter Glaser. Similar to the settlers of Kunešov, those who settled in Sklené were also granted the freedoms of Žilina. The charter stated that, among other rights, the new villagers received the freedom to build their own mills.[310]

Another village established around that time was settled by a certain John (*Johannes*), as reflected in the settlement's is unknown, but it likely occurred around the middle of the fourteenth century. John Göldner, a burgher of Kremnica, acquired the settlement in 1361. In addition to the right to build on the water, he gained ownership of a four-wheeled mill that had been constructed in the settlement.[311] When Göldner acquired the village, he also inherited the rights of Žilina for his property.

One further settlement foundation in the neighborhood of Kremnica occurred in 1364. The nobles in the aforementioned settlement of Felsőmutna sought to establish villages in the forests belonging to their rather extensive estate. The Mutna Forest, or by its Árpádian-period name, *Chernakou*, lay in the southernmost part of historic Turóc County.[312] It was divided into two sections by the river sharing the same name as the county (Turiec by its modern name).[313] The river proved to be a significant asset in settling the area, as its runoff greatly surpassed that of the rivers bordering Kremnica. This fact and the region's richly forested land were the primary factors in its settlement. The settlement process took considerable time: in 1263, the land of the Mutna Forest was allocated to new settlers during the reign of King Béla IV in exchange for the fishing and hunting services they provided to the king.[314] In the middle of the fourteenth century, plans to settle the area were finalized. The settlement was organized under the leadership of a man called Nicholas, son of Matthew, with the intent to establish a village or villages to exploit the resources of the uninhabited forest regions. Nicholas, as *advocatus*, received significant benefits, including two empty plots and a site suitable for a mill *(locus molendini, see* Chapter 3.1)—that is, the right to build a mill—in the territory of the forest, as well as a place for a mill outside of the forest by the River Turiec, the latter also including a plot for the miller's house. This time, the settlement was distinct because the *hospites* were granted the rights of the mining town of Krupina.[315] However, the *advocatus* and his heirs held a special right to receive every sixth denar of the taxes and gifts the settlers paid to the landlords. Since this latter stipulation is not included in the original privileges of Krupina, it is referred to in the *hospes*-privilege as a custom of Žilina. Accordingly, while in general, the charter refers to the rights of the settlers as being the same as for the inhabitants of Krupina, in one specific instance, the charter still regards Žilina's privileges as authoritative.[316]

It is no surprise that the nobles of Mutna made significant efforts to settle the area during that period. The rapid growth of nearby Kremnica must have brought previously unprecedented economic benefits to the forest's settlers. However, proximity to Kremnica was not only an economic advantage but also a considerable threat in the long run, as the wealthy town, from the beginning of the fifteenth century onward, made numerous attempts to acquire ownership of the villages in the area.[317] Kremnica not only expanded its land holdings in the early fifteenth century but continued to do so in the following century as well. By the late fifteenth century, the landlords of Mutna—referred to as *predium* (perhaps meaning, but not necessarily, a lost settlement[318])—claimed that the burghers of Kremnica had hired mercenaries to take over their territories.[319] Not much later, in the 1520s, the authority of Turóc County prohibited Kremnica from seizing any part of the village of Felsőmutna.[320]

From the perspective of mills, one additional *hospes*-settlement worth mentioning, the village of Dolný Turček, also founded in the Mutna Forest. This settlement, located a few kilometers north of Kremnica, had significant mills as the aforementioned River Turiec flowed through the village. According to a document dated 1371, the settlement had two mills by then, one operating with as many as eight wheels and another with six.[321] Aside from providing details about one of the largest milling complexes in fourteenth-century Hungary, the document also contains an exceptional clause: it annuls all prior documents issued by Kremnica concerning property rights in the river. This passage is essential for two reasons. First, it strongly contradicts contemporary customary law, which, as discussed in Chapter 3.1, recognizes earlier-acquired rights as superior to those acquired later in matters of water rights and other aspects of daily life. Second, the document suggests that Kremnica made earlier decisions regarding the rights to water in Dolný Turček; these documents have survived to this day. This latter issue will be revisited in the conclusion of this subchapter, as this passage may be crucial to the present inquiry. The last *hospes*-village worth mentioning is Horná Štubňa, then known as Újlehotka. The name of the settlement is significant, as it is a compound word consisting of "új" (meaning "new" in Hungarian) and "lehotka" (a word of Slavic origin meaning forest clearance, which is a loanword in Hungarian).[322] In 1390, the settlers were granted the rights of Krupina, and the leader of the settlement process, Bartos and his heirs, had the right to establish mills freely.[323]

As the list of settlements discussed above shows, the burghers of Kremnica acquired significant properties in the vicinity of the mining town and had several mills constructed on these estates. However, the mills within Kremnica's borders were not solely in the possession of the burghers of Kremnica; many were owned by burghers from other mining towns. For instance, Nicholas Smyt, a burgher from Nová Baňa, came into possession of a four-wheeled water mill that had previously belonged to Kadold.[324] Soon afterward, Smyt sold it to a burgher from the town of Louny in Bohemia.[325] He was not the only outsider involved in the milling industry within Kremnica's borders. Burghers from Buda, as well as the archbishopric of Esztergom, owned numerous mills in the mining town.[326] The latter institution acquired a large mill in the city in the 1460s, which it then leased back to Kremnica. They purchased the mill for 4000 florins and leased it back to the town for as much as 300 golden florins a year,[327] making it the most significant single source of income from mills in the entire archbishopric.[328]

Aside from the fact that several mills were owned by individuals other than the burghers of Kremica, including the largest one in town, there were also more mills in the burghers' possession for grinding grain and crushing the locally mined ore. The well-preserved fifteenth-century account books of the city precisely recorded the mills and their owners (*see* Figure 3.15), also noting which of the two aforementioned streams (in the Soler and Collner valleys) the mill operated by. While a detailed analysis of the individual mills would shed light on the town's historical topography, it is less relevant to this subchapter. However, their very existence is undoubtedly vital for understanding water rights in general.

Some lessons can be drawn from the data presented above. First, since the town's inception, the burghers of Kremnica have possessed several water mills.

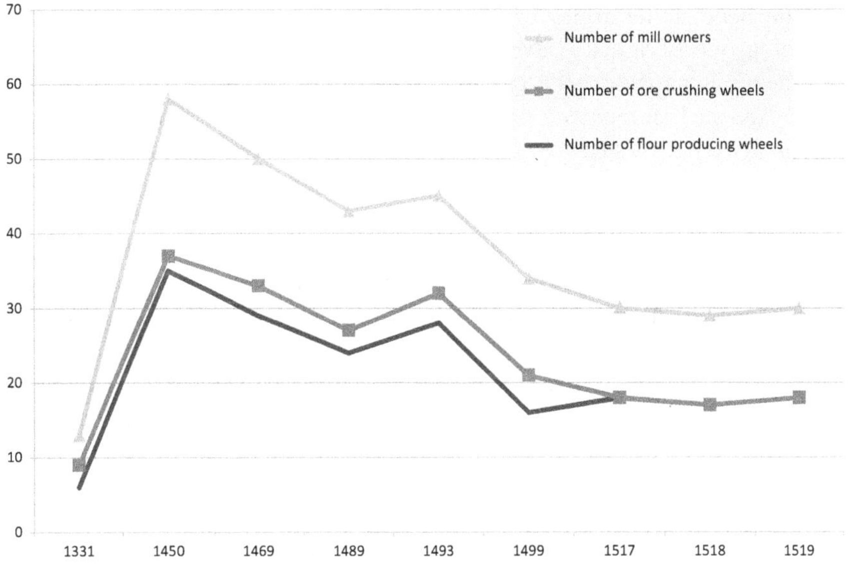

FIGURE 3.15 The mills of Kremnica and their ownership (1331–1519).

Source: Fejérpataky, *Magyarországi városok*, 625–626 (1442–1443), 625–636 (1450), MNL OL DF 250 101 (1469), MNL OL DF 250 088 (1489), MNL OL DF 250 090 (1499), MNL OL DF 250 102 (1517), MNL OL DF 250 076 (1518), MNL OL DF 250 077 (1519).

Given the 1331 pledge contract, it is worth noting that, even if the privileges granted to Kremnica in 1328 do not explicitly mention this privilege, the Germans and settlers from other areas enjoyed this freedom. The existence of the right to use waterways for constructing mills is supported by the fact that *hospes* villages founded under the leadership of Kremnica were also granted this freedom. Otherwise, the *advocati* from the various *hospes* villages would have gained more freedoms than the burghers who commissioned them to lead the settling of the villages. The aforementioned 1371 document regarding Dolný Turček is crucial here, as it shows that the town of Kremnica had already exercised its water rights before the end of the fourteenth century. To illustrate the significance of securing these rights in the fourteenth century for mining settlements, the following short subchapter offers an overview of the history of the water rights related to the establishment of another mining town, Nová Baňa.

3.3.4 The Foundation of Nová Baňa and Its Water Mills

Unlike the water mills of Kremnica, which, despite the relative abundance of available resources, have not been extensively studied, the mills of Nová Baňa have been the subject of thorough analyses. Lacking the foundation charter, the most important source on the early history of this mining town is the lawsuit that unfolded between the settlement as it developed and the Benedictine abbey of Hronský Beňadik concerning water and, in general, property rights.[329] Even if the founding charter has not survived, the document's issuance can be precisely dated to either August or September 1345.[330] As such, it represents a relatively close parallel to Kremnica, founded less than two decades earlier. Nová Baňa had probably begun to grow in the late 1330s, when the burghers of Pukanec began to settle the area to exploit the newly discovered ore.

With the onset of mining activity in the area, the establishment of the settlement right on the border of the royal estate and the lands of the Benedictines of Hronský Beňadik also began. The primary issue was that the most crucial water resources for the new town originated from the Benedictines' estate, and as demonstrated by the example of Kremnica, water resources were vital for founding a mining settlement. It appears that the Benedictines initially had no objection to the miners' presence, as the abbey anticipated income from a portion of their estate, which had not been financially viable until then. The abbey's expectations seemed to have been justified, as not long before 1340,[331] they successfully leased a mill site by the Büksavnica Stream to two burghers of Pukanec who wanted to use it for the newly opened mine on the nearby royal estate. As highlighted in earlier research, the lease contract contains an essential clause stating that the plot and the mill to be constructed were leased, not owned, because they were not located within the territory of the royal forest of Savnic. Therefore, the abbey asserted that the customary law of the royal settlers held no legal significance in the area, nor would the future privileges granted to the mining town and its burghers.[332] This clause is intriguing, as it suggests that the settlers may enjoy specific water-use rights in this instance. These rights could hardly have been anything other than the freedom to exploit water resources—that is, establishing a mill—within the confines of their settlement. In addition to the leases granted to the burghers of Pukanec from 1345 onward, the abbey began systematically leasing its lands suitable for the construction of mills. Initially, the abbey provided its *provisor*, a certain Peter, a place for a mill by the Büksavnica Stream for an annual payment of one florin.[333] A few weeks later, Abbot Siegfried leased another mill site to a certain Nicholas, son of Leo and Dytel dicto Moder, by the nearby Verence Stream,[334] and finally, within the same timeframe, the aforementioned Kadold *urburarius* obtained a lease for a mill site by the Büksavnica.[335]

The question of the border between the abbey's lands and the king's soon became a source of conflict. In 1345, King Louis I already warned Nová Baňa to respect the monastery's property rights. The mills were likely completed by 1345, and to resolve the dispute in 1347, the king ordered the leaseholders to pay one florin of rent to the abbey of Hronský Beňadik for each mill built on the Benedictines' estate. The burghers of Nová Baňa, despite their repeated promises, did not pay the rent regularly. The most relevant question for this subchapter is why the burghers of Nová Baňa may have chosen not to pay the one-florin rent. The territory where the mills stood, as discussed above, lay within the borders of the Benedictine abbey. However, after Nová Baňa was populated, the mills' leaseholders, likely lacking detailed perambulations, treated them as the town's property. In response to such claims, King Louis I repeatedly warned the burghers not to infringe on the abbey's property rights. However, if the mills were within the borders of Nová Baňa (and naturally, the lessees argued that in their interests), they would not have been required to pay this rent.[336] If the mills belonged to Nová Baňa, the settlers could have operated them independently, without incurring any associated duties, which was likely the customary law in similar cases.

3.3.5 *Mills in Mining Towns: Who Was Entitled to What?*

In the previous pages, drawing on examples from medieval Hungarian mining towns established in the thirteenth and fourteenth centuries, the book aimed to demonstrate the importance of the rights and legal disputes surrounding the construction of water mills and the general use of water for various purposes. As shown by the examples of Kremnica and Nová Baňa, constructing mills in the settlements shortly after their founding to grind grain, process mined metal ore, and prepare timber for building the mine shafts was already a crucial issue. Accordingly, although the charters of privilege for the Upper Hungarian and Transylvanian mining towns rarely mention the right to construct mills and utilize water resources in general, these questions are discussed with striking frequency in the early legal documentation of these settlements. This is also true for the early pragmatic written documents from Kremnica and Nová Baňa mentioned above, in which at least half of the documents issued by the authorities of the two towns address related issues.[337]

Previous researchers have attempted to identify the economic and other privileges associated with mining and royal towns, as recorded in foundation charters, to illustrate the freedoms certain communities acquired in medieval Hungary. Although the charters granted to mining settlements rarely included provisions regarding the right to build mills and to exploit water resources in general, the preceding subchapters demonstrate that, at settlements established in the thirteenth and fourteenth centuries, settlers effectively exercised this right. Therefore, there may be valid grounds to accept the hypothesis suggested in the introduction to the subchapter. One additional issue requires clarification: to what extent can the foundation charters of these settlements be used to understand the economic privileges the settlers received? Apart from the specific references within them, which should not be underestimated, these documents likely encompassed only some of the freedoms and economic benefits that burghers possessed in a mining town.

3.4 Rivers as Borders: Another Minefield?

> This river [Tiber] moreover circles that splendid mountain on which the city of Perugia is situated and while flowing a great distance through its district, the river itself is bordered by plains, hills and similar places (...) when I was resting from my lecturing and to relax, was travelling towards a certain villa situated near Perugia above the Tiber, I began to contemplate the bends of the Tiber, its alluvion, the islands arising in the river, the changes of the river-bed as well as a host of unanswered questions which I had come across in practice. (...) I began to consider in various ways what the legal position was, not believing that I would take it any further (...) while I slept that night, I had a vision near dawn that a certain man, whose countenance I found gentle, came to me and he said the following: "Write down what you have begun to think about and since there is a need for illustration, provide mathematical diagrams."[338]

The quotation above is from the prologue of a treatise by Bartolus of Sassoferrato, one of the most celebrated jurists of the fourteenth century. Bartolus, a most influential and, by the time the text was written in 1355, busy professor at the University of Perugia, decided to take some time off.[339] As mentioned in the text, he headed to a villa overlooking the valley of the Tiber, the location of which has not been identified by historians. There, he began considering the question of to whom the islands emerging in the river belonged and what happened to the ownership of a certain piece of land when the river, which constitutes its border, starts to meander along a different path, eroding parts of one's property and adding to the far bank. Although this subchapter extends the problem to (semi)artificial riverbed changes, it addresses the same question. What happened to the ownership of these lands, and to what degree was that a source of conflict in the medieval Kingdom of Hungary? Was there a customary law related to adjudicating similar cases, or was the problem so rare that no stable norm formed in the Middle Ages? As argued in previous chapters, the process of building mills and creating river obstructions can be traced back to the Árpádian period. However, these issues were far less evident with border and riverbed changes.

As suggested by the prologue filled with various tropes, Bartolus initially thought it was an eccentric idea to discuss an issue like this in a treatise until a mysterious man urged him to do so. Although the anecdotal story behind the inspiration for the treatise known as *Tractatus de fluminibus seu Tyberiadis*, often referred to simply as *Tyberiadis*, may not have had much connection to what occurred, it is hard to believe that the entire treatise was completed in just a few weeks, as Bartolus claims. Existing scholarship on the treatise places major importance on the circumstances surrounding the work's formation, particularly the possible holiday and the location of the villa mentioned.[340] These questions are relevant to the study of Bartolus but are likely less significant with respect to the issue of rivers as borders. More interestingly, a major authority on Roman law at this time was writing a work dedicated to this topic. It appears from the preceding quote that he did not consider the question worthy of similar treatment (*Et circa multa dubia que de facto ocurrerant et alia ego ipse ex aspectu fluminis excibatam, quid iuris esset, cepi aliqualiter intueri, non tamen credens ultra procedere, ne recreationem propter quam accesseram inpedirem*). The prologue is somewhat controversial, as Bartolus also notes that he encountered similar problems during his legal practice.

Bartolus's work consists of three parts and addresses two questions: who owns the land when an island emerges from a river, and, when the river changes course, how should the borders of connected estates be demarcated? One of the most important features of the treatise is that it combines legal reasoning and geometry in its argumentation. Bartolus drew numerous geometric figures, using them to meticulously describe how newly emerging islands or lands connected to existing land should be divided. Tied to the manuscript tradition, these drawings exhibit several variants; however, an autograph fragment from the *Tyberiadis* preserves several figures that can be associated with Bartolus himself.[341]

Although historians have tended to focus on the circumstances surrounding the work's creation and to address the text itself less, this is not the only legal source that addresses the problem of shifting boundaries. A series of law codes, customary law collections, and documents related to individual lawsuits testify to its practicality. Furthermore, the very fact that Bartolus engaged in writing such a work suggests that changes in estate borders due to shifting riverbeds were not as rare as one might think. To what extent were the questions he raised significant as theoretical matters? To what extent did he aim to provide an example, with this text, of the potential to combine geometry and law rather than simply addressing a legal problem? Does the Hungarian source material provide insights into similar issues? How were such cases resolved in medieval Europe and Hungary? In this upcoming subchapter, these matters are discussed through a series of land-related disputes from the Kingdom of Hungary that preserve such cases.

Although Bartolus discusses in depth in his work what happens when a piece of land is created by alluvial deposits on the banks of a river, he partially overlooks a difficulty that appears to be a recurring theme in the Hungarian source material.[342] He never considered what precisely occurs if the piece of land in question belonged to the landlord on the other bank of the river; that is, it did not emerge from the river but was separated from one person's land and was attached by the river to land belonging to someone else. In the cases examined by Bartolus, the principles of geometry can be applied with the aid of geometric diagrams.[343]

The general principle suggested by the famous jurist is straightforward: it aligns the ownership of the islands with the parallel landownership on the nearest bank. In Bartolus' treatise, the same applies to lands that emerge from the river due to years of alluvial accumulation. This phenomenon in Roman law is commonly referred to as *accretio* (accretion), and it was adjudicated in the ancient legal tradition in a manner similar to Bartolus's approach.[344] This issue was also significant in the case of lands emerging along seashores.[345] In the medieval Kingdom of Hungary and, likely, in most of Europe except the Low Countries and some parts of England and Germany, the formation of new lands was primarily a consequence of shifts in river flow rather than of seawater movements. This process in legal tradition is typically referred to as *avulsio* (avulsion), a special form of accretion. Bartolus did not propose a definitive solution to this question, and it's unclear why he partly omitted this issue. It may be related to the fact that the primary motivation for his work was to demonstrate the possibilities of integrating geometry into law. The geometric principle Bartolus presents is well-suited to classical cases of accretion, but for avulsion, the legal principle takes precedence, leaving little room for geometry. Even if Bartolus overlooked this particular problem, it clearly emerged as a recurrent issue in medieval court cases.

Consequently, scholars have long devoted attention to the problem. It has been discussed across various distinct fields, ranging from legal history to physical geography and geomorphology. Historians have analyzed emblematic local-level events and examined how the processes in question affected societies, economies, and political structures. The most important example is a major hydrological change in the watershed of the River Po. The event, usually referred to as Rotta della Cucca (Cucca breach), first attracted the attention of Italian historians in the eighteenth century. Ludovico Antonio Muratori, one of the pioneers in the study of medieval Italy, was the first historian to discuss the Cucca breach.[346] The basis for the assessment of the event is a chapter in Paul the Deacon's *Historia Langobardorum*, in which the cleric describes the floods of 589. As he notes, several floods occurred along rivers in Italy this year, with the most dangerous being that of the River Adige.[347] Scholarship, strongly building on the work of Muratori, has viewed the event as a driving force in the transformation of the riverine landscapes and the courses of the Rivers Po and Adige. It created extensive marshlands around the mouths of the two rivers and transformed property boundaries in Veneto. In recent decades, historians have criticized this oversimplified view of landscape change in the region and have tended to regard it as a long process in which the channel system maintained by the Roman administration and subsequently abandoned by the Lombards was deliberately neglected. The marshes that developed as a result of the cessation of maintenance proved useful as a form of frontier defense against the Exarchate of Ravenna.[348] In explaining the involved processes, the change in landholdings was only a marginal issue, but it nonetheless drew scholars' attention to the fact that riverine landscapes were not nearly as fixed in the Middle Ages as they became in the nineteenth–twenty-first centuries, by which time most of the rivers in densely populated areas of Europe had undergone extensive regulation through canalization.

In the nineteenth century, research on the changing waterscapes of the Low Countries also generated interest among Dutch and Flemish historians. As in the Italian case, the key element of the geographical processes that captured the attention of historians from the nineteenth century onward was not primarily the change in individual estate borders, but rather the significant transformation of extensive stretches of seashore. Administrative and cartographic documents were the first sources utilized by historians in the Low Countries. Over the past fifty years, research conducted by geographers and geomorphologists has greatly expanded opportunities to study the hydrological processes of the Holocene, in this case, the Late Holocene. Geographers, often collaborating with historians, have demonstrated the potential of researching the avulsion histories of specific rivers in the Low Countries,[349] southern France,[350] and other areas of Europe.[351]

The combination of written and cartographic evidence with geomorphology not only led to studies on changes in riverine landscapes and, concurrently, the related landholding structures in Western Europe, but also produced research addressing the issue in Central Europe, particularly in Hungary. Several studies have examined transformations in the waterscape around Vienna during the late medieval and Early Modern periods, resulting in numerous lawsuits among landowners along the River Danube.[352] Research also revealed significant changes in historical riverine landscapes, including the Danube, Tisza, Rába, Mura, and Dráva river valleys.[353] However, most of the studies mentioned above focused on alterations in riverbeds and the alluvial development of major rivers in Central Europe.

Much less attention has been given to smaller rivers and streams, despite their potential relevance to these boundary issues. There is ample source material, including perambulation documents and cartographic data, partly because streams formed boundaries of estates even more frequently than major rivers. The Hungarian source material, especially up to the late fourteenth century, is somewhat exceptional in this regard, as perambulations account for probably seven to eight percent of the entire medieval legal source material.[354] When legal disputes concerning changes in riverbeds arose, this group of sources became crucial to the discussion, as perambulations were often conducted as stages in legal disputes. One recurrent type of dispute involved alterations to riverbeds. These legal matters concerned not only natural changes in hydrological conditions but also artificial modifications to mill races, fishponds, irrigation channels, and similar features, which, in some cases, could affect estate boundaries. Alongside historians and geomorphologists, legal writers also focused on the problems posed by shifting riverbeds in areas where rivers served as boundaries. These scholars primarily contributed by analyzing collections of Roman law and examining the contemporary implications of avulsion.[355]

All of the approaches listed above are important, as they often help contextualize the results using the Hungarian source material. In the following pages, drawing on source material from the medieval Kingdom of Hungary, it is argued that the problem raised by Bartolus is anything but theoretical, and that there was an increasingly crystallized customary law regarding how to resolve similar disputes, even if that law was not based on the principles he advocated.

3.4.1 The Border between Ľubotín and Orlov: What Can a Single Case Reveal?

In 1349, one of the landlords in Sáros County, Rikalf son of Rikalf, wrote a supplication to King Louis I, stating that the River Poprad had detached a piece of land from his village, Ľubotín. Although the river had previously demarcated his land from the king's village, Orlov, its changing course began to flow through his estate. It physically separated a large piece of land that had belonged to Ľubotín (*see* Figure 3.16).[356] Owning Ľubotín was fraught with difficulties. Approximately twenty years earlier, the family had to appeal to Louis's father, King Charles I, when Phillip Druget, the influential Italian nobleman and member of the king's entourage, attempted to acquire some of the lands belonging to Ľubotín. The efforts of the Drugets may not have been surprising, as this was the period when the family rapidly extended its power in the region. While the Drugets might have been anticipated as a problem, losing land due to the River Poprad's shifting riverbed may not have been one of the potential issues Rikalf had envisioned.[357]

In 1349, King Louis ordered Sáros County to investigate the case in response to the appeal. It was led by Tivadar, a noble county judge (*iudex nobilium* or *szolgabíró* in Hungarian), and a delegate (*homo provinciae*),

FIGURE 3.16 The surroundings of Ľubotín and Orlov on the sheet of the First Military Survey.

Source: Vienna, Österreichische Staatsarchiv, Kriegsarchiv KPS Kartensammlung B IXa 527. XXI.02.

a man named Jacob, son of Sükösd.[358] During the investigation, they interrogated local nobility and tenant peasants, particularly those living in two nearby settlements: Plaveč (Palocsa) and Gerlavágása (meaning "Gerla's clearance," a vanished settlers' village near the three previously mentioned settlements). The investigation concluded that the supplicant was right and that the River Poprad indeed had detached lands from Rikalf's property.[359] According to the surviving documents, the changes in the riverbed were not due to human intervention but were likely caused by floods that altered the area's hydrography. However, this was never explicitly stated in the related charters.[360]

Despite the verification of Rikalf's claim, the dispute likely remained unresolved at that time. Ten years later, in 1359, the case was brought back to court by Rikalf's family. Based on documents issued in 1359, it can be inferred that during the ten-year period between the cases, the lands in question were used by the king's tenant peasants and were never returned to Rikalf and his family. Clearly, in 1349, Rikalf intended to reclaim the disputed lands. The investigation concluded that these lands had originally been part of his possessions. This suggests that, in such cases, detached land was presumed to belong to the original owners, in this instance, Rikalf. However, it took ten years to recover them in response to the complaints from Rikalf and Peter, son of Ladislaus, both of whom belonged to the same family. Ultimately, the lands were reassigned not solely because they had belonged to Rikalf's family, but also due to the family's service to King Louis I.[361] This time, the donation referred to the reason for the alteration in the riverbed's management. As indicated in the 1349 documents, the main flow of the River Poprad had not been artificially changed; rather, it resulted from the river's swift current. The case was settled by reinstating the previous owners to the lands in question in the presence of a deputy of the chapter of Spišská Kapitula (Szepes) and a *homo regius*. As usual, a new perambulation of the disputed land was conducted, during which part of the land was described as the "former bed of the River Poprad."

Even if, in 1359, the land appears to have been reassigned to Rikalf and his family, the issue remained unresolved for the rest of the Middle Ages. Nearly fifty years later, in 1405, at the joint diet of Abaúj and Sáros Counties held in Košice, the noble judges and *alispánok* (*vicecomites*) of the latter county testified that the same lands that had been disputed in 1349 and 1359 originally belonged to Ľubotín and, therefore, rightfully belonged to Rikalf's successors.[362]

The set of documents relating to the lands beside the River Poprad between Ľubotín and Orlov reveals several important issues. In 1349, according to surviving sources, Rikalf attempted to prove that the River Poprad had changed course, as demonstrating this would have allowed him to continue using the lands, even though by then they lay on the opposite bank. The existence of this legal case suggests that even if the changes in the course of the River Poprad were part of a natural process, land ownership may still not have been affected. However, the situation becomes less clear in light of the lawsuit of 1359, in which the lands in question were (re)instituted in the Rikalf sons in exchange for services rendered to the crown, rather than because they had originally belonged to the family. Nonetheless, the fact that the same lands were returned to the Rikalf family indicates that the issue was also connected to their previous ownership. Given these seemingly contradictory documents, it is certainly worth considering whether customary law existed in medieval Hungary that applied to similar cases.

3.4.2 Riverbed Changes and Estate Borders: Was There a Medieval Customary Law in Hungary?

As previously noted, István Tringli not only discussed mill construction but also addressed the problem at hand. He suggested that changes in estate boundaries resulting from riverbed alterations may have been a minor issue in medieval Hungary, and that these concerns were certainly less significant than lawsuits concerning water mills. The number of lawsuits associated with the two problems indicates that disputes over milling rights were more common than those related to riverbed changes. As discussed in Chapter 3.1, by the late thirteenth and fourteenth centuries, the number of mills in Hungary had grown sufficiently large that they sometimes hindered each other. The establishment of a defined customary law governing the use of water and the construction of water mills was the obvious outcome. However, the ongoing struggle for land ownership between Ľubotín and Orlov suggests that, regarding riverbed changes, the norm was ambiguous, or that the tenants of Orlov, the king's land, attempted to exploit their favorable political position against Ľubotín. A single case cannot clarify this issue as effectively as the results from multiple legal disputes. Before presenting some of the representative cases, this subchapter examines the most evident source that one can reference when discussing whether customary law existed concerning a specific question: Stephen Werbőczy's *Tripartitum*. Werbőczy addresses the issue in a surprisingly extensive manner, given its seemingly minor significance:

> Then, as the boundaries and borders of many free cities, villages, estates and many towns and deserted lands are set and defined by rivers and streams; and by the flood and force of these waters often large pieces of land, meadows and woods are separated, carried away and attached to the area of another neighboring city, town or estate; since the river, driven by vehement flood often strays and spills from its usual course, flow and bed into a new bed; so some people think and believe that the

lands, meadows and woods that were annexed and attached to the area of another neighboring free city, town or village due to a change in the flow, course or bed of the river ought to belong to and come into the possession of that free city, town or village; arguing and stating that their boundaries are set by the flow, course and bed of the river. But this opinion is not correct.

[1] For this way many frauds could be committed, and the waters and rivers—with hidden canals, and sometimes by making shallow dikes, or raising dams or filling up the bed—could be driven into a new course and bed in any direction, according to will; thus someone could easily usurp another's lands, woods and meadows.

[2] Therefore, the opposite opinion shall be accepted as correct (…)[363]

These points in Werbőczy's work were first discussed in the nineteenth century. In his examination of this section of the *Tripartitum*, Rezső Dell'Adami, a professor of civil law, suggested that, unlike much of this work, in this specific case, Werbőczy applied the principles of Roman law rather than Hungarian customary law to land boundary disputes linked to changes in the course of streams.[364] In the 1930s, the aforementioned Alajos Degré revisited the issue. He also highlighted that Hungarian customary law may have differed from the Roman law that Werbőczy applied in this instance. Unlike Dell'Adami, Degré collected several documents that support this argument.[365] Following Degré's lead, Tringli accepted that the origins of this part of Werbőczy's work stem from Roman law. However, Roman law and the Digest (*Digesta*) itself are not as clear-cut on this issue as Werbőczy's text or what has been proposed by subsequent scholars.[366] The *Sachsenspiegel* (Saxon Mirror), compiled in the early thirteenth century, presents a similar solution to the issues addressed in the *Tripartitum*. Nevertheless, research on the German customary law compilation has revealed that the influence of the Digest was more pronounced.[367] When discussing alterations to the waterscape, all three legal sources rest on the principle that, when a river's course changes rapidly, the detachment of a portion of land from one person's property and its reattachment to another person's property should not affect land ownership. However, while the two medieval customary law collections did not incorporate any exceptions to this principle, the Digest did include a significant contrary circumstance: what happens if the shift in the river's flow proves permanent.[368]

From the perspective of this subchapter, the most important question is whether Werbőczy's points reflected the practices of the late medieval period. The picture that emerges from an analysis of court cases from the period up to the early sixteenth century is not straightforward. Based on the example from the Poprad River valley, it appears that Werbőczy summarized existing customs. However, the quantitative analysis indicates that the chief justice applied a different legal principle from that prevailing in his time. Although the available secondary literature references only a handful of cases, the analysis of the cartularies discussed in Chapter 1.2.2 identified more than fifty relevant cases (see the list in Appendix 3). It does not make sense to discuss each case individually, primarily because little information is available about the background of the cases. Aside from some trends to be examined, certain cases provide a better understanding of the different legal norms or illustrate shifts in practice. Two questions arose from these cases. First, did (natural) changes in the course of the river mean that land boundaries also changed in medieval Hungarian legal practice? Secondly, how were natural and artificial changes in riverbeds distinguished from one another? While the answer to the latter appears obvious at first glance, this question, as evidenced by the sources, lies at the heart of numerous legal disputes.

The earliest known lawsuit illustrating the application of customary law, as described by Werbőczy, dates back to 1338. The document indicates that a certain Ivánka, son of John Turóci, submitted a complaint to the *ispán* of Zólyom County, Master Doncs. Ivánka alleged that his interests were harmed by a land transaction involving Galovany, an estate adjacent to his own. Provost Paul, son of Gele, and Gál, son of Jacob, son of Albert, agreed to exchange specific parcels of land. According to Ivánka's complaint, Paul took possession of a plot bordering his own. While Ivánka cultivated crops, Paul raised livestock, and Ivánka claimed that Paul's animals damaged his plowlands and meadows. Paul, however, asserted that Ivánka had established boundary markers on fields which he (Paul) received from *ispán* Doncs himself. Ivánka maintained that Paul had set up these markers. Doncs summoned Paul and fifty witnesses to appear at the convent of Turóc (Kláštor pod Znievom) and testify that it was Ivánka who erected the boundary markers. The case was further complicated by the fact that Paul was also the notary for *ispán* Doncs. It was revealed that Paul wrote the charter and that Ivánka had indeed placed the markers. This aspect of the case became somewhat irrelevant as the boundaries later shifted when the Porouathka Stream altered its course, detaching a section of land from Ivánka's estate and attaching it to Paul's.[369] This strongly suggests that, regardless of the boundary markers, the boundary between the two properties shifted with changes in the streambed's location.

Another case from the following year suggests that similar cases were not consistently treated during the Angevin period. In 1339, a dispute arose over the border between two estates, Čoltovo and Lekeňa (now part of Bohúňovo), in Gömör County, not far from the lands involved in the previous case. The Hablucapataka stream marked the boundary between the two estates in one section until it flowed into the Sajó River (*see* Figure 3.17). The litigants were Paul, son of Gallus, who owned Čoltovo, and Nicholas Forgách's sons, Andrew and Nicholas, who owned Lekeňa. Both parties claimed that

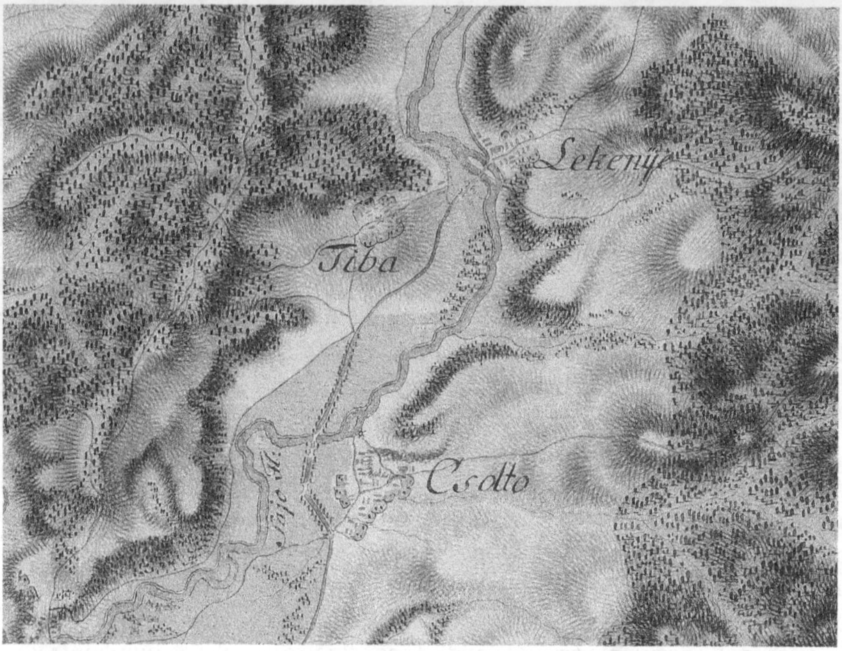

FIGURE 3.17 Čoltovo and Lekeňa on the sheet of the First Military Survey.
Source: Vienna, Österreichische Staatsarchiv, Kriegsarchiv, KPS Kartensammlung B IXa 527. XIX.09.

the other had taken possession of lands that rightfully belonged to them. According to Andrew, Paul had artificially redirected the stream into a new riverbed. However, witnesses testified that the river's course had changed as a result of flooding. The alluvion carried by the floods silted up the former streambed, causing the water to shift its path. The importance of determining whether the change in the riverbed was artificial or natural highlights that the two cases were evaluated differently. During this period, a natural alteration of the riverbed was recognized as an illegitimate boundary shift, while an artificial modification was not. Although the change in the riverbed should have implied a shift in land ownership, this did not occur. It was revealed during the trial that the entire contested area fell within the lands of Andrew Forgách. Consequently, the disputed territory remained under his control.[370]

From the same decade, however, there is a case suggesting that, for the parties involved, even natural changes in a river's course did not alter land ownership. For instance, in 1340, when the boundary between Szentmárton and Kóród was demarcated (today both settlements belong to Coroisânmărtin), a section of the boundary was formed by the Holt-Küküllő River (a branch or backwater of the Târnava Mică River). Likely due to the undefined riverbed, the two parties decided to erect boundary markers together in a dry section of the bed. They did this to secure the boundary between the lands in case floods washed away the riverbed.[371] This indicates that even natural changes to the boundaries were not associated with a change in ownership, or at least that parties sometimes reached agreements that deviated from such a custom. It was probably in the interests of both parties to exercise caution, as it may not have been clear which course the river would take if the old riverbed were filled in. None of the landowners would have known whether they would gain or lose territory. It is difficult to determine the exact reason for this type of agreement because, at the time of the First Military Survey (and thus the first accurate mapping of the area), this branch of the Târnava Mică had disappeared entirely (*see* Figure 3.18).

Even clearer evidence of the not-yet-crystallized customary law in these cases is provided by a boundary dispute from 1347. The parties involved were the bishopric of Eger and Stephen, son of Paul of Ónod. One section of the boundary between Ónod and Hídvég (present-day Sajóhídvég) was the bed of the River Sajó. However, over time, the hydrological conditions of the area changed, leading to the emergence of two islands with usable lands (*duabus angulationibus vulgo Zygeth vocatis*) and places for fishing (*loca piscaturarum*) on the Ónod side of the river. As a result of this hydromorphological change, the main course of the River Sajó began to flow within the borders of Hídvég. The First Military Survey sheet does not help to identify the former bed of the Sajó River between Hídvég and Ónod, but a detailed mid-nineteenth-century manuscript map of the area does (*see* Figure 3.19).[372] The bishopric of Eger attempted to claim the aforementioned land, worth 13 marks, but Stephen objected to the bishopric's claim. Paul Nagymartoni, the chief justice of Hungary, required Stephen and thirteen other nobles to swear an oath that the land in question had belonged to him. As Stephen took the oath along with the necessary noble witnesses, the lands in question were returned to him. Furthermore, because he had made a false claim, Nicholas Dörögdi, the bishop of Eger, was fined 13 marks.[373] This indicates that, according to the chief justice, the case was clear and that ownership of the land was not altered by the relocation of the riverbed and the shifting of some land from one bank of the river to the other. The fact that the bishop had to pay a fine indicates that Nagymartoni considered this the norm. In light of the few cases discussed above, this issue was not as clear as suggested by the chief justice's decision. Rather, this case may have been part of an effort to establish a custom for evaluating similar cases.

Analyzing every case involving similar issues would be unnecessary, as the considerations at issue are largely the same as those in the disputes discussed above. The mid-fourteenth-century cases discussed above demonstrate, in relatively good detail, that this was when rights related to water-mill construction were clarified and when various parties involved in waterscapes began to emphasize putting these rights in writing to secure their access to

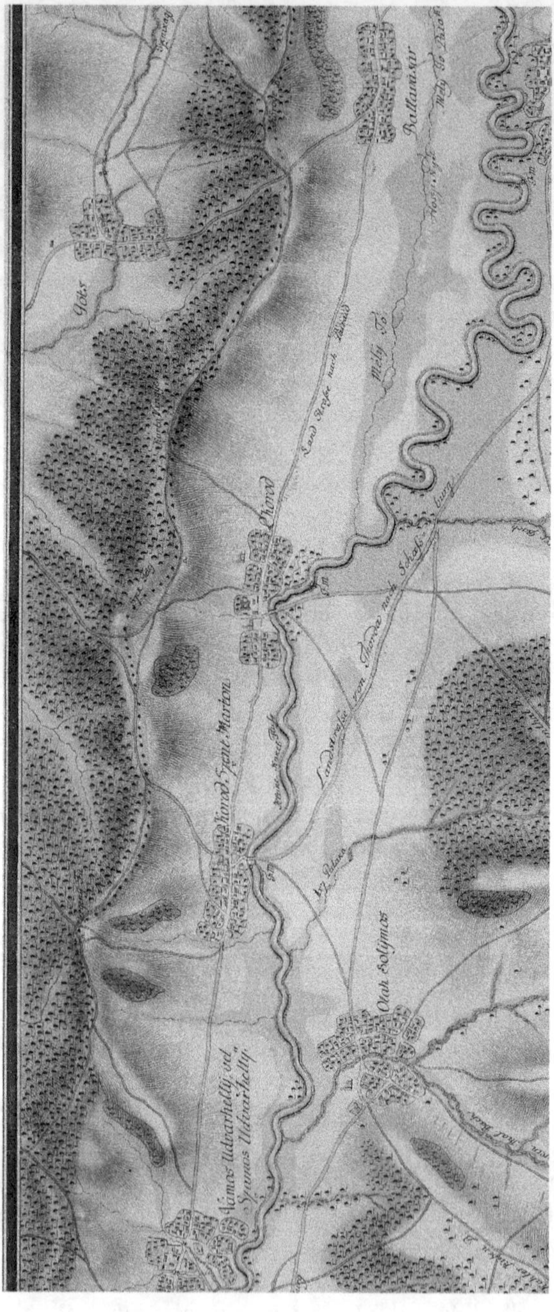

FIGURE 3.18 Szentmárton and Kóród on a sheet of the First Military Survey.

Source: Vienna, Österreichische Staatsarchiv, Kriegsarchiv, KPS Kartensammlung B IXa 715. 158.

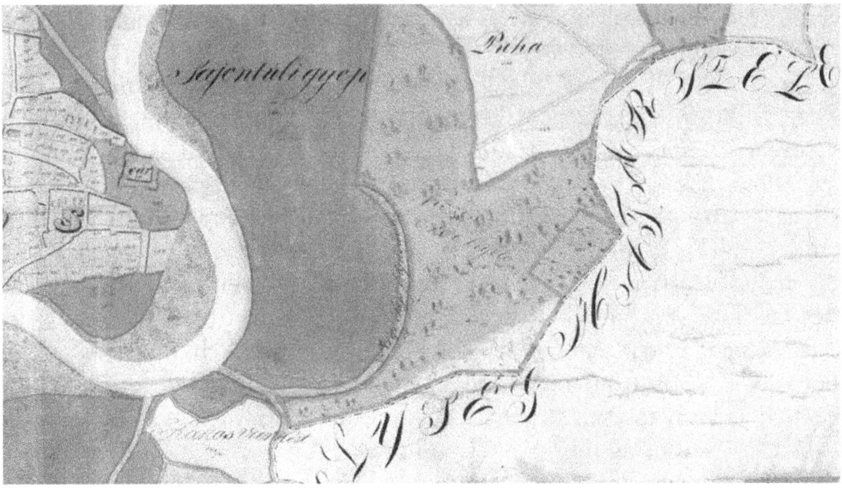

FIGURE 3.19 The border between Ónod and Hídvég in the mid-nineteenth century, showing the location of the former bed of the River Sajó in the center.[374]

Source: MNL OL S 73. no. 102. Online: https://maps.hungaricana.hu/hu/MOLTerkeptar/11395/.

water. The difficulty in identifying comparable established legal customs regarding riverbed changes and land ownership is attributable to the relative infrequency of such lawsuits compared to cases concerning the construction of new mills.[375]

From the late fourteenth century onward, however, there were only a few cases where a change in the riverbed did not result in a change in the ownership of the lands involved.[376] Of course, this does not mean that similar riverbed changes did not lead to lawsuits. Likely, a few hundred similar cases are preserved in the late-medieval legal records of Hungary. In all likelihood, these lawsuits point to the same process identified here and would provide a stronger foundation for the conclusions suggested. Some of the examples discussed above, in addition to highlighting practices that differ from those in late medieval cases, are exceptional from another perspective as well. Among the more than fifty cases collected, only about a dozen can be linked to natural riverbed changes. In the majority (about three-fourths) of these legal disputes, human intervention contributed to the formation of new riverbeds, which is unsurprising given that, by the late medieval period, rivers and streams had been transformed by human activities. Most of the related arguments revolved around the nature of the riverbed modifications. Sometimes, the distinction between artificial and natural modification was unclear, which likely facilitated illegal actions. Due to the unregulated flow of rivers in the Carpathian Basin, their beds changed much more frequently than in the nineteenth and twentieth centuries, when significant regulation works were implemented.

This is not only true for smaller river branches but also for major rivers. Shifts in the riverbeds of smaller rivers and streams were likely everyday occurrences, especially in the lowlands and neighboring hilly areas of the Carpathian Basin. The primary concern was often not the change itself, but rather whether the river would return to its old bed and continue flowing through it. This is probably why Ulpian's On the Edict (*Ad edictum*), which was included in Justinian's Digest, prohibited any intervention that would alter the flow of a "public river" (as he and Roman authors generally refer to permanent waters) after a flood or under any other circumstances. This perspective stems from the belief that rivers that changed course would eventually revert to their original beds, presumably when water levels were at their lowest during the summer months.[377] This is why the Digest set out different principles for adjudicating short- and long-term modifications of riverbeds, which also defined estate boundaries.

Riverbeds were often formed through earthwork construction, causing rivers to shift to a new bed. Sometimes these alterations are difficult to recognize, as evidenced by legal documents from the Middle Ages.[378] In this way, sizable pieces of land became attached to others.[379] Due to the anticipated significant income from the disputed lands, the riverbanks were destroyed, and channels were excavated to redirect the water. Because many of the lawsuits focused on how a river's flow had been altered, it appears that the two instances were treated differently in late medieval legal practice. Specifically, while a change in a river's flow due to natural hydromorphological processes, without direct human intervention, resulted in a change in ownership, in cases where direct human intervention in a river's flow was demonstrated in court, the ownership of the lands in question remained unchanged.

These deliberate, man-made alterations to the flow of rivers were not always intended to gain possession of someone else's land. A case from the early fifteenth century illustrates the extent to which some interventions in a river had unexpected and, more importantly, unwanted consequences, even for those who carried out the works. In 1405, two sons of Pető of Gerse (the eponym of the prominent Gersei Pethő family), John and Thomas, submitted a complaint regarding the construction of a new channel by the Sárvíz Stream between their Gerse and Sármelléke estates (now both part of Gersekarát) and the land holdings of the nobles of Telekes. The Pető sons had excavated the new channel, intended to supply water to a mill they had built. The stream probably had a small discharge, so its entire flow was diverted into this artificial channel. It is reasonable to assume that the original natural stream was small, since the present-day valley stream in question contains only intermediate flow and fills only after rainfall. The construction of mills by similarly sized streams was not unique to premodern times. Later, these mills were often referred to in Hungarian as *pokolidő* (meaning "storm time") or *felhőt kiáltó* ("sky squalling") mills.[380] Due to the diversion of water, the old riverbed, which from that time likely received no water for most of the year, began to silt up. The nobles of Telekes exploited this change. They began to consider the channel as the new riverbed and border between Telekes and the Pető estates. The nobles started treating the meadows between the two branches of the stream as their own. Even if the new streambed was meant to serve the interests of the Pető family, it ultimately resulted in the detachment of part of their estate and the occupation of these areas by the nobles of Telekes.[381]

3.4.3 Riverbeds and Borders: Some Conclusions

The previous subchapter provided an overview of medieval Hungarian legal customs concerning a particular legal circumstance. The issue of shifting river courses and estate boundaries raised by Bartolus in his treatise, however trivial it may seem, likely had real practical significance in everyday practice. When the humanist Peter Lossai wrote the first treatise on geodesy in the Kingdom of Hungary in the mid-fifteenth century, he attributed the emergence of this profession to the annual flooding of the Nile in ancient times. According to his work, these floods altered land boundaries annually, necessitating remeasurement of the plots.[382] In addition to this unique source, other legal evidence also suggests that the problem was reasonably practical. Although the more than fifty cases identified and partially presented here are far from comprehensive, they provide a sample that enables the identification of a variety of practices and customs. A more systematic study of a larger proportion of the available source material would likely have yielded similar results, although it might have been possible to present the formation of legal customs in a more nuanced manner.

The sources discussed above nevertheless suggest that, from the Angevin period onward, changes in riverbed location led to recurrent property disputes. While court cases in the fourteenth century do not show a clear pattern, from the fifteenth century onward, changes in a riverbed's location typically coincided with alterations in the ownership of the adjacent land. By the fifteenth century, a largely settled customary law had been established, under which similar cases were adjudicated. For the preceding period, the sources also indicate that most of these riverbed changes were not solely the result of natural hydromorphological processes but rather stemmed from deliberate interventions in river flow. The legal customs mentioned here did not apply in these cases.

Identifying such human interventions was often tricky, especially because, as shown above, these processes were sometimes partly artificial and partly natural. At times, these changes were not even intended by those who ordered earthwork or construction along a river. Although none of the lawsuits discussed above suggest this issue *per se*, changing riverbeds may have been caused indirectly by interventions at entirely different sections of the same waterway. The ambiguous nature of these changes was recognized as early as the Middle Ages, which is likely why Werbőczy attempted to change the existing legal customs in his *Tripartitum*. To some extent, he applied Roman law by building on specific points from Justinian's Digest. However, he did not fully adhere to the Roman legal tradition; instead, he modified it to clarify similar situations as much as possible. By stabilizing estate boundaries, even in cases involving changes to the bed of the border river, he likely hoped to resolve such disputes.

Although this book does not focus on the Early Modern period, it is worth considering the relevance of the conclusions drawn to similar legal procedures in the sixteenth and seventeenth centuries. The body of lawsuits collected for this later period remains incomplete, so it would be unwise to generalize at this point. However, one point is clear from the limited sources the existing literature has drawn upon. First, in the century following the publication of the *Tripartitum* and the fall of the medieval Kingdom of Hungary to the Ottomans, Werbőczy's legal principles were not systematically applied; instead, *consuetudo* prevailed.[383] Nonetheless, almost 150 (!) years after the completion of Werbőczy's work, the principles he established in the *Tripartitum* were enacted as law. In 1655, Act 81 referenced the relevant passages of the *Tripartitum*: "About lands which by [flash] floods or floods that happened or happen slowly were carried away from a land and were attached to another, Book 1 Title 87 of the Tripartitum has to be applied."[384] Although by this time many points of the *Tripartitum* had become standard references in lawsuits and many had also been enacted as laws, the aforementioned point of Werbőczy's work was not among them.

Here are two additional sources discussed that support the latter point. However, as the current collection is far from comprehensive for the Early Modern period, these sources can only serve as indications that other cases may be identified through a more systematic analysis of similar sources. The first of the two cases mentioned is a letter from 1655, the year in which Act 81 was passed by King Ferdinand III, suggesting that the issue remained unresolved. While the most significant threats in the area from the beginning of the sixteenth century were the Ottomans' advances and raids, hostilities also occurred in other directions. Vas County has marked the western border of Hungary since the Árpádian period. This border was as contested as the Kingdom of Hungary's southern border. As discussed in detail recently, from the thirteenth century onward, there were recurring border disputes between the Holy Roman Empire, the Habsburg territories, and the Kingdom of Hungary. This unrest, despite some peace treaties, led during the Angevin period to the creation of border committees that met regularly.[385] Nobles on both sides of the border actively engaged in plundering and, periodically, occupied lands owned by landlords on the opposing side. In December 1655, the provisor of one of the most influential noble families in Transdanubia, the Batthyány family, was involved in a lawsuit in Dobersdorf (now part of Rudersdorf) concerning plots of land beside the River Lafnitz. The letter reveals that Magdalene, the sister of Adam Batthyány (1610–1659), the landlord of the family's vast estate complex, informed him of the property dispute. According to the letter, the family insisted that even if the river changed its course, their lands would not change hands.[386] The opposing party, however, held a different perspective, perhaps because, in this section, the River Lafnitz marked the border between the Kingdom of Hungary and the Austrian duchies, where legal practices also differed.[387] In such cases, there was no basis for prioritizing the Hungarian legal system.

Another similar case is preserved from a slightly later time. In 1680, the nobles of Vas County, as usual, gathered in Murska Sobota for their annual meeting to discuss legal matters and the state of defense against the Ottomans. This time, the debates focused on territories along the River Mura, which, in certain sections, formed the border between the Kingdom of Hungary and Styria from the Árpádian period onward. In our modern perspective, the River Mura may have effectively served as a state and estate border; however, this area frequently experienced disputes, not only because the landlords were eager for plunder but also due to the river's susceptibility to extremely rapid hydrological changes. The regular floods of this alpine river and the characteristics of the area along the Styrian–Hungarian border led to significant alterations in the location of riverbeds and floodplains on several occasions during the Middle Ages and the Early Modern period. The changing nature of this landscape is well illustrated by numerous eighteenth- and early nineteenth-century manuscript maps that depict this section of the River Mura. These maps were created to clarify which islands and peninsulas belonged to which estate and eventually state. In many instances, the maps not only indicate the river's state at the time of their creation but also note which estate various pieces of land previously belonged to.[388]

The case from 1680 involved two estates: Felső-Petánc (part of present-day Petanjci) on the left bank, which belonged to the Kingdom of Hungary, and the Styrian village of Szatóc (Šratovci [Schrottendorf] in the borders of present-day Radenci) on the right bank of the River Mura. The investigation conducted by the authority of Vas County focused on an island approximately a hundred acres in size. The land originally belonged to the village of Felső-Petánc, but thirty or forty years before the dispute arose, the frequent floods of the River Mura had torn it from the lands of Felső-Petánc and attached it to the village of Szatóc. It seems that from the 1640s or 1650s until 1679, the landlords, the Nádasdy family, took no measures to reclaim their lands. During these decades, the inhabitants of the Styrian village treated the land as their own. Then, in 1679, Andrew Nádasdy (a member of the less affluent Darabosi branch of the family) decided to reclaim the disputed lands. Austrian and German territorial law, like Hungarian, was likely influenced by Roman law by the mid-seventeenth century.[389] Therefore, similar to Hungarian legislation discussed above, the law of the land would have supported Nádasdy's claim. Andrew Nádasdy, however, was likely either unaware of the Styrian customs in this context or unwilling to pursue a lengthy lawsuit against the landlords of Szatóc, the Eggenbergs, then one of the wealthiest families in Styria. It may have seemed more straightforward to attempt to reoccupy his lands by force. His retinue did not stop after occupying the lands; they also severed the ferry ropes operated by the Styrians between the two banks of the Mura. The noble judge (*juratus assessor*) of Vas County, Peter Körmendi, inquired about the original status of the land, suggesting he was aware of the legal principles to be followed in such cases, but Nádasdy was probably not.

These two brief episodes are not unique along river borders in the Middle Ages and the Early Modern period. What stands out and likely deserves consideration, however, is the Batthyány family's awareness, in contrast to Andrew Nádasdy's ignorance, of the judicial options in such cases. Even though twenty-five years had passed since the law was enacted, it may not yet have been widely understood by 1680. This further illustrates the earlier point that, in infrequent cases, even legal experts, such as noble judges, sometimes lacked a solid grasp of the norms. In both the Dobersdorf case and that of Felső-Petánc, the issues were compounded by the fact that the lands not only changed owners and estates but also their governing state. Despite these jurisdictional complications, the manner in which the procedure was handled in either case does not appear to have been affected. Both sources should remain simple snapshots of a larger picture, and it's possible that the Styrian party eventually appealed against the occupation of the lands if their customs differed.

3.5 Conclusions: Rivers as Conflict Environments

The previous main chapter addressed the book's central questions, specifically how the use of water bodies by different actors in the rural and urban landscapes of the Kingdom of Hungary during the Middle Ages gave rise to conflicts and to negotiated resolutions. The organization of the various subchapters reflects resource-use conflicts in both rural and urban contexts surrounding water used for purposes such as milling, fishing, defense, and navigation along major rivers and small streams. The presented case studies illustrate how these conflicts were resolved and, notably, which legal principles were upheld by the courts that rendered judgments in these cases.

Whether a landowner or farmer sowed oats or barley on a specific estate was a private decision with little impact on surrounding estates. Of course, where herd animals were kept, it was always the owner's responsibility to keep them within the limits of a designated pasture.[390] Felling trees on one's property might have eventually altered the hydrological conditions and erosion rates in areas outside the estate. Still, this process resulted in only a few regulations or lawsuits in Central Europe during the premodern period. The exploitation of forest resources also remained within the realm of property rights throughout the Middle Ages.[391]

Since the Roman legal tradition, waterways, especially navigable ones, have been treated differently from plowed lands or forests. These entities have never been considered private property; instead, they have belonged to the sovereign. Rulers, of course, granted certain related benefits to institutions, groups, or individuals, such as fishing rights along the Danube in the territory of present-day Budapest, or the rights of settlers to use the waters for fishing or for building mills in some of the mining towns of Upper Hungary and Transylvania. However, the ownership of minor, non-navigable rivers and streams also differed from plowed lands, meadows, or forests. The various types of interventions in river flow, as demonstrated in Chapter 3.1, led to changes, amongst others, in water levels, hydrological regimes, siltation, fish catch, and navigation possibilities.[392] If a person intentionally modified the flow of a water body, it could lead to unintended consequences. Of course, many intended changes directly harmed the interests of other individuals, as shown in the examples in this chapter. Because of the nature of waterways, these consequences could affect large areas and sometimes territories quite distant from the location of the obstruction or diversion introduced into a river.

Accordingly, the organization of water use along both major waterways and minor rivers and streams needed to be negotiated among various stakeholders interested in these water bodies. Despite the numerous lawsuits related to such conflicts, very few medieval laws addressed water-related issues. Nevertheless, a variety of lawsuits indicate that established norms regarding lawful water manipulation existed. The absence of direct regulations, of course, led to numerous lawsuits from the onset of legal literacy in Hungary. The preserved cases provide insight into how water-related conflicts were resolved. In many instances, as shown in the previous subchapters, only fragments of the lawsuits have survived. Often, there is little information on the implications of these decisions. Nonetheless, the glimpses provided by these legal debates illustrate the importance of securing rights to water and associated benefits for different individuals and institutions. The increasing number of disputes over water use indicates that the waters of the Carpathian Basin were utilized with greater intensity over time.

Can patterns be observed in the numerous legal cases discussed in the previous subchapters? Were there regions or periods with more conflicts, or did regional differences influence the resolution of such disputes? Certain areas of conflict are overrepresented, but these coincide with regions that either have more surviving documentary evidence or are simply those where sources have been more systematically published in earlier centuries, particularly in cartularies. These areas include urban areas notably the broader environment of free royal and mining towns on one hand, and regions with a denser network of ecclesiastical institutions, most importantly monasteries and chapters, on the other (*see* Figure 3.20).

Areas with fewer conflicts overlap with parts of the country that produced a smaller body of written evidence, either because lower population density led to fewer legal disputes or because the relevant sources were more heavily destroyed than others. Examples include mountainous regions such as the Carpathians and the Bihor Mountains. These areas were less populated than the rest of the Carpathian Basin, and many parts of these mountains were frontier regions until the end of the Árpádian period.[393] Another area significantly underrepresented in legal sources is the Great Hungarian Plain. The lack of sources warrants further explanation. First, the population density in this area, especially during the late medieval period, from which most surviving documents originate, was certainly lower than in other parts of the central basin, particularly Transdanubia.[394] Second, without free royal towns and with only a few ecclesiastical institutions permitted to issue legally binding documents, the region preserves only a relatively scant amount of documentary evidence. This scarcity was likely exacerbated by the area's heavy Ottoman occupation, which likely influenced the survival of a limited number of documents produced there. Third, and likely most importantly, the area's hydrography differed significantly from the rest of the Carpathian Basin. The few major rivers in the region, including the Danube and the Tisza, as well as some streams in the Körös River watershed, had relatively slow currents and were of limited utility for operating mills. Outside of these waterways, there were extensive marshlands unfit for constructing mills. These water bodies were primarily used for fishing, and, due to clearer associated rights, this led to fewer lawsuits. Therefore, rather than highlighting the importance of specific geographical areas in water-use conflicts, it is more accurate to assert that water bodies were more intensively contested resources in urban settings and in hilly areas of the countryside.

The picture becomes less clear when examining the temporal distribution of the cases analyzed here. Periods during which the sources are more common also produce higher numbers of legal disputes. More importantly, the distribution of sources edited in the various cartularies studied and discussed in Chapter 1.2.2 is highly uneven. The majority of sources from

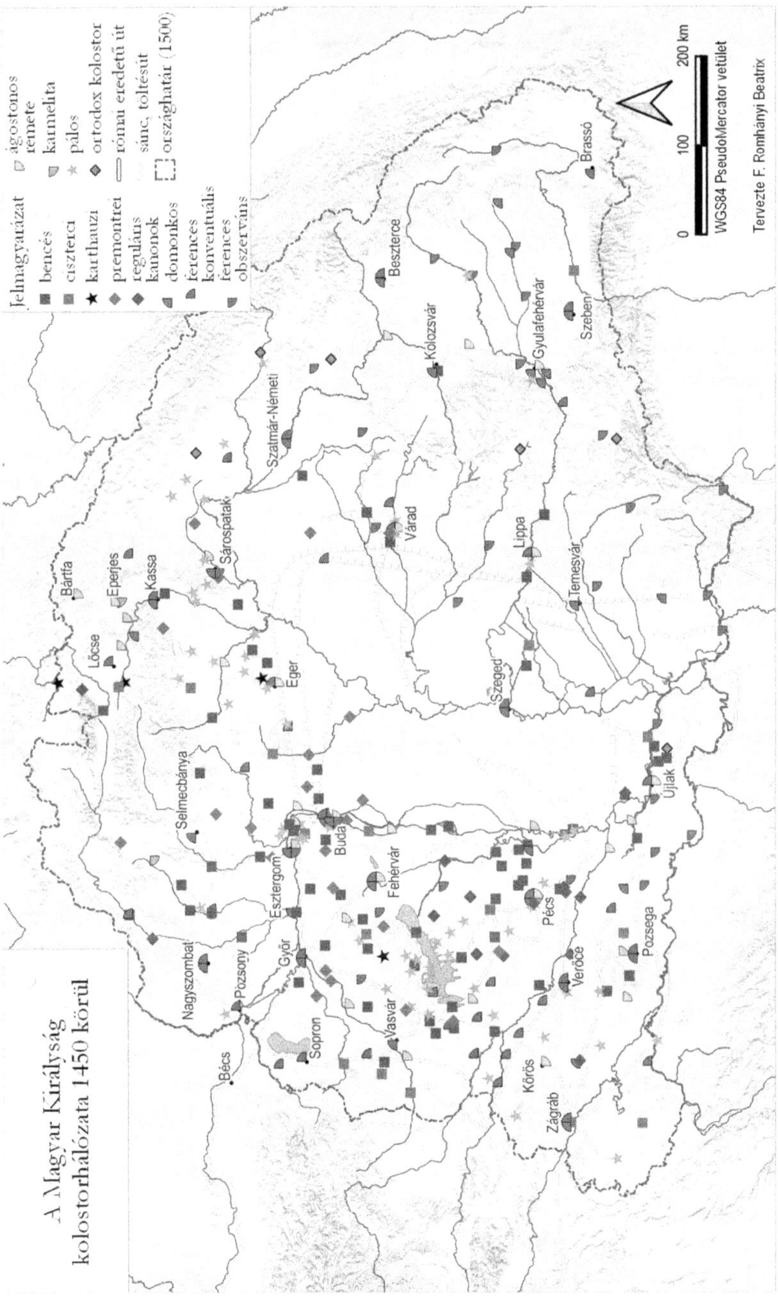

FIGURE 3.20 Monasteries and chapters in the Kingdom of Hungary in the Late Middle Ages (after Beatrix Romhányi).

the Árpádian period were either published in extenso or summarized. The same is true for the first two-thirds of the Angevin period. However, the situation differs for the period from around 1365 to 1387, for which sources have been collected only partially in cartularies. Thanks to a research group's enormous efforts, the first forty years of Sigismund's reign are now better covered. Consequently, more lawsuits concerning water rights have come to light. However, the period from the late 1420s onward is less systematically covered in the various cartularies. The surviving sources from King Matthias' reign are notably under-edited, except for those concerning the Kingdom of Hungary's diplomatic relations. This uneven distribution may lead to false conclusions. Even if the majority of the cases discussed in the previous subchapters fall between the late thirteenth and the mid-fifteenth century, this does not imply that this period saw more water-related lawsuits. Instead, it indicates that legal documents for this period are more accessible. Nevertheless, from the fourteenth century onward, there was not only a steep rise in the number of surviving records, but also concomitant attempts by various institutions and individuals to secure their rights regarding water use.

Identifying spatial and temporal patterns in the development of water-related disputes and customary law in the Kingdom of Hungary during the Middle Ages can be challenging. However, these conflicts have become increasingly evident in the legal awareness and daily lives of people and institutions that depend on, or anticipate income from, farming, industrial, and transport activities involving water bodies. The numerous water-related cases in the sections of the summary of medieval Hungarian customary law, the *Tripartitum*, clearly indicate the significant role of water in the economic life of medieval Hungary.

The duration for which this customary law has been upheld remains to be discussed. There is insufficient space to delve into the complex legal developments in Hungary and its political disintegration from the mid-sixteenth century onward. Still, a single case will be briefly presented to illustrate the enduring nature of these medieval norms. The case concerns the Nádasdy family, previously mentioned in connection with violent trespass in Styria during the 1680s. This time, however, the file records a dispute involving the more prominent heirs of this significant Early Modern family. The case concerns an appeal by Francis Nádasdy (1708–1783) addressed to Veszprém County on September 20, 1770. Nádasdy's appeal concerned a new mill constructed by the village of Berhida, then owned by the cathedral chapter of Veszprém. According to Nádasdy's claim, the newly built mill caused losses to his water mills in Kiskovácsi (part of present-day Berhida). The aristocrat contended that the dam of the new mill affected the mills in Kiskovácsi by altering the flow of water reaching them, which were located downstream along the Séd Stream.[395] In response to the appeal, a committee was dispatched to investigate whether there were valid grounds for the complaint. The response affirmed his claims, leading Nádasdy to demand, in the aforementioned document, the demolition of the chapter's mill. This case parallels dozens, if not hundreds, of other cases from the early modern and modern times. The source, had it been in Latin rather than Hungarian, which by the late eighteenth century had become a commonly used language in administrative and legal matters under the jurisdiction of the counties, could easily have been a late-medieval document as well. The manner in which such cases were handled remained virtually unchanged until modern times, when watermills gradually became obsolete. Until that point, however, practitioners who possessed practical legal knowledge played a crucial role in addressing such legal issues. These individuals were those who worked with water or owned water-related infrastructure. The next chapter focuses on them.

Notes

1 See: Katalin Szende, "Stadt und Naturlandschaft im ungarischen Donauraum des Mittelalters," in *Europäische Städte im Mittelalter* (Forschungen und Beiträge zur Wiener Stadtgeschichte, 52), eds. Ferdinand Opll, and Christoph Sonnlechner (Vienna, Bozen, and Innsbruck: Studien Verlag, 2010), 365–397.

2 The subchapter includes some results published already in: András Vadas, "Some Remarks on the Legal Regulations and Practice of Mill Constructions in Medieval Hungary," in *Wasser in der mittelalterlichen Kultur: Gebrauch – Wahrnehmung – Symbolik / Water in Medieval Culture: Uses, Perceptions, and Symbolism* (Das Mittelalter. Beihefte, 4), eds. Gerlinde Huber-Rebenich, Christian Rohr, and Michael Stolz (Berlin: de Gruyter, 2017), 291–314.

3 For the most important exceptions, see Ingo Herbert Kropač, "Mühlen und Mühlenrecht in der Steiermark während des Mittelalters," (PhD-diss. Universität zu Graz, 1982), 40–41, Reinhard Härtel, "Von Wasser- und Mühlenrecht im Hochmittelalter." in *Recht und Geschichte. Festschrift Hermann Baltl zum 70. Geburtstag*, ed. Helfried Valentinitsch (Graz: Leykam, 1988), 221, Mathieu Arnoux, "Les moulins à eau en Europe occidentale (IXe–XIIe siècle). Aux origines d'une économie institutionnelle de l'énergie hydraulique," in *L'acqua nei secoli altomedievali: Spoleto, 12–17 aprile 2007,* 2 vols (Settimane di Studio del Centro Italiano di Studi sull'Alto Medioevo, 55) (Spoleto: Presso La sede della Fondazione, 2008), vol. 2, 715–720. See also: Ryszard Kiersnowski, "Znaki graniczne w Polsce średniowiecznej," [Border signs in medieval Poland] *Archeologia Polski* 5 (1960): 267 and Alexandra Chira, "Documentary Evidence on the Uses of Water-Mills in Medieval Transylvania. A Case Study of the River Someş," in *Usus aquarum: Interdisziplinäre Studien zur Nutzung und Bedeutung von Gewässern im Mittelalter* (Forschungen zur Geschichte und Kultur des östlichen Mitteleuropa, 54), eds. Christoph Mielzarek, and Christian Zschieschang (Vienna: Böhlau, 2019), 65–67. https://doi.org/10.7788/9783412500894.51

4 István Tringli, "A magyar szokásjog a malomépítésről," [Hungarian customary law of mill construction] in *Tanulmányok a középkorról* [Studies on medieval history] (Analecta Medievalia, 1), ed. Tibor Neumann (Budapest and Piliscsaba: PPKE, 2001), 251–267.

5 E.g., László Szabolcs Gulyás, "A mezővárosi önkormányzat funkciói és társadalmi háttere a középkori Hegyalján," [Functions of the autonomy of market towns and its social background in the medieval Hegyalja region] (PhD-diss., Debreceni

Egyetem, 2008), 144, Kristóf Keglevich, "A garamszentbenedeki bencések és Újbánya polgárai a 14. század közepén," [The Benedictines of Hronský Beňadik and the burghers of Nová Baňa in the mid-14th century] in *Tiszteletkör. Történeti tanulmányok Draskóczy István egyetemi tanár 60. születésnapjára* [Lap of honor. Studies in honor of the 60th birthday of Professor István Draskóczy], eds. Gábor Mikó, Bence Péterfi, and András Vadas (Budapest: ELTE Eötvös Kiadó, 2012), 144, and László Ferenczi, "'Molendinum ad aquas calidas.' A pilisi ciszterciek az állítólagos 'Fehéregyházán'. Történeti topográfiai és tájrégészeti kutatás a pilisi apátság birtokán," [The Cistercian of Pilis at the alleged 'Fehéregyháza'. Historical topographic and landscape archaeological research at the estates of the Pilis abbey] *Studia Comitatensia* [NS] 1 (2014): 146.

6 Ecclesiastical bodies entrusted with issuing legally binding documents in medieval Hungary, with similar functions as public notaries in Western Europe. See: László Solymosi, "Die glaubwürdigen Orte (loca credibilia) Ungarns im 14.–15. Jahrhundert," *Archiv für Diplomatik* 55 (2009): 175–190. https://doi.org/10.7788/afd.2009.55.jg.175 and Katalin Szende, *Trust, Authority, and the Written Word in the Royal Towns of Medieval Hungary* (Utrecht Studies in Medieval Literacy, 41) (Turnhout: Brepols, 2018), 72–84.

7 Tamás Kőfalvi, "Malmok és malomhelyek a pécsváradi konvent hiteleshelyi gyakorlatában." [Mills and mill sites in the practice of the credible place of the convent of Pécsvárad] in *Studia Professoris – Professor Studiorum. Tanulmányok Érszegi Géza hatvanadik születésnapjára* [Studies in honor of the 60th birthday of Géza Érszegi], eds. Tibor Almási, István Draskóczy, and Éva Jancsó (Budapest: Magyar Országos Levéltár, 2005), 158.

8 József Holub, *Zala megye középkori vízrajza* [Medieval hydrography of Zala County] (A Göcseji Múzeum közleményei, 23) (Zalaegerszeg: Göcseji Múzeum, 1963), 48–49 and Pavol Maliniak, "Mlynárstvo na strednom Pohroní v stredoveku a na začiatku novoveku," [Milling in central Pohronský region in the Middle Ages and in the beginning of the Modern Age] in *Z histórie technicko-hospodárskeho využitia vodných tokov na strednom Pohroní* [From the history of technical and economic use of watercourses in the central Pohronský region], eds. Ján Žilák, and Pavel Hrončok (Banská Bystrica: Centrum vedy a výskumu UMB: Fakulta humanitných vied UMB, 2011), 38.

9 Alajos Degré, *Magyar halászati jog a középkorban* [Fishing law in medieval Hungary] (Budapest: Sárkány Nyomda, 1939), 83 note 48.

10 Holub, *Zala megye*, 48–49. On mill roads, see: Magdolna Szilágyi, *On the Road: The History and Archaeology of Medieval Communication Networks in East-Central Europe* (Series Minor, 35) (Budapest: Archaeolingua, 2014), 152–154.

11 Holub, *Zala megye,* 48–49. See for a similar assumption: Dénes Jankovich B., "Adatok a Körösvidék középkori vízrajzához és a vizek hasznosításához," [Data to the medieval hydrography and water use of the Körös region] *Békés Megyei Múzeumok Közleményei* 16 (1996): 340.

12 Tamás Vajda, "Újabb adatok Árpád-kori vízimalmainkról," [New data on Árpádian-age water mills] in *Középkortörténeti tanulmányok 5. Az V. Medievisztikai PhD-konferencia* [Studies in medieval history, 5. Papers read at the 5th conference of medievalists] eds. Éva Révész Éva, and Miklós Halmágyi (Szeged: Szegedi Középkorász Műhely, 2007), 211–245.

13 Tamás Vajda, "Hazai vízimalmaink 1301–1325 közötti okleveles adatai," [Documentary evidence on water mills in Hungary from 1301 to 1325] in *Középkortörténeti tanulmányok. A III. Medievisztikai PhD-konferencia (Szeged, 2003. május 8–9.) előadásai* [Studies in medieval history. Paper presented at the 3rd PhD-conference in medieval studies, Szeged, 8–9 May 2003], ed. Boglárka Weisz (Szeged: Szegedi Középkorász Műhely, 2003), 212–213.

14 András K. Németh, "Vizek és vízgazdálkodás a középkori Tolna megyében I. Vízimalmok." [Waters and water management in medieval Tolna County, I. Water mills] *A Wosinsky Mór Múzeum Évkönyve* 35 (2013) [2015]: 123.

15 Karl S. Bader stat., "Kollektaneen zu Geschichte und Streuung eines rechtstopographischen Begriffs," *Blätter für deutsche Landesgeschichte* 101 (1965): 56–57; Härtel, "Von Wasser- und Mühlenrecht," 221 and Arnoux, "Les moulins à eau," 715–720.

16 Emil Cernea, "Dreptul asupra vadului de moară" [Legal rights on mill dams] *Analele Universității București* 40 (1991): 93–95; Livia Magina argued that mill sites were simply plots suitable for putting up mills: eadem, "'Molendinis, Piscinis et Piscaturis': The Utilisation of Water Resources in the Banat in the Medieval and Early Modern Periods," in *Government and Law in Medieval Moldavia, Transylvania and Wallachia* (Studies in Russia and Eastern Europe, 11), eds. Martyn Rady, and Alexandru Simon (London: UCL School of Slavonic and East European Studies, 2013), 65. See also: Chira, "Documentary Evidence," 65–67.

17 See e.g., MNL OL DL 70 718. Discussed in: Tringli, "A magyar szokásjog a malomépítésről."

18 See e.g., MNL OL DL 14 969. For similar cases in Italy, see: Gerrit Jasper Schenk, "Managing Natural Hazards: Environment, Society, and Politics in Tuscany and the Upper Rhine Valley in the Renaissance (1270–1570)," in *Historical Disasters in Context: Science, Religion, and Politics* (Routledge Studies in Cultural History, 15), eds. Andrea Janku, Gerrit Jasper Schenk, and Franz Mauelshagen (New York: Routledge, 2012), 34–36.

19 E.g., MNL OL DL 31 222, 70 170, 15 070, 15 098, 15 133, 15 344, 103 566, etc.

20 Moldenhauer, "Mühlen und Mühlenrecht in Mecklenburg," and *Zbior praw litewskich od roku 1389 do roku 1529 tudzież rozprawy sejmowe o tychże prawach od roku 1544 do roku 1563* [Collection of Lithuanian laws from 1389 to 1529 and parliamentary hearings on these rights from 1544 to 1563], ed. Adam Tytus Działyński (Poznań: W drukarni na Garbarach, 1841), 213 [*De inundacione prati, vel stagni ex stagno alterius*].

21 Werbőczy, *Tripartitum*. On the work, see the volume: *A magyar jog fejlődésének fél évezrede. Werbőczy és a Hármaskönyv 500 év múltán* [Half a millennium of the development of the Hungarian legal system. Werbőczy and the Tripartitum after 500 years], ed. Gábor Máthé (Budapest: Nemzeti Közszolgálati Egyetem, 2014) (esp. the contribution by István Tringli) and Martyn Rady, *Customary Law in Hungary: Courts, Texts and the Tripartitum* (Oxford: Oxford University Press, 2015). https://doi.org/10.1093/acprof:oso/9780198743910.001.0001.

22 *Primo ratione temporis prout dictum est immediate. Quia prius emanavit privilegium alterius quod non potest tolli per posterius etiam cum clausula derogatoria.* Werbőczy, *Tripartitum*, Book 2. Ch. 11. 1.

23 MNL OL DL 15 540 (January 9, 1461). See the edition of the document in the appendix (1. no. 4). The charter is mentioned in Csilla Zatykó, "Adalékok egy

1460-as Zala megyei birtokmegosztó oklevelhez," [Contributions to an estate division charter from Zala County from 1460] in *"Es tu scholaris." Ünnepi tanulmányok Kubinyi András 75. születésnapjára* [Studies in honor of the 75th birthday of András Kubinyi] (Monumenta historica Budapestinensia, 13), eds. Beatrix F. Romhányi et al. (Budapest: Budapesti Történeti Múzeum, 2004), 128.

24 Cf. the first reference to water mills in from the Czech lands where mill sites are referred to as *loco ad molendina edificanda* discussed in Chapter 2.2.

25 Kropač, "Mühlen und Mühlenrecht," 40–41 and Härtel, "Von Wasser- und Mühlenrecht," 221. For the period of Arnulf: Berent Schwineköper, "'Cum aquis aquarumqve decursibus' – Zu den Pertinenzformeln der Herrscherurkunden bis sur Zeit Ottos I," in *Festschrift für Helmut Beumann zum 65. Geburtstag*, ed. Kurt-Ulrich Jäschke (Thorbecke: Sigmaringen, 1977), 37 and 56.

26 DHA, vol. 1. 170–174 no. 50/II. For criticism of the document, see Bernát Kumorovitz L., "A zselicszentjakabi alapítólevél 1061-ből, 'Pest' legkorábbi említése," [The foundation charter of Zelicsszentjakab from 1061 and the earliest mention of 'Pest'] *Tanulmányok Budapest Múltjából* 16 (1964): 43-83.

27 Kumorovitz, "A zselicszentjakabi," 52.

28 Kumorovitz, "A zselicszentjakabi," 47–50. See also: Zsoldos, *Archontológia*, 15, 191, and 340.

29 DHA, vol. 1. 173. The text that has been added to the charter also includes reference to a mill site: *Idem bonae memoriae vir dedit Sancto Jacobo alium locum Jana nominatum cum saltu lignorum et cum terra ad unum aratrum pertinente, quatuor mansiones servorum, locum molae cum piscina.* The part in question is at 173–174.

30 Tamás Vajda, "Okleveles adatok Árpád-kori vízimalmainkról," [Documentary data on Árpádian-age water mills] in *Medievisztikai tanulmányok. A IV. Medievisztikai PhD-konferencia (Szeged, 2005. június 9–10.) előadásai* [Studies in medieval history. Papers read at the 4th conference of medievalists, Szeged, 9–10 June 2005], eds. Szabolcs Marton, and Éva Teiszler (Szeged: Szegedi Középkorász Műhely, 2005), 196.

31 *[XL.] In predio Karalka (...) habent terram ad tria aratra, preter prata et feneta et pascua peccorum communiter cum villa et congruum locum aptum ad molendinum. A pannonhalmi Szent-Benedek-Rend története*, 12 vols [History of the Benedictine order of Pannonhalma], eds. László Erdélyi, and Pongrácz Sörös (Budapest: Stephaneum, 1902–1916), vol. 1. 771–787 no. 185. The relevant part, here at 781.

32 MNL OL DL 40 059. Edited in: Ferencz Kubinyi, Jr., *Árpádkori oklevelek 1095–1301 / Codex Diplomaticus Arpadianus* (Monumenta Hungariae historica, 1) (Pest: [Osterlamm], 1867), 46–50 no. 58–60.

33 For its best edition, see: DHA, vol. 1. 228–240 no. 81.

34 Balázs Péter Karlinszky, "A veszprémi káptalan a középkorban. A veszprémi székeskáptalan középkori birtokai," [The chapter of Veszprém in the Middle Ages. The medieval estates of the chapter] (PhD-diss., Pázmány Péter Katolikus Egyetem, 2013), 17–24.

35 *In predio Betereg, quod aliter Soyl nominatur, habet ecclesia terram ad IIII aratra, V servos, III aratores, IIII vinitores; IX iobagiones, III molendina et unum vacuum locum molendini, prata eciam atque silvam (...) In villa Vaschul habet ecclesia terram ad IX aratra, vineas X, servos VIII, yobag primos II, hospites*

iobag VII, aratores IIII^{or}, molendina III et I locum molendini. MNL OL DF 200 611. For its critical edition and diplomatic analysis of the charter: DHA, vol. 1. 228–242 no. 81. For the text, see here: 237–238.

36 Dénes Szabó suggest that the village was close to the Danube or the River Žitava in the vicinity of present-day Moča. Dénes Szabó, *A dömösi adománylevél hely- és vízrajza* [Topography and hydrography of the donation charter of Dömös] (A Magyar Nyelvtudományi Társaság Kiadványai, 85) (Budapest: Akadémiai Kiadó, 1954), 44–45.

37 MNL OL DF 238 281. For its best edition, see: Dénes Szabó, "A dömösi prépostság adománylevele (1138/1329). I–III.," [Donation charter of the provostry of Dömös (1138/1329)] *Magyar Nyelv* 32 (1936): 54–57, 130–135, and 203–206. For the research history of the important document, see Gábor Thoroczkay, "A dömösi prépostság története az alapítástól I. Károly uralkodásának végéig," [The history of the provostry of Dömös from its foundation to the end of the rule of King Charles I] *Fons* 19 (2012): 414–417.

38 For their locations, see: ÁMTF, vol. 1. 420, 428–429, 474, and Kristóf Keglevich, *A garamszentbenedeki apátság története az Árpád- és az Anjou-korban, 1075–1403* [The history of the abbey of Hronský Beňadik in the Árpádian and Angevin period, 1075–1403] (Capitulum, 8) (Szeged: Szegedi Tudományegyetem Törteneti Intézete, 2012), 65–66, 178–179, and 193. For the River Kompa (Compa), see Melinda Szőke, *A garamszentbenedeki apátság alapítólevelének nyelvtörténeti vizsgálata* [Linguistic historical study of the foundation charter of the abbey of Hronský Beňadik] (A Magyar Névarchívum Kiadványai, 33) (Debrecen: Debreceni Egyetemi Kiadó, 2015), 179.

39 *Item aliam possessionem Zanto in vicinato adiacentem cum LXXII mansionibus hospitum, et terram ad XL aratra, cum fenetis et aliis pertinenciis suis, et locum molendini super fluvium Cumpa, qui transit ipsas possessiones.* For the better edition of the charters, see: MES, vol. 1. 116 no 89. and CDES, vol. 1. 79–80 no. 82, for its summary: RA, vol. 1/1. 31 no. 90. For the will: Keglevich, *A garamszentbenedeki apátság története*, 65–66, 178–179, and 193.

40 *Est Bolotyn, de quo egreditur fluvius, qui vocatur Foc; et ab egressu totus est ecclesie. Tendit autem ad meridiem et venit ad villam Lustah, circuitque insulam, que tota est prefate ecclesie, facitque loca molendinis apta.* Edited in: *A pannonhalmi Szent-Benedek-Rend*, vol. 10. 502–517 no. 8. The relevant part, here: 516. For the mills at Fok and its surrounding, see: Andrea Kiss, " 'Rivulus namque, qui dicitur Fuk, fluens de prefato lacu' – Fok, Sár, Foksár," in *Antropogén ökológiai változások a Kárpát-medencében* [Anthropogenic ecological changes in the Carpathian Basin], eds. Bertalan Andrásfalvy, and Gábor Vargyas ([Pécs] and Budapest: PTE Néprajz–Kulturális Antropológia Tanszék–L'Harmattan, 2009), 49-64. For Lusta estate, see here: 53 and 59, and Éva Kovács, *A Tihanyi összeírás mint helynévtörténeti forrás* [The conscription of Tihany as a source of historical toponymy] (Magyar Névarchívum kiadványai, 34) (Debrecen: Debreceni Egyetemi Kiadó, 2015), 190–191.

41 *Iuxta villam [Mager etiam] Wodazow dedit duo territoria cum duobus locis molendinorum.* CD, vol. 3/1. 153–163. For the diplomatic analysis of the document, and its more recent edition: Géza Érszegi, "A leleszi monostor alapítóleveléről, 1214," [On the foundation charter of the abbey of Leles] *Szabolcs-Szatmár-Beregi Levéltári Évkönyv* 16 (2003): 13–28, for the relevant part, see: 24. See also for the summary of the document: RA, vol. 1/1. 96–97 no. 295.

42 *Predium Megehid nomine cum duobus locis molendinorum (...) Habet insuper duo loca molendinorum super aquam Raba et piscinam unam cum duabus vineis in exitu porte civitatis, iuxta viam, qua pergitur ad Sanctum Gothardum.* Edited in AÚO, vol. 6. 383–385 no. 234. For its more recent edition, see: UB, vol. 1. 72–74 no. 105. Its summary: RA, vol. 1/1. 105–106 no. 322. See also: Vajda, "Okleveles adatok Árpád-kori," 197 and 209.

43 *Contulit eciam in Zumuld terram arabilem et tria aratra et pratum, quod situm est iuxta grangyam, quam predicti fratres construxerunt iuxta pomerium et molendinum quod positum est iuxta grangyam abbatis de Thata et alium locum molendini ibidem et silvam, que vocatur Rugus, et vineam in Stanch.* Edited in: UB, vol. 1. 102–109 no. 143, here: 106. See also: Lajos Kiss, "Középkori földrajzi nevek magyarázata," [Explanations of medieval place names] *Magyar Nyelv* 86 (1990): 169. [Rögös]. For the estates of the abbey of Klostermarienberg, see: László Ferenczi, "Management of Monastic Landscapes. A Spatial Analysis of the Economy of Cistercian Monasteries in Medieval Hungary." (PhD-diss., Central European University, 2018).

44 *Terram Sag, ad Castrum Hunth pertinentem, que potest sufficere usui duorum aratrorum et dimidii, et locum molendini super aquam Ipul et fenetum.* Edited in: ÁÚO, vol. 7. 32–33 no. 20, its summary: RA, vol. 1/2. 192 no. 627. See also: ÁMTF, vol. 3. 240–242, and Vajda, "Okleveles adatok Árpád-kori," 210. For the confirmation of the donation of the estate from 1239: ÁÚO, vol. 7. 66–67 no. 45., its summary: RA, vol. 1/2. 199–200 no. 654.

45 *Preterea Ecclesia habet quandam terram, que vocatur Inka, circumquaque metatam, fenetum sufficientem, loca apta ad stagna et ad molendina, silvas satis, de quantitate certificari non potitimus, quia incolis caret. In villa Kulked XX mansiones, terram ad XII aratra, fenetum cum arundineto quasi XL iugera, stagnum in I loco, et locum ad molendinum, silvam quasi L iugera.* MNL OL DF 206 903. Edited in: ÁÚO, vol. 6. 466–477 no. 295, and: *A pannonhalmi Szent-Benedek-Rend*, vol. 1. 694–696 no. 107, its summary: RA, vol. 1/1. 145 no. 451.

46 Cf. Károly Tagányi, "Gyepü és gyepüelve, I-IV.," [*Gyepű* and *gyepűelve*] *Magyar Nyelv* 9 (1913): 97–104, 145–152, 201–206, and 254–266, and Attila Zsoldos, "A vasi várispánság felbomlása," [Dissolution of the royal castle district of Vas] *Vasi Szemle* 54, no. 1 (2000): 27–46, idem, "Confinium és marchia. Az Árpád-kori határvédelem néhány intézményéről," [*Confinium* and *marchia*: on some institutions of Árpádian-age frontier protection] *Századok* 134 (2000): 99–116, and István Herényi, *A nyugati gyepü erődítményei: erődítmények, várak Nyugat-Magyarországon* [Constructions of the Western *gyepű*: fortifications and castles in Western-Hungary] (Budapest: Heraldika, 2007).

47 *Et si ipsam terram a Pousa et Laurencio recipere non possent, nichilominus darent Ponetho locum molendini cum loco curie apto ad statum molendini, de suo iure proprio (...) idem Mathias et N[icholaus] statuisse dixerunt et dedisse Poneto locum molendini cum loco curie apto ad molendinum in superiori parte insule Mathie et Nicholai in angulatione fluvii Zomus.* MNL OL DF 252 590. For its low-quality edition: CD, vol. 4/1. 167–169, somewhat better is: ÁÚO, vol. 11. 309–310 no. 217, its summary: *Az Árpád-kori nádorok és helyetteseik okleveleinek kritikai jegyzéke* [Critical register of the charters of Árpádian-age palatines and their deputies] (A Magyar Országos Levéltár kiadványai, II. Forráskiadványok, 51), ed. Tibor Szőcs (Budapest: Magyar Országos Levéltár, 2012), 75–76 no. 62.

48 ÁUO, vol. 6. 383–386 no. 234. See also: Vajda, "Földesúri jogok," 87; HO, vol. 4. 20–22 no. 7. For Megyehíd, see: Dezső Csánki, *Magyarország történelmi földrajza a Hunyadiak korában*, 5 vols [Historical geography of Hungary in the age of the Hunyadi family] (Budapest: MTA, 1890–1914), vol. 2. 775.

49 ÁÚO, vol. 7. 308–309 no. 215, its summary: RA, vol. 1/2. 289 no. 934. For the forged nature of the document, see: János Karácsonyi, *A hamis, hibáskeltű és keltezetlen oklevelek jegyzéke 1400-ig: a Történelmi Tár 1908 évi számában megjelent „pótlások..."-kal kiegészítve* [Register of forged, wrongly dated and undated charters up to 1400, complemented by the additions published in the *Történelmi Tár* in 1908] (Aetas Könyvek, 1) (Szeged: Aetas, 1988), 22 no. 105. This forged charter is transcribed in the similarly forged charter of King Ladislas IV from 1272: RA, vol. 2/1. 147 no. 2189, then these are transcribed in the forged charter of Ladislas from 1278: RA, vol. 2/2–3. 214 no. 2881.

50 MNL OL DL 49 548., its editions: ÁÚO, vol. 9. 60–62 no. 36, UB, vol. 2. 57–58 no. 79.

51 *Inde reflectitur versus septemtrionem et descendit ad paludem super qua est vetus locus molendini...* MNL OL DL 78 961. (before September 10, 1256). Edited in: Zichy, vol. 1. 9–10 no. 12, its summary: Tamás Kőfalvi, *A pécsváradi konvent hiteleshelyi oklevéltára, 1254–1526* [Cartulary of the credible place of the convent of Pécsvárad, 1254–1526] (Szeged: [SZTE], 2006), 55–56 no. 2.

52 *Et a parte septemtrionali possessioni ecclesie antedicte et in eodem fluvio cu[rrendo] pervenit ad locum molendini Magistri Magii, et ibi exit de fluvio, et penes eundem fluvium a parte meridionali est una meta terrea, in qua meta pro signo est fracta pars lapidis molaris, qua separat a possessione Koxo Martini filii Magii, et a possessione Koxo Benedicti filii Corneli...* MNL OL DL 84 029, its edition: ÁÚO, vol. 8. 37–41 no. 26.

53 For early modern borders marked by mill stones: Lajos Takács, *Határjelek, határjárás a feudális kor végén Magyarországon* [Boundary signs, paermabulation at the end of feudal age in Hungary] (Budapest: Akadémiai, 1987), 102.

54 Karácsonyi, *A hamis*, 26 no. 122 and Vajda, "Okleveles adatok Árpád-kori vízimalmainkról," 213 considers the document authentic, while György Györffy suggested it is a fourteenth-century forgery: ÁMTF, vol. 1. 113.

55 For instance, there is a royal charter by King Andrew II in which he donates the estate of Tomaj (*Tumoy*) in 1216. According to Imre Szentpétery, the authenticity of the document, based on its structure and the land ownership, is dubious. The reference to the mill site may be another reason to suggest that the document was forged. *Sed compulsus Pete filius Leustah ob timorem suae possessionis quia Ochz similiter fuit adversarius, emit xl marcis argenti partem, quae in Tumoy Zaladiensi Tibam praenominatum contingebat et locum unius molendini et decem libertinos (...) necnon V vineas sitas in Wyta.* MNL OL DL 91 101, edited in HO, vol. 6. 11–12 no. 8, its summary: RA vol. 1/1. 102 no. 310. For the estate structure, see: MES, vol. 1. 214–215 no. 220.

56 E.g., *Deinde in meatu seu cursu ipsius putei venit ad fluvium Ipol, et descendit in ipsum fluvium ex inferiori parte molendini predicti comitis Michaelis et filiorum suorum, et deinde sursum in eodem fluvio Ipol eundo venit ad locum antiqui molendini, abhinc vadit ab locum molendini Mark filii Johannis prenotati.* MNL OL DL 1617 and 66 685, edited in: AO, vol. 1. 10–13 no. 8; *Et de illa transitur penes quoddam stagnum lateris a parte orientali adiacentis usque ad locum*

cuiusdam molendini antiqui, ab hinc transit in decursu aque ipsius molendinarii et atingit terram Pustawzud alio nomine Agagzug vocatam. MNL OL DL 77 809, edited in: Zichy, vol. 4. 178–185 no. 153, *Ab occidente est una terra, que dicitur Kisvar et ibi est unus locus molendini desolatus et est molendinum idem Iohannis sartoris et Georgi et Egidi fili Pauli Ianka, sic quod idem molendinum vendiderat avus uxoris Georgi, qui vocabatur Iacobus, vinea tamen pertinebat ad ecclesiam de Beel.* MNL OL DF 207 754, edited at: http://monasterium.net/mom/HU-PBFL/PannHOSB/1433_V_18/charter.

57 *Quandam possessionem suam Fonchol vocatam in comitatu Abawyuariensi existentem inter possessiones Bakta Popy Fay Deuecher Toka et inter eandem possessionem Forrow adiacentem et eisdem undique* [...] *vicinantem simul cum uno loco molendini sui in quo edificium ad ipsum molendinum aptum et tres rothe lapides molares [...] situari possunt seu edificari, super fluvio Hornadvyze a plaga occidentali scilicet a parte dicte ville Forow adiacente cum omnibus utilitatibus et pertinenciis suis ac sub eisdem metis antiquis et limitacionibus quibus ipsam sui predecessores et* [...] *[possi]disset, dedisset, donasset et contulisset.* MNL OL DL 2798, edited in: AO, vol. 3. 55–57 no. 43. For a similar case, see: MNL OL DL 43 182.

58 *Firmitatem vinculum quendam locum pro molendino aptum cum canali in superficie possessionis ipsorum Wgralregd vocate in fluvio Crisii ex opposito molendini eorum a parte possessionis eiusdem Ladislai Peterd predicte existentem pro fabrica molendini eidem Ladislao filio Jacovi dedissent donassent et contulissent immo dederunt donaverunt et contulerunt.* MNL OL DL 7521. See also: Jankovich, "Adatok a Körösvidék középkori vízrajzához," 337. *Intra metas dicte possessionis eorum Eghazasgesen super fluvio Zerethwa aut predicto ryvulo Holknecheche locum molendini, ubi unus finis clausure ipsius molendini ad metas ipsorum Nicolai et Ladislai filii sui protendi posset et deberet et sine eo ipsum molendinum construi et edificari nequiret, ibi et in tali loco iidem Albertus et filii Ewdyn molendinum construi et edificari faciendi haberent facultatem contradictione ipsorum Nicolai et Ladislai filii sui in eo, quod finis clausure huiusmodi molendini ad metas eorum protenderetur, obviare eisdemque molendino in eodem portionem de regni consuetudine vendicare non valentibus.* MNL OL DF 275 823 and *Modo simili tradiderunt, contulerunt assignarunt antelato Nicolao filio Benedicti terram ipsorum cum loco molendini super possesionem Gargow in valle post quam vadit ductus aque seu fossatum ad molendinum cum cunctis utilitatibus et pertinentiis* [...]. MNL OL DL 63 858.

59 *Primo praefati frater Jacobus vicarius et conventus dicti claustri Beatae Mariae Virginis quandam totalem sylvam antefati claustri Morichzewge vocatam in comitatu Zempleminiensi circa metas possessionis Kak peaedictae adiacentem, quam ab orientali antiquus meatus fluvii Hernad, ab occidentali vero partibus modernus meatus dicti fluvii Hernad circumiacerem, simulcum locis molendinorum super eidem meatibus existentem aliisque universis utilitatibus eiusdem silvae.* MNL OL DL 16 955.

60 *Ad supplicationem et instantiam fratrum heremitarum de eadem inclinati, quoddam molendinum nostrum quodammodo patiens ruinam et desolatum, in rivulo Kekyspataka vocato, in territorio ville nostre regalis Zenthendre vocate habitum, et locum alterius molendini in eodem rivulo supra predictum molendinum existentem, ab omni iurisdictione et potestate castri Wyssegradiensis

et castellanorum eiusdem prorsus et per omnia excepta et exempta, cum omnibus ipsorum utilitatibus et pertinentiis universis, auctoritate regalis excellentie eidem ecclesie sancti regis Ladislai et fratribus heremitis in ea deo famulantibus, elemosinalem et dotalitiam provisionem facientes, duximus concedenda, imo damus et conferimus pleno iure perpetuo et irrevocabiliter possidenda tenenda pariter et habenda. MNL OL DL 7121, edited in: AO, vol 7. 70–71 no. 40. Mentioned in: Romhányi, *Pauline Economy,* 71. See also for abandoned mills: Maliniak "Mlynárstvo na strednom Pohroní v stredoveku," 38.

61 MNL OL DL 15 680, edited in: AO, vol. 6. 244–249 no. 160. For naming of mills, see: Christian Zschieschang, "Zur Benennung von Mühlen im Mittelalter," in *Wassermühlen und Wassernutzung,* 193–232 http://doi.org/10.1515/9783110547023-004 and Christian Zschieschang, "Beobachtungen zur Entstehung von Mühlennamen westlich der Oder," in *Usus aquarum,* 87–162. https://doi.org/10.7788/9783412500894.87.

62 The charter from 1216: MNL OL DL 91 101. Other examples include MNL OL DL 98 120 and 67 620.

63 MNL OL DF 265 871.

64 *In loco alterius deserti molendini similiter duarum rotharum Zekemolna nuncupato.* MNL OL DL 58 253.

65 MNL OL DL 2553 (*reparari facere*); 43 505 (*restaurandum*) or 17 428 (*reformare*).

66 Cf. Werbőczy, *Tripartitum,* Book 1. Ch. 133. 136–137.

67 Werbőczy, *Tripartitum,* Book 1. Ch. 133. 136–137. Cf. György Bónis, *Középkori jogunk elemei. Római jog, kánonjog, szokásjog* [Elements of Hungarian medieval law. Roman law, canon low, customary law] (Budapest: Közgazdasági és Jogi Könyvkiadó, 1972), 167–168 and Tamás Vajda, "Árpád- és Anjou-kori vízimalmaink tájalakító hatása," [Water mills as driving forces of landscape change in the Árpádian and Angevin Periods] in *Micae mediaevales II. Fiatal történészek dolgozatai a középkori Magyarországról és Európáról* [Studies of young historians on medieval Hungary and Europe], eds. Péter Jakab et al. (Budapest: ELTE BTK Történelemtudományok Doktori Iskola, 2012), 59–75.

68 E.g., *In villa Woyanouch duas sessiones iobagionales populosas et insuper tres desertas, edificia domorum habentes, necnon unum locum molendini subtus pellentis, aqua eiusdem nonquam deficiente (...) in villa Mathewsouch decem sessiones iobagionales populosas et insuper quatuor desertas, necnon septem molendina subtus pellentia, aquis eorundem nonquam deficientibus et tria loca molendinorum deserta et edificia non habentia (...) in villa Zorouch tres sessiones iobagionales populosas et unam desertam ac unum molendinum (...) in Boysinch duas sessiones iobagionales populosas et unam desertam ac locum unius molendini desertum subtus pellentis, aqua eiusdem nonquam deficiente.* MNL OL DL 262 183. (February 6, 1418), for its summary, with the edition of the relevant part: ZsO, vol. 6. 397–400 no. 1416 or *Coram eodem conventu personaliter constitutus quendam locum molendini in fine et intra metas possesionis Bandayka in fluvio Thorna appellato aliis decurrens subtus pellente nunc edificiis destitutum apud manus suas pro centum florenis auri titulo pignoris habitum dudum dicto condam Petro filio condam Pauli patri suo a Johanni filii Johannis de dicta Ayka impignoratum prefato Andree doctori et vicario dicti ecclesie Wespriminensi pro eisdem centum florenis auri puri plene ut dixisset ab eo accepisset simulcum uno*

loco sessionali in quo molendinator ipsius molendini moratus fuisset... MNL OL DL 66 292 (September 1, 1467).

69 *Item in terra Kochou Magister Etthuruh habet eandem partem, quam prius habuit metis distinctam et Nicolaus habet similiter eandem porcionem, quam prius habebat metis separatam et molendinum, quod est in Kochou iuxta ecclesiam cessit Magistro Etthuruh; aliud vero molendinum quod est ibidem in fine ville inferiori cessit magistro Nicolao et in locis molendinorum adhuc vacuis equalem partes habebunt porcionem.* MNL OL DL 67 264, edited in: ÁÚO, vol. 12. 267–268 no. 225.

70 As noted above Vajda suggested that the term *locus molendini* is used to describe pieces of land that had different owners than the surrounding areas. He may have come to this conclusion based on such lease contracts.

71 *Concessimus seu condonavimus Jacobo, famulo Kaboldi urbararii domini regis, civi de Bakabania, super fluvium Byksewniche, intra metas possessionis nostri monasterii habitum unum fundum, seu locum molendini nonum decimum ab inferiori parte ipsius rivuli sursum numerando.* For the latest edition of the charter, see: Kristóf Keglevich, *A garamszentbenedeki apátság 13. és 14. századi oklevelei (1225–1403)* [13th–14th-century charters of the abbey of Hronský Beňadik (1223–1403] (Capitulum, 9) (Szeged: Szegedi Tudományegyetem Történeti Intézete, 2014), 336–337 no. 8. For fairly similar lease contracts, see: see here 334–335 nos. 5 and 6. On the regulations of distances between mills, see: Moldenhauer, "Mühlen und Mühlenrecht," 221 and Francesco Salvestrini, *Libera città su fiume regale. Firenze e l'Arno dall'Antichità al Quattrocento* (Florence: Nardini, 2005).

72 For the latest edition, see: Keglevich, *A garamszentbenedeki apátság*, 336 no. 7.

73 The relevant sheets of the military surveys are available online: http://mapire.eu.

74 See for instance: András K. Németh, and Gábor Máté. "Szempontok és példák a középkori eredetű malmok és malomhelyek folytonosságának vizsgálatához," [Notes and examples to the study of the continuity of medieval mills and mill sites] in *Középkori elemek a mai magyar anyagi kultúrában* [Medieval elements in the present-day Hungarian material culture] (Tanulmányok a Kárpát-medence anyagi kultúrája köréből, 1), eds. Anikó Báti, and Zsigmond Csoma (Budapest: Agroinform, 2014), 47–68.

75 See e.g., the abbey of Pécsvárad: MNL OL DL 3683. With some omissions edited in: AO, vol. 5. 267–268 no. 138.

76 *Quoddam locum molendini ipsorum intra metas eiusdem possesionis Zakachy in fluvio Harahalya vocato habitum necnon duo iugera terrarum arabilium inter ipsum locum ipsius molendini in quibus alveum aque ipsius molendini edificandi preparari deberet necnon cursi antiquo meatu ipsius aque ac quoddam parvum pratum ibidem existens* MNL OL DL 8856.

77 *Nos magistri Nicolaus et Stephanus vicecomites de Zaboch et quatuor iudices de eodem significamus quibus expedit universis praesentium per tenorem quod magistri Dominicus et Ladislaus filii Stephani filii Keleed nobiles de Berench accedentes ad nostram personalem praesentiam et nobis uno ore equalique voce et concorditer atque sana mente significare curaverunt quod in universis eorum possessionis* [sic!] *ubique existentibus talem celebrassent in consequens divisionem quod recta dimidietas dicte possessionis Berench vocate a parte orientali dicto magistro Dominico simul cum suo proprio loco sessionali cessisset praeter locum cuiusdam molendini a parte meridionali in rivulo Sebuscher habitam* [sic!] *quod*

locum magistro Ladislao antedicto successisset et similiter dimidietas eiusdem possessionis dicto magistro Ladislao scilicet a parte occidentali suo loco similiter proprio sessionali succedissent. Item idem dominus magister Dominicus unum locum pro molendino ubi voluerit liberam habebit occupandi facultatem, residuas vero tam villam loca molendina, quam alias utilitates quovismodo vocitatas communiter possidebunt. MNL OL DL 96 258.

78 *Ex uberiori gracia nostra eidem Nicolao unum locum molendini prope dictam possessionem [Kochuba] in fluvio Fekethe Keres vocato gratiose concedimus ita ut ipse in dicto loco quem elegerit.* MNL OL DL 44 405.

79 *Idem Iohannes et Georgius pro ipso molendino per dictum Ladislaum de Zewles facto pro se modo simili unum locum molendini in dicto fluvio Tharna eligere et sibi ipsis aliud molendinum construi facere.* MNL OL DL 18 035. For another similar case, see: MNL OL DL 64 552.

80 Tringli, " A magyar szokásjog," 255.

81 *Tandem magister Mykse, filius Reynoldi, de genere Ratholdi, ibidem in instanti coram nobis personaliter comparens dixit protestando quod predictum molendinum filiorum Bagar olim inter duo molendina sua super eundem fluvium ab utraque parte tum superiori tum inferiori decurencia, in preiudicium eorum, per facti potenciam, in indebito et inconvenienti loco fundatum ac hedificatum extitisset et ne per constructionem molendini in loco molendini filiorum Bagar supradicto, sibi in posterum preiudicium valeat generari, neque dicta loca molendinorum suorum propriorum, et terre ad eadem pertinentes, in divisionem huiusmodi inmiscendo, occupentur et alienentur ab ipso, a rehedificacione ipsius molendini, et occupacione terrarum et locorum molendinorum suorum predietarum, fratres nostros capitulum et nobiles antedictos, perinodum protestacionis prohiberet et prohibuit coram nobis.* MNL OL DF 281 805, edited in: *Oklevelek Hontvármegyei magán-levéltárakból I. 1256–1399* [Charters from private archives of Hont County, 1256–1399], ed. Ferenc Kubinyi, Jr. (Budapest: [no publisher], 1888), 61–62 no. 42.

82 MNL OL DF 201 008 and 201 009.

83 Arnoux, "Les moulins à eau," 715–720.

84 *The Development of Literate Mentalities in East Central Europe* (Utrecht Studies in Medieval Literacy, 9), eds. Anna Adamska and Marco Mostert (Turnhout: Brepols, 2004). https://doi.org/10.1484/M.USML-EB.6.09070802050003050104040909.

85 See: Richard C. Hoffmann, "Footprint Metaphor and Metabolic Realities Environmental Impacts of Medieval European Cities," in *Natures Past: The Environment and Human History* (Comparative Studies in Society and History), ed. Paolo Squatriti (Ann Arbor, MI: University of Michigan Press, 2007), 288–325. See also: James A. Galloway, Derek Keene, and Margaret Murphy, "Fuelling the City: Production and Distribution of Firewood and Fuel in London's Region, 1290–1400," *The Economic History Review* NS 49 (1996): 447–472. https://doi.org/10.1111/j.1468-0289.1996.tb00577.x, and James A. Galloway, "Metropolitan food and fuel supply in medieval England: regional and international contexts," in *Food supply, Demand and Trade: Aspects of the Economic Relationship between Town and Countryside (Middle Ages – 19th century)* (Comparative Rural History Network – Publications, 14), eds. E. Thoen, and P. Van Cruyningen (Turnhout: Brepols, 2012), 7–18. https://doi.org/10.1484/M.CORN-EB.4.00053.

86 Judit Klement, *Hazai vállalkozók a hőskorban: A budapesti gőzmalomipar vállalkozói a 19. század második felében* [Hungarian entrepreneurs in the heroic age: the entrepreneurs of the steam mill industry in the second half of the 19th century] (Budapest: Eötvös Kiadó, 2012), 56–69.

87 See the studies in the volume: *Montagnes, fleuves, forêts dans l'histoire. Barrières ou lignes de convergence? (Berge, Plusse, Walder in der Geschichte. Hindernisse oder Begegnungsräume?)*, ed. Jean-François Bergier (St. Katharienen: Scripta Mercaturae, 1989).

88 The three settlements were completely independent of each other in a legal sense until modern times (up to 1873), with different privileges and characters. Cf. *Medieval Buda in Context* (Brill's Companions for European History, 10), eds. Balázs Nagy et al. (Leiden and Boston: Brill, 2016). https://doi.org/10.1163/9789004307674.

89 For a recent translation of the work, see: Prokopios, *The Wars of Justinian*, trans. by H. B. Dewing, revised by Anthony Kaldellis (Indianapolis, IN: Hackett, 2014), Book V/20. 298. See. Robert J. Forbes, *Studies in Ancient Technology, Volume 2. Irrigation and Drainage, Power [Water- and Windmills], Land Transport and Road Building, The Coming of the Camel* (Leiden: Brill, 1993), 105–106, https://doi.org/10.1163/9789004450547 and Örjan Wikander, "The Water-Mill," in *Handbook of Ancient Water Technology* (Technology and Change in History, 2), ed. Örjan Wikander (Leiden: Brill, 2000), 383. https://doi.org/10.1163/9789004473829_019.

90 On ship mills in general, see: Daniela Gräf, *Boat Mills in Europe from Early Medieval to Modern Times* (Veröffentlichungen des Landesamtes für Archäologie mit Landesmuseum für Vorgeschichte, 51 = Bibliotheca molinologica, 19) (Dresden: Landesamt für Archäologie mit Landesmuseum für Vorgeschichte, 2006). For early medieval ship mills, see: Martin Schoebel, "Die Mainzer Rheinmühlen im Mittelalter: Ein Beitrag zur Geschichte der rheinischen Schiffsmühle," *Blätter für deutsche Landesgeschichte* 129 (1993): 79–115, and Dietrich Lohrmann, "Schiffsmühlen auf dem Tiber in Rom nach Papsturkunden des 10.–11. Jahrhunderts," in *Ex ipsis rerum documentis: Beiträge zur Mediävistik. Festschrift für Harald Zimmermann zum 65. Geburtstag*, eds. Klaus Herbers, Hans-Henning Kortüm, and Carlo Servatius (Sigmaringen: Jan Thorbecke, 1991), 277–286. On ship mills along the Danube in the Viennese section: Christina Spitzbart-Glasl, "Feste Wassermühlen und Schiffsmühlen als Bestandteil der Wiener Gewässerlandschaft," in *Donau-Stadt-Landschaften. Budapest – Wien. Beiträge der Tagungen in Wien (16.04.2014) und Budapest (23.–24.01.2015)*, eds. Máté Tamáska, and Csaba Szabó (Berlin: Lit Verlag, 2016), 263–78.

91 Germain Sicard, *The Origins of Corporations. The Mills of Toulouse in the Middle Ages* (New Haven, CT: Yale University Press, 2015).

92 Walter Endrei, "A magyar hajómalom," [Ship mills in Hungary] *Technikatörténeti Szemle* 21 (1995): 10, Vajda, "Árpád- és Anjou-kori vízimalmaink," 70, Balázs, "Vízimalmok, szárazmalmok, szélmalmok," Zsófia Vajkai, "Malomtípusok és a molnár mesterség a XIX. századi Magyarországon II.," [Mill types and millers in 19th-century Hungary] *A Magyar Mezőgazdasági Múzeum Közleményei 1981–1983*, 355–363, and Anon., "Hajómalmok," [Ship mills] in *Magyar néprajz*, vol. 3. http://mek.oszk.hu/02100/02152/html/03/23.html.

93 For Poland, see: Urszula Sowina, *Water, Towns and People. Polish Lands against a European Background until the Mid-16th Century* (Polish Studies – Transdisciplinary Perspectives, 17) (Frankfurt am Main: Peter Lang, 2016), 84–85.

94 BTOE, vol. 1. 279–281 no. 259 (1292) and AO, vol. 1. 102–103 no. 96 (its summary: AOklt, vol. 2. 371–372 no. 775).

95 Judit Benda, "A kereskedelem épületei a középkori Budán, 3. Malmok, pékek és kenyérszékek a későközépkori Budán," [Trade buildings in medieval Buda, 3. Mills, bakeries, bread shops at late medieval Buda] *Tanulmányok Budapest Múltjából* 38 (2013): 11–12 with reference to the Early Modern situation. See for the post-Ottoman period: Katalin Simon, "A Duna és a hajómalmok Pest-Budán a 18. században és a 19. század első felében," [The Danube and ship mills at Pest-Buda in the 18th and the first half of the 19th century] *Urbs. Magyar Várostörténeti Évkönyv* 10–11 (2017): 301–321.

96 MNL OL DL 1691, edited in: AO, vol. 1. 102–103 no. 96 (for its summary: AOklt, vol. 2. 371–372 no. 775).

97 For the problem of carpenters and millers, see Chapter 4.

98 *Iidem index et iurati cives fluvium Danobii inter calidas aquas superiores Budenses et porcionem scilicet possessionariam dictarum dominarum priorisse et sanctimonialium in margine eiusdem Danobii adiacens et inter possessionem Jenew similiter earundem dominarum sanctimonialium, ubi scilicet semper et ab antiquo dicti prepositus et capitulum ecclesie Budensis ad dicte domine sanctimoniales portum seu wadum navalem habuissent, haberentque de presenti, tam ex parte calidarum aquarum quam ex parte dicte possessionis Jenew vocate, quo scilicet naves hincinde ad portum ipsius Danobii remigio duci solite fuissent, nonnulla molendina ligari et locari, ac multos palos et ligna wlgo zelep vocata ad transitum dictarum navium in ipsum fluvium Danobii incuti et locari fecissent, per hocque dictum wadum adeo impedivissent et in cessanter impedirent, ut naves in ipso wado hincinde transire nequirent.* PMTOE 231 no. 882 (MNL OL DL 15 787).

99 ÁMTF, vol. 4. 502 and 584.

100 On the hospital, see: Katalin Irásné Melis, "Pest kialakulása és a középkori város építési periódusai," [The formation of Pest and its medieval building periods] *Budapest Régiségei* 47 (2014): 135 and 155. (with erroneously identifying George Révész as the master of the hospital and dating the document once to 1460 and then some pages later to 1476).

101 *Tamen nunc quidam Georgius Rewez, Paulus Aurifaber et magister hospitalis S. Nicolai confessoris de Pesth in medio meatus ipsius piscature prope murum Beechwara, vbi numquam alias molendina fuissent, senticulis suis firmiter in profundo Danobii fixis, molendina sua locassent perque recia et piscatores ipsos ita turbarent, vt vix medietatem piscium, quam antea prendidissent, prendere possent.* Nándor Knauz, "A budai káptalan regestái 1148–1649," [Summary of the documents of the chapter of Buda] *Magyar Történelmi Tár* 12 (1863): 23–24 and 41. See also: PMTOE 248 no. 949 and László Bártfai Szabó, *Óbuda egyházi intézményei a középkorban* [Church institutions of Óbuda in the Middle Ages] (Budapest: [Author's edition], 1935), 107 no. 142. For sturgeon fishing, see Sándor Takáts, "A komáromi vizahalászat a XVI. században I–II.," [Sturgeon fishing at Komárno in the 16th century] *Magyar Gazdaságtörténelmi Szemle* 4 (1897): 425–445 and 485–509 and László Bartosiewicz et al., "Animal

Exploitation in Medieval Hungary," in *Economy of Medieval Hungary* (East Central and Eastern Europe in the Middle Ages, 450–1450, 49), eds. József Laszlovszky et al. (Leiden and Boston: Brill, 2018), 138–142. https://doi.org/10.1163/9789004363908_007. For the problem of fish, ferries and mills, see: Richard C. Hoffmann and Christoph Sonnlechner, "Vom Archivobjekt zum Umweltschutz: Maximilians Patent von 1506," *Studien zur Wiener Geschichte. Jahrbuch des Vereins für Geschichte der Stadt Wien* 62–63 (2006–2007) [2011]: 79–133.

102 Bártfai Szabó, *Óbuda egyházi intézményei*, 105 no. 135. *Historia critica regum Hungariae. Vols. 8–19. Stirpis mixtae: 1301–1526*, 12 vols (Budae: Weingand & Koepf, 1788–1794), vol. 13. 718. On Szentfalva (or Szenterzsébetfalva), see: András Kubinyi, "Budapest története a későbbi középkorban Buda elestéig," [History of Budapest in the later Middle Ages, to the fall of Buda] in *Budapest története*, 5 vols [History of Budapest], ed. László Gerevich (Budapest: Budapest Főváros Tanácsa, 1975–1980), vol. 2. 15 and 41 note 27; István Tringli, "Pest megye a késő középkorban," [Pest County in the late medieval period] in *Pest megye monográfiája*, vol. 1/2 [Monography of Pest County], eds. Domokos Kosáry, Attila Zsoldos, and István Torma (Budapest: Pest Megye Monográfia Közalapítvány, 2001), passim (both as Szenterzsébetfalva and Szentfalva), and ÁMTF, vol. 4. 557 (Szenterzsébet).

103 *Quod in fluvio Danubii in territoriis possessionis nostre Zenthffalwa vocate in comitatu Pestiensi existentis habite, ubi habilius et utilius videbitur, unum locum molendini integri, quod vulgariter 'zelep' appellatur, eligere et illud molendinum iuxta cursum temporis, estivalis scilicet et hiemalis in loco, ubi voluerit, construi facere et locare fructusque et utilitates eiusdem ipse Michael Bodo ac sui heredes et successores universi absque solutione census soliti, qui vulgariter 'ferthon' vocatur, et etiam aliarum taxarum nostrarum ordinariarum et extraordinariarum perpetue, pacifice, quiete et sine contradictione quorumcumque percipere possint et valeant. Quocirca vobis, iudicibus, iuratis ceterisque populis et iobagionibus nostris in dicta possessione nostra Zenthffalwa constitutis et existentibus, item cunctis operariis et factoribus nostris in officiolatu eiusdem possessionis Zenthffalwa qualitercunque existentibus, presentibus atque futuris harum serie firmiter committimus et mandamus, quatenus dictum locum molendini annotato Michaeli Bodo in dicto fluvio Danubii, in territoriis eiusdem possessionis Zenthffalwa eligere et illud sibi pacifice construi facere et eidem suisque heredibus et posteritatibus universis uti, tenere fructusque eiusdem percipere et percipi facere permittatis et permitti facere debeatis, et aliud nullo modo facere presumpmatis.* MNL OL DL 46 153. edited in *A Szapolyai család oklevéltára / Documenta Szapolyaiana I. Levelek és oklevelek / Epistulae et litterae (1458–1526)*, ed. Tibor Neumann (Budapest: MTA Bölcsészettudományi Kutatóközpont Történettudományi Intézet, 2012), 188–189 no. 198. See also: András Kubinyi, "Budai és pesti polgárok családi összeköttetései a Jagelló-korban," [Family contacts of the burghers of Buda and Pest in the Jagiello period] *Levéltári Közlemények* 37 (1966): 266 (later published under the same title, in *Tanulmányok Budapest középkori történetéről*, 2 vols [Várostörténeti tanulmányok, 10], eds. István Kenyeres, Péter Kis, and Csaba Sasfi [Budapest: Budapest Főváros Levéltára, 2009], vol. 2. 513–570), and Benda, "A kereskedelem épületei a középkori Budán, 3.," 11–12.

104 See: András Vadas, and János M. Bak, "Diets and Synods in Buda and Its Environs," and András Kubinyi, "Buda, Medieval Capital of Hungary," in *Medieval Buda in Context*, 322–344 https://doi.org/10.1163/9789004307674_015 and 366–385 https://doi.org/10.1163/9789004307674_017 respectively.

105 *In dieta pro festo beati Georgii martyris campo Rákos iuxta contenta decreti celebrata per dominum Joannem Bornemyzza, regium thesaurarium, in medium regnicolarum regestum istud missum est. Post cuius quidem diete finem, tertia videlicet die, feria utputa sexta post Stanislai, egregii et memeria perpetua digni viri Ambrosius Vydfy de Mohorra, Georgius de Kalonda ac magister Nicolaus Vydfy de Woykoch, regni Sclavonie prothono-tarius unacum Ladislao Keletfy de Stwinack, Anbrosio* [sic!] *Erdewgh de Wragowcz, et ceteris notabilibus viris, navi in Danubio ad Pesth de Bwda navigantes tactu cuiusdam molendini submersa pariter occubuerunt suffocatique aquis decesserunt. Eadem in dieta dominus Emericus de Peren in palatinum ante castrum Budense est creatus per magistrum Stephanum de Verbewcz tunc prothonotarium presente manu sua.* Lajos Thallóczy, *Jajca (bánság, vár és város) története* [History of (the banate, castle and town of] Jajce (Magyarország melléktartományainak oklevéltára. Codex diplomaticus partium regno Hungariae adnexarum, 4) (Budapest: MTA, 1915), 182–183 no. 114.

106 *1504–1566, memoria rerum. A Magyarországon legutóbbi László király fiának legutóbbi Lajos királynak születése óta esett dolgok emlékezete. Verancsics-évkönyv* (Magyar Helikon), ed. József Bessenyei (Budapest: Magyar Helikon, 1981), 17–18. Referred to in Benda, "A kereskedelem épületei a középkori Budán, 3.," 11.

107 K. Németh, "Vizek és vízgazdálkodás, I.," passim, Béla Czigány, "Adatok a Győr megyei hajósmolnárok életéhez, I–V.," [Data to the life of ship millers in Győr County] *Arrabona* 2 (1960): 133–139, 4 (1962): 97–116, 5 (1963): 207–228, 7 (1965): 413–428, 9 (1967): 149–178, 10 (1968): 123–140, and 12 (1970): 187–211 and János Attila Tóth, "Adatok a dunai hajómalmok régészetéhez," [Data on the archaeology of Danube ship mills] *Archaeologiai Értesítő* 148 (2023): 301–323 https://doi.org/10.1556/0208.2023.00059. See also Simon, "A Duna és a hajómalmok."

108 András Vadas and László Ferenczi, "Small Urban Waters and Environmental Pressure before Industrialization: The Case of Hungary," *Journal of Historical Geography* 82 (2023): 98–109. https://doi.org/10.1016/j.jhg.2023.09.008.

109 Marija Karbić, and Bruno Škreblin, "Grad na rijeci ili pored nje: srednjovjekovni Zagreb i rijeka Sava," [The Town on or Close to the River: Medieval Zagreb and the Sava River] in *Rijeka Sava u povijesti* [River Sava in history] (Bibliotheca Croatica: Slavonica, Sirmiensia et Baranyansia Posebna izdanja, 15), ed. Branko Ostajmer (Slavonski Brod: Hrvatski Institut za Povijest, 2015), 163–180.

110 *Potok u srcu Zagreba: uz potok Medveščak-od izvora do ušća / A Brook in the Heart of Zagreb: Along the Medveščak Brook from Source to Mouth: Muzej grada Zagreba, spanj-studeni 2005*, ed. Nada Premerl (Zagreb: Muzej grada Zagreba, 2005), and Ivka Kljajić, and San Mikulec, "Cartographic-Historical Analyses of the Medveščak Stream Regulation and Quantum GIS," *Kartografija i geoinformacije – Journal Cartography and Geoinformation* 12, no. 19 (2013): 30–43.

111 Andrija-Ljubomir Lisac, "Mlinarstvo Zagreba od najstarijih vremena do početka XX stoljeća," [Mills of Zagreb from the earliest times to the beginning of the 20th

century] *Zbornik Historijskog zavoda Jugoslavenske akademije* 8 (1977): 220 note 2.

112 See: Nada Klaić, *Zagreb u srednjem vijeku* [Zagreb in the Middle Ages] (Zagreb: Liber, 1982), Neven Budak, "Budući da smo htjeli u Zagrebu na brdu Gradecu sagraditi slobodni grad..." [As we wanted to build a free city in Zagreb on the Gradec Hill ...] in *Zlatna bula 1242–1992* [Golden bull], ed. Zlatko Stublić (Zagreb: Muzej grada Zagreba, 1992), 21. For the mills of Zagreb, see: Lisac, "Mlinarstvo Zagreba," 217–290.

113 For the water management of the cathedral chapter's estates, see: Petra Vručina, "Kako gospodariti vodom u srednjemu vijeku? Primjer Zagrebačkoga kaptola i njegovih vlastelinstava u 14. i 15. stoljeću / Water Management in the Middle Ages: The Example of the Zagreb Chapter in the 14th and 15th Centuries," *Miscellanea Hadriatica et Mediterranea* 8 (2021): 33–64 https://doi.org/10.15291/misc.3612.

114 *Petrus habuisset tria molendina a patre castri Grech, in aqua que dividit terram castri Grech a terra fratrum Zagrabiensium, que quidem archidiaconus sepedictus ab antiquo possederat titulo empcionis. Verum pro bono pacis, mediante bano Stephano, exstitit ordinatum ut duo molendina de tribus, scilicet unum melius et aliud deterius, deberet habere archidiaconus antedictus, tercium vero molendinum, melius deteriori, deberent habere cives de Grech, suis usibus aplicatum.* MCZ, vol. 1. 26–27 no. 32. For the royal confirmation of the ban's decision, see MCZ, vol. 1, 29–30 no. 37. For Stephen: Zsoldos, *Archontológia*, 45.

115 Cf. Engel, *Archontológia*, vol. 1. 16.

116 *Preterea idem Nicolaus banus possessiones quorumdam hominum, ut dicebant, infidclium domini regis, videlicet quoddam balneum in inferiori parte civitatis Zagrabiensis existens, cum omnibus domibus, utilitatibus et edificiis suis ac fundo curie in quo idem balneum existit et molendinum eorumdem infidelium in eadem civitate in fluvio de suburbio castri Medue currenti situm.* MCZ, vol. 1. 169–171 no. 194. For Mikcs of Ákos: Engel, *Archontológia*, vol. 1. 16.

117 *Accione eiusdem capituli ecclesie Zagrabiensis in facto possessionario contestata ad octavas festi Penthecostes tunc affuturas, prorogata, in quibus una accione ipsius capituli Zagrabiensis super facto cuiusdam fossati taliter proposita, quod ipsi hospites sub monte et in metis dicte civitatis de Grech quoddam fossatum propria eorum auctoritate et potencia fodere incipiendo et finem eius ad terram dicti capituli deducendo, quedam aqua per ipsum fossatum deducta terram ipsius capituli destruxisset. In quorum contrarium, preallegatum fossatum et terram ad quam finis eius fossati perveniret, intra metas dicte civitatis procuratores eorumdem hospitum adiacere allegarant.* MCZ, vol. 1. 445–452 [Addenda] no. 4.

118 *Ecce dominus Jacobus, dicte zagrabiensis ecclesie prepositus. tam pro se personaliter, quam cum procuratoriis litteris dicti capituli pro eodem capitulo comparendo, se et ipsum capitulum cum preallegatis hospitibus super facto possessionario in causa fuisse et iam diu ventilatam et ad presentes octavas Epiphaniarum domini esse deductam, in qua eciam ipsum capitulum deberet instrumenta exhibere, astruendo universas litteras tam causales quam privilegiales tempore combustionis civitatis de Wysegrad (in con)servator\o domini Tbatamerii, Albensis ecclesie prepositi, in quo per magistrum Demetrium arcbydiaconum Zzagrabiensem, tunc procuratorem ipsius capituli reposite*

fuerant ad servandum, omnino fuissent concremate et amisse. MCZ, vol. 1. 447 [Addenda] no. 4.

119 Cf. Chapter 3.1.

120 For Nicholas Angielini, see Géza Pálffy, *Die Anfänge der Militärkartographie in der Habsburgermonarchie. Die regelmäßige kartographische Tätigkeit der Burgbaumeisterfamilie Angielini an den kroatisch-slawonischen und den ungarischen Grenzen in den Jahren 1560–1570 / A haditérképészet kezdetei a Habsburg Monarchiában. Az Angielini várépítész-família rendszeres térképészeti tevékenysége a horvát–szlavón és a magyarországi határvidéken az 1560–1570-es években* (Budapest: Ungarisches Nationalarchiv / Magyar Országos Levéltár, 2011).

121 *Ad locu m molendini super eodem rivulo sancti Emerici prope puteum, deinde per eundem rivulum sursum ad duo molendina abbatis de Zagrabia super eodem rivulo, que molendina ad metas capituli includuntur, deinde super ipsum rivulum ad tria molendina capituli, et dehinc ad molendinum damini episcopi, adhinc iterum ad tria molendina capituli, semper cum dicto rivulo in totum cum molendinis ad capitulum scilicet* [?] *Zagrabiensem ecclesiam pertinente, dehinc per verum meatum ipsius rivuli versus septentrionem pervenit ad duo molendina Grecensia, que aqua de dicto rivulo recepta ducuntur. Abhinc tendit per antiquum meatum sepedicti rivuli ad molendinum Heremitarum, deinde per antiquum meatum ipsius rivuli transit quedam molendina Gracensia in vicinitate predii Nicolai*" MCZ, vol. 1. 129–130 no. 14. Cf. Igor Karaman, *Tisućljetni Zagreb* [Zagreb through a thousand years] (Zagreb: Školska knjiga, 1994) (map 20).

122 *Documenta Artis Paulinorum*, 3 vols, eds. Béla Gyéressy, and Ferenc Hervay (Budapest: MTA Művészettörténeti Kutató Csoportja, 1975–1978), vol. 3. 320–321, and 325. See also: Romhányi¸ *Kolostorok és társaskáptalanok*, 55.

123 Cf. Nada Klaić, *Povijest Zagreba. 1. Zagreb u srednjem vijeku* [History of Zagreb, vol. 1. Zagreb in the Middle Ages] (Zagreb: SNL, 1982), 477.

124 *Molendina vero domini abbatis claustri beate Marie virginis de dicta Zagrabia inter predictas metas et signa metalia existencia, nos et ipsi cives in suo statu pacifico relinquimus et commisimus; (…) Hoc non pretermisso, quod incipiendo a predictis duabus metis prope dictum molendinum filiorum Chupor erectis usque predictarn villam Gracchan prenominati cives a parte occidentali in ipsorum territorio absque jmpedimento et destruccione rnolendini sancti Johannis Nove ville in porcione seu territorio ipsius civitatis existentis et habiti, molendina et alia edificia quotquot volunt, edificandi et construendi et aquam seu decursum aque ipsius rivuli de vero cursu in locis necessariis ad fossata excipiendi liberam habebunt facultatem.* MCZ, vol. 1. 341–345 no. 362.

125 MCZ, vol. 1. 250–254 no. 262.

126 MCZ, vol. 2. 28 no. 30, 28 no. 31, 27 no. 29, 29 no. 31, 29 no. 31, and 28 no. 31 (ZsO, vol. 9. 171 no. 531, 172 no. 533, 221 no. 691, 235 no. 733, 237 no. 739, and 253 no. 795).

127 *Fluvium Chirkvenik de suo cursu per idem fossatum fluere permisissent* MCZ, vol. 2, 44 no. 36 (the original of the document is lost).

128 *Venerunt ad nos cives civitatis zagrabiensis in modo gravis querele, quomodo aquas ipsis ab antiquo spectatas violenti manu occuparetis, quod nostra credulitate de vobis non fuissemus spectaturi, cum et alia colloquia habuimus simul, in quibus talia minime fuerunt perpetrata, sicuti domino Georgio de Bexin, bene notum est, quod nulla parcium alteri parti absque quod aliquod*

preiudicium aut dampnum inferret, quod in nostri absencia minime observatum est, ex quo utique aliud non vultis, quam serenissime regie maiestatis civitatem juribus privare et ipsos condempnare. MCZ, vol. 2. 364–365 no. 295.

129 Karbić, and Škreblin, "Grad na rijeci."

130 Adrienn Papp, "Oszmán-török fürdők Budán – A Rudas és Rác fürdő régészeti kutatása (2004–2006)," [Ottoman bathing houses at Buda – archaeological research of th Rudas and Rác bathing houses, 2004–2006] in *A középkor és a kora újkor régészete Magyarországon*, 2 vols [Archaeology of the Middle Ages and the Early Modern Period in Hungary], eds. Elek Benkő, and Gyöngyi Kovács (Budapest: MTA Régészeti Intézet, 2010), vol. 1. 207–220, Adrienn Papp, "Fürdők a török kori Budán." [Bathing houses at Ottoman-period Buda] (PhD-diss., Eötvös Loránd Tudományegyetem, 2014).

131 Paula Zsidi, "A Duna szerepe Aquincum topográfiájában," [The role of the Danube in the topography of Aquincum] *Budapest Régiségei* 41 (2008): 57–83.

132 Nicolas Marcel de La Vigne, *Bvdae ac Pesti ichonographica descriptio tvm etia(m) earv(m) vicinarv(m), castrorv(m), aggerorv(m), fossarv(m), prospectvv(m), nvda, ac recta scenographica delineatio, aliarvmve dictae Bvdae obsidionis particvlaritatv(m) à sac(ratissimi) caes(arei) mai(estatis) exercitv, eivsqve praecelsorvm foederatorvm copiis factae* ([Vienna] [?], 1686). Held at: Budapesti Történeti Múzeum [Budapest History Museum], Metszettár [Collection of prints]. György Rózsa, *Budapest régi látképei. 1493–1800. Alte ansichten von Budapest* (2nd ed. Budapest: Új Művészet Kiadó, 1999), Cat. 27, 121, 150, and H. Bredekow, *Geometrischer Grundt-Riß der Neunmahl Bekriegten Undt Blutig Besiegten Königlichen Ungarischen Resitends Stadt Buda vel Offen* [Vienna] [?)] 1686). Held at: Hessisches Staatsarchiv Marburg, Wilhelmshöher Kriegskarten, vol. 6. no. 20. See for their recent editions at: András Végh, *Buda I. kötet, 1686-ig* [Buda, I. Until 1686] (Magyar Várostörténeti Atlasz, 4) (Budapest: Budapesti Történeti Múzeum, Pázmány Péter Katolikus Egyetem Bölcsészet- és Társadalomtudományi Kar Régészeti Tanszék, and Archaeolingua Alapítvány, 2015). For these representations, see also: Adrienn Papp, "Török kori lőpormalom Budán," [Ottoman-period gunpowder mill at Buda] *Keletkutatás 2015* (Tavasz [Spring]): 105–113.

133 Cf. István Kenyeres, "The Fate of the Medieval Archives of Buda and Pest," in *Medieval Buda in Context*, 52–68. https://doi.org/10.1163/9789004307674_004.

134 András Kubinyi. "Budafelhévíz topográfiája és gazdasági fejlődése," [Topography and economic development of Budafelhévíz] *Tanulmányok Budapest Múltjából* 16 (1964): 85–180 (published under the same title in *Tanulmányok Budapest középkori történetéről*, vol. 1. 115–182).

135 Benda, "A kereskedelem épületei a középkori Budán, 3," Katalin Szende, "Mills and Towns: Textual Evidence and Cartographic Conjectures from Hungarian Towns in the Pre-Industrial Period," in *Extra muros. Vorstädtische Räume in Spätmittelalter und Früher Neuzeit. Espaces suburbains au bas Moyen Âge et à l'époque moderne* (Städteforschung, A 91), eds. Guy Thewes, and Martin Uhrmacher (Cologne and Vienna: Böhlau, 2019), 485–516. https://doi.org/10.7788/9783412515164.485, and Végh, *Buda*.

136 For hospitals in medieval Hungary, see Judit Majorossy, and Katalin Szende, "Hospitals in Medieval and Early Modern Hungary," in *Europäisches Spitalwesen. Institutionelle Fürsorge in Mittelalter und Früher* (Neuzeit Quelleneditionen des

Instituts für Österreichische Geschichtsforschung, 5), eds. Martin Scheutz et al. (Vienna and Munich: Oldenbourg, 2008), 409–454. https://doi.org/10.7767/boehlau.9783205160885.409. For the Holy Spirit Hospital, see: Lajos Pásztor, *A magyarság vallásos élete a Jagellók korában* [Religious life of the Hungarians in the Jagiello age] (Budapest: [no publisher], 1940), 53 and 56, Zoltán Somogyi, *A középkori Magyarország szegényügye* [Taking care of poors in medieval Hungary] (Palaestra Calasanctiana, 37) (Budapest: Stephaneum, 1941), 37–48, and Kubinyi, "Budafelhévíz," 123–129. More recently, see: Judit Lászay, and Adrienn Papp, "Investigations in the Budafelhévíz Baths," in *Régészeti kutatások Magyarországon 2007 / Archaeological Investigation in Hungary 2007*, ed. Júlia Kisfaludi (Budapest: Kulturális Örökségvédelmi Hivatal and Magyar Nemzeti Múzeum, 2008), 63–84.

137 MES, vol. 3. 157–158. Végh, *Buda*, 49. See also: Kubinyi, "Budafelhévíz," 123–124.

138 BTOE, vol. 1. 151, 181, 241–242, and 246. Kubinyi, "Budafelhévíz," 130 note 336, raised that in the latter confirmation, mills are referred to in the plural. He suggested that the other mills may have been elsewhere, but this is probably only the traditional plural that is prevailing in similar lists of benefits and belongings. Cf. Chapter 2.

139 CD, vol. 8/4. 369–372 no. 182. For the first reference of a standing mill there, see: CD, vol. 8/4. 292–294 no. 134.

140 CD, vol. 8/4. 292–294 no. 134, and CD, vol. 8/4. 369–372 no. 182.

141 Herta Bertalan, "Das Klarissinnenkloster von Obuda aus dem 14. Jahrhundert," *Acta archaeologica Academiae Scientiarum Hungaricae* 34 (1982): 151–176, and Brian McEntee, "Queen Elizabeth of Hungary (1320-1380) and Óbuda: Patronage, Personality and Place," in *La diplomatie des états Angevins: aux XIIIe et XIVe siècles / Diplomacy in the countries of the Angevin Dynasty in the thirteenth – fourteenth centuries: actes du colloque international de Szeged, Visegrád, Budapest: 13–16 septembre 2007*, eds. Zoltán Kordé, and István Petrovics (Rome and Szeged: Dipart. di Storia Ungh. del Medioevo e della prima Età Moderna Univ. degli Studi di Szeged and Accademia d'Ungheria in Roma Istituto Storico "Fraknói", 2010), 209–218 (with reference to the prolific literature in the topic).

142 *Eiusdem molendini in Calidis Aquis prope Ecclesiam hospitalis Sancti Spiritus Budae inter duo molendina in ipsis Calidis Aquis existentia habitum.* MNL OL DL 5289. See also: MNL OL DL 6370, 6625, 6755 with locating it to the same place. Cf. Végh, *Buda*, 55 (locating it to the latter place).

143 *Molendinum in qualidis aquis existens circa sanctum Spiritum.* MNL OL DL 7564, edited in: BTOE, vol. 3/1. 38–39 no. 91.

144 *Quoddam molendinum in fluvio Calidarum aquarum superiorum a parte Danubii habitum religiosis dominabus sanctimonialibus claustri beate virginis de Veteribuda.* MNL OL DL 10 956, edited in BTOE, vol. 3/2. 50–51 no. 771.

145 *Molendini in superioribus Calidis aquis, in fine eiusdem aque prope balneum religiosarum dominarum sanctimonialium de Insula ieporum.*" MNL OL DL 12 317. (erroneously indicated as MNL OL DL 12 327 at Kubinyi, "Budafelhévíz," 131 note 339). BTOE, vol. 3/2. 161 no. 978 and MNL OL DL 12 325, edited in: BTOE, vol. 3/2. 163 no. 982.

146 MNL OL DL 15119, edited in CD, vol 11. 70–73 no. 20, and BTOE, vol. 3/2. 298 no. 1189. Cf. Kubinyi, "Budafelhévíz," 131.

147 MNL OL DL 13179, edited in BTOE, vol. 3/2. 299. no. 1192.

148 MNL OL DL 13 188, edited in BTOE, vol. 3/2. 300 no. 1194. Katalin Holl-Gyürky, "The Use of Glass in Medieval Hungary," *Journal of Glass Studies* 28 (1986): 77–78.

149 MNL OL DL 13 207, edited in BTOE, vol. 3/2. 302–303 no. 1199. The lease contract: MNL OL DL 13 299, edited in BTOE, vo. 3/3. 309–310 no. 1214.

150 Kubinyi, "Budafelhévíz," 131.

151 His grant of arms from 1476: MNL OL DL 50 534. Edited in *Monumenta Hungariae Heraldica. Magyar czimeres emlékek*, 3 vols, eds. Antal Áldásy, and László Fejérpataky (Budapest: Franklin, 1901–1926), vol. 2. 61–62. Cf. Dániel Kálmán, "A középkori magyar armálisok címerei 1437–1490 között," [Coats of arms of the medieval Hungarian grants of arms between 1437 and 1490] in *Micae mediaevales V. Fiatal történészek dolgozatai a középkori Magyarországról és Európáról* [Studies of young historians on medieval Hungary and Europe] (ELTE BTK Történelemtudományok Doktori Iskola Tanulmányok – Konferenciák, 9), eds. Laura Fábián et al. (Budapest: ELTE BTK Történelemtudományok Doktori Iskola 2016), 149–150.

152 MNL OL DL 22 546.

153 For millers, see Chapter 4.

154 MNL OL DL 26 131. Kubinyi, "Budafelhévíz," 135.

155 MNL OL 22 014, edited with omissions in: Knauz, "Budai regeszták," 50–52 no. 77.

156 Kubinyi, "Budafelhévíz," 135 note 355. See also: Knauz, "Budai regeszták," 72 no. 94.

157 E.g., the case of the mills along the River Tapolca around Pápa: István Wöller, *A Tapolca vízimalmai Pápán és környékén* [The water mills of the River Tapolca at Pápa and its surroundings] (Fejezetek Pápa város történetéből, 1) (Pápa: Pápa Város Önkormányzata, 1994), or by the River Hévfő at Oradea: Renáta Skorka, "A váradi ferences rendi kolostorok topográfiája," [Topography of the Franciscan friaries of Oradea] in *Nagyvárad és Bihar az Anjou-korban* [Oradea and Bihor in the Angevin period] (Tanulmányok Biharország történetéről, 5), ed. Attila Zsoldos (Oradea: Varadinum Kulturális Alapítvány, 2018), 175–181 and 281 (map).

158 Roberta Magnusson, and Paolo Squatriti, "The Technologies of Water in Medieval Italy," in *Working with Water in Medieval Europe* (Technology and Change in History, 3), ed. Paolo Squatriti (Leiden, Boston, and Cologne: Brill, 2000), 253–254. https://doi.org/10.1163/9789047400110_010.

159 This subchapter of the book partially relies on an article published earlier: András Vadas, "Economic Exploitation of Urban Moats in Medieval Hungary with Special Regard to the Town of Prešov," *Mesto a Dejiny* 6 no. 2 (2017): 6–21.

160 E.g., *Die Befestigung der mittelalterlichen Stadt* (Städteforschung, A 45), eds. Gabriele Isenberg, and Scholkmann, Barbara (Cologne, Weimar, and Vienna: Böhlau, 1997), Oliver Creighton, and Robert Higham, *Medieval Town Walls. An Archaeology and Social History of Urban Defence* (Stroud: Tempus, 2005), or *City Walls: the Urban Enceinte in Global Perspective*. ed. James D. Tracy (New York: Cambridge University Press, 2000). For East Central Europe, see the overview of Teodor Octavian Gheorgiu, *Cetatile oraselor: apararea urbana în centrul si estul Europei în Evul Mediu* [Castle towns: urban defense in Central and Eastern Europe in the Middle Ages] (Bucureşti: Editura Simetria, 2000).

161 Richard Holt, "Medieval England's Water-Related Technologies," in *Working with Water*, 79–81. https://doi.org/10.1163/9789047400110_006
162 Mark Gardiner, "Hythes, Small Ports, and Other Landing Places in Later Medieval England," in *Waterways and Canal-Building in Medieval England*, ed. John Blair (Oxford: Oxford University Press, 2007), 90. https://doi.org/10.1093/acprof:osobl/9780198723134.003.0005
163 Guillerme, *The Age of Water.*
164 Charles L. H. Coulson, *Castles in Medieval Society: Fortresses in England, France, and Ireland in the Central Middle Ages* (Oxford: Oxford University Press, 2003). https://doi.org/10.1093/acprof:oso/9780198208242.001.0001.
165 For the problem, in a mostly German context, see Matthias Untermann, "Erscheinungsformen der Stadtbefestigung," in *Die Befestigung der mittelalterlichen*, 3–25.
166 On the series, see: *Lords and Towns in Medieval Europe: The European Historic Towns Atlas Project*, eds. Anngret Simms, and Howard B. Clarke (Farnham: Ashgate, 2015), and Nick Millea, Keith Parry, and Adrian Phillips, "The European Historic Towns Atlas Project," *IMCoS Journal* no. 153 (2018): 26–36. For the atlases published so far: www.historictownsatlas.org.uk/content/bibliography.
167 See for exceptions, the example of Freistadt in Upper-Austria: Elisabeth Gruber, Elisabeth Gruber: *"Raittung und außgab zum gepew." Kommunale Rechnungspraxis im oberösterreichischen Freistadt Edition und Kommentar der Stadtgrabenrechnung (1389–1392)* (Quelleneditionen des Instituts für Österreichische Geschichtsforschung, 14) (Vienna: Böhlau, 2015), and on Vienna: Ferdinand Opll, Heike Krause, and Christoph Sonnlechner, *Wien als Festungsstadt im 16. Jahrhundert. Zum kartografischen Werk der Mailänder Familie Angielini* (Vienna, Cologne, and Weimar: Böhlau, 2017), esp. 176–177. https://doi.org/110.26530/oapen_626456.
168 Adrian Andrei Rusu, *Castelarea Carpatică. Fortificații și cetăți din Transilvania și teritoriile învecinate (sec. XIII–XIV)* [Carpathian Castelaria. Fortifications and citadels in Transylvania and the neighbouring territories] (Cluj-Napoca: Editura Mega, 2005), and Ákos Karczag, and Tibor Szabó, *Erdély, Partium és a Bánság erődített helyei* [Fortified places in Transylvania, Partium and Banate] (2nd ed. Budapest: Semmelweis, 2012).
169 Endre Marosi, *XVI. századi váraink, 1521–1606* [Castles in 16th-century Hungary] (Miskolc: Hungária-Európa Alapítvány and Borsod-Abaúj-Zemplén Megyei Levéltár, 1991).
170 For an overview of the results: Katalin Szende, "A Kárpát-medence középkori városainak régészeti kutatása az elmúlt két évtizedben," [Archaeological research into medieval towns in the Carpathian Basin since 1990] in *A középkor és a kora újkor régészete Magyarországon*, vol. 1. 149–150. For earlier research see also: Imre Holl, "Feuerwaffen und Stadtmauern. Angaben zur Entwicklung der Wehrarchitektur des 15. Jahrhunderts," *Acta Archaeologica Academiae Scientiarum Hungaricae* 33 (1981): 201–243.
171 Imre Holl, "Sopron középkori városfalai," [Town walls in medieval Sopron] *Archaeologiai Értesítő* 94 (1967): 155–183, 95 (1968): 188–205, 98 (1971): 24–44, 100 (1973): 180–207. The most important from the point of view of this chapter is 95 (1968): 198–204. On the fortification of the towns in Slovakia with

reference to moats: Martin Miňo, "Niekoľko postrehov k fortifikačným prvkom miest stredoslovenskej banskej oblasti," [Some notes on the fortification elements of central Slovakian mining towns] *Archaeologia historica* 36 (2011): 289–302. On Banská Bystrica: Igor Graus, "Vývoj fortifikácie Banskej Bystrice do konca 16. storočia v podmienkach špecifík pasívnej obrany stredoslovenských banských miest," [Development of the fortifications of the town of Banská Bystrica until the end of the 16th century under the specific conditions of passive defense of the Middle-Slovakian mining towns] *Vojenská história* 4 (2000): 3–21.

172 Péter Rabb, "Városi védművek a középkori Magyarországon, 2 vols," [Urban fortifications in medieval Hungary] (PhD-diss., Budapesti Műszaki Egyetem, 2005), on fortification moats, see vol.1. 47–48.

173 See: Guillerme, *The Age of Water.*

174 *Filarete's Treatise on Architecture: Being the Treatise by Antonio di Piero Averlino, Known as Filarete,* 2 vols, ed. John R. Spencer (New Haven, CT: Yale University Press, 1965), vol. 1. 52–53.

175 Norbert Mátyus, "Filarete és Bonfini. Egy humanista építészeti tankönyv Itáliában és Magyarországon," [Filarete and Bonini. A humanist architectural treatise in Italy and Hungary] in *Techné. A szakképzés módszerei az ókorban és a középkorban* [Techné. Forms of crafts' knowledge transfer in the Ancient Times and the Middle Ages], eds. Márta Munding, Kornél Szovák, and László Takács (Piliscsaba: Avicenna Közel-Kelet Kutatások Intézete, 2018), 345–370.

176 Holt, "Medieval England's," 79–82.

177 *Castrum suum nobile Varziacum, quod, propter loci immunita, prius frequentibus patebat depredantium incursibus, muros reparando antiquos et fossas turribus ingentibus circumsepsit. Domos episcopales juxta ecclesiam beate Eugenie muro solido, magnificis edificiis, turribus ac propugnaculis a primis condidit fundamentis. Presidium quoque episcopale ulterius muro solidiori, turribus munitioribus et antemuralibus in tutam et ad expugnandum diflicilem munitionem provexit, et fossis mire amplitudinis vallavit in girum, in quibus, grata aquarum exuberantia ex fontibus circumfluentibus eas replens, non modicum facit ad munitionis augmentum; et tum ex piscium multitudine qui ibi nutriuntur optimi, tum ex molendinis que edificavit ibidem, tum ex aliis utilitatibus suis, episcopo totique municipio multe commoditatis existunt.* Louis M. Duru, *Bibliothèque historique de l'Yonne; ou, Collection de légendes, chroniques et documents divers pour servir à l'histoire des différentes contrées qui forment aujord'hui ce département*, vol. 1 (Auxerre: Perriquet, 1801), 431–450, and Coulson, *Castles in Medieval Society*, 70.

178 For similar data from Hungary, see: Jankovich, "Adatok a Körösvidék," 323.

179 Otto Brunner, *Die Finanzen der Stadt Wien: von den Anfängen bis ins 16. Jahrhundert* (Vienna: Dt. Verl. für Jugend u. Volk. 1929).

180 András Kubinyi, "Városhálózat a késő középkori Kárpát-medencében," [Urban network in the late medieval Carpathian Basin]. *Történelmi Szemle* 46 (2004): 17–18, and see the detailed overview of the relavant research history: Bálint Lakatos, *Mezővárosi oklevelek: Települési önkormányzat és írásbeliség a késő középkori Magyarországon, 1301–1526* [Market town charters. Urban autonomy and literacy in late medieval Hungary] (Budapest: Bölcsészettudományi Kutatóközpont Történettudományi Intézet, 2019), 11–16. See also: Rabb, "Városi védművek," esp. its appendix. Many of its data have to be modified

in light of the recent work: Karczag, and Szabó, *Erdély, Partium és a Bánság*. See most recently: Beatrix F. Romhányi and Katalin Szende, "Cities, Towns and Market Towns in the Context of the Regional Disparities of the Kingdom of Hungary around 1500," *Mesto a dejiny* 14, no. 1 (2025): 6–32. https://doi.org/10.33542/CAH2025-1-01.

181 Jankó, Kücsán, and Szende, *Sopron*, Gazetteer: 2 (legal status), 7 (population), 26 (mills). References to archival data on mills can be found in this publication.

182 Katalin Szende, "Neighbourhoods, Suburbs and Ethnic Quarters in the Hungarian Towns of the Thirteenth to Fifteenth Centuries," in *Cities – Coins – Commerce: Essays Presented to Ian Blanchard on the Occasion of his 70th Birthday*, ed. Philipp Robinson Rösener (Stuttgart: Franz Steiner Verlag, 2012), 43–64.

183 Jankó, Kücsán, and Szende, *Sopron*.

184 Szende, "Mills and Towns," 496–497, Jankó, Kücsán, and Szende, *Sopron*, passim and 82 (Gazeteer: Mills).

185 Imre Holl, "Sopron (Ödenburg) im Mittelalter," *Acta Archaeologica Academiae Scientiarum Hungaricae* 31 (1979): 110–119.

186 MNL OL DF 201 756, January 16, 1340. Házi, *Sopron*, vol. 1/1. 76–77 no. 136.

187 *Item dem Vlreich zimermann hab ich geben 4 tagion per 20, das er dy rinnen zu der mul bey dem voderntor hat gemacht, facit LXXX den.* Házi, *Sopron*, vol. 2/3. 83 no. 10.

188 *Item Thomas Zenikel vnd Johannes Zeniekel geprueder von gurmennt rugent an stat ir selbs auf die nachgeschriben stucken vnd gueter auf halben tail [von wegen Stephan des Zeniekel purger zu Odenburg vnd fr(awen) Margareten die dann sein hausfr(aw) gewesen ist] von erst auf ain mayerhoff gelegen am Korenmarkt vor der Stat Odenburg zenachst der spitalmül. Gerichtsbuch. Bírósági Könyv 1423–1531* (Sopron Város Történeti Forrásai, A/2), eds. Jenő Házi, and János Németh (Sopron: Győr-Moson-Sopron Megye Soproni Levéltára, 2005), 107–108 no. 29 (1438). See also *Gerichtsbuch*, 117 no. 43 (1440). *Item desselben tags hab ich geben dem Hanns mulner, das er an der mül im graben geczimert hat, dem hab ich beczalt IX taglon per XX, facit VI sol. den.* Házi, *Sopron*, vol. 2/3. 345 no. 23 (1440–1441), and the tax paid by the miller, Hans: Házi, *Sopron*, vol. 2/3. 345 no. 27 (1442).

189 On the fulling mills see: István Dávidházy, "A soproni posztókallók," [The fulling mills of Sopron] *Soproni Szemle* 43 (1989): 193–205. This is taken over by Jankó, Kücsán, and Szende, *Sopron*, 27 Fig. 10. For the exact site of the mill, see recently: József Kücsán, "A Várkerület belső ívének beépítése, 1766–1801," [The building up of the inner line of the Grabenrunde, 1766–1801] *Soproni Szemle* 69 (2015): 329, and 331. Fig. 5.

190 Dávidházy, "A soproni posztókallók," passim.

191 *Anno domini M CCCC Im LXXXXVI-ten Jare Am mitichen Goczleichnambsabent habenn die herrenn den Teicht Jm Attengräben so etwan Cristoff härb seliger gehabt hat Widerumb von dem Erberen Jacob pilsner eingenomen vnd Jm <vnd seinen Erben,> all versessen Zynns so gemainer stat da von gefallen werdenn ganncz nachgelassen, Auch da für gebenn das allt Od gemeir hinder des Khörnmarkh gelegenn mit garten vnd annder aller zugehorung des Nachper ist des Jacob Joachim seligen einsetz, vnd an der annderen seitten Wolfganng Osterperger Einseczs ledig vnd frey Actum vtsupra. Gedenkbuch. Feljegyzési*

könyv 1492–1543 (Sopron Város Történeti Forrásai, A/3), eds. Károly Mollay, and Károly Goda (Sopron: Győr-Moson-Sopron Megyei Levéltár Soproni Levéltára, 2006), 60 no. 61 (June 1, 1496). See also: Dávidházy, "A soproni posztókallók," Fig. 10, and Jankó, Kücsán, and Szende, *Sopron*, 27 Fig. 10, and 82 (Fishponds). For the fish keeper see also: Károly Mollay, "A Tómalom középkori előzményei (Fejezet a soproni határ történetéből)," [Medieval predecessors of the Tómalom (An episode from the history of the borders of Sopron)] *Sopron Szemle* 46 (1992): 154.

192 Anita D. Szakács, *Sopron város tanácsa bírósági jegyzőkönyveinek regesztái III. 1570–1579* [Summaries of the protocols of the court of law of the town of Sopron, 1] (Sopron: Győr-Moson-Sopron Megye Soproni Levéltára, 2011), 21–22 no. 2 (MNL GyMSML SL IV. A. 1002. a. III. 309–310. p.).

193 *Eodem anno piscinulam, e regione turris novae coeno refertam, denuo fructuose et utiliter expurgari fecit, circumquaque palis et sepibus bene munivit. Lackner Kristófnak, mindkét jog doktorának rövid önéletrajza. Vitae Christophori Lackhner I. U. D. Hominis, brevis consignatio* (Sopron Város Történeti Forrásai, C/5), ed. Gergely Tóth (Sopron: Győr-Moson-Sopron Megyei Levéltár Soproni Levéltára, 2008): 138–139 cap. 38. See also: Gergely Tóth, "Előre megírt emlékezet: Lackner Kristóf különös önéletrajza," [The memory written untimely: the curious autobiography of Kristóf Lackner] in *Lackner Kristófnak, mindkét jog doktorának rövid önéletrajza* 16, and 31.

194 Mátyás Bél, *Sopron vármegye leírása / Descriptio Comitatus Semproniensis*, 3 vols (Sopron Város Történeti Forrásai, C/4), ed. Katalin Mária Kincses (Sopron: Soproni Levéltár, 2006), vol. 3. 172–173 and 186–187. Cf. Jankó, Kücsán, and Szende, *Sopron*, 82.

195 MNL Győr-Moson-Sopron Vármegye Soproni Levéltára (Győr-Moson-Sopron County Archives, Sopron Town Archive) IV.1003.a. Tanácsülési jegyzőkönyvek (Rathsprotocoll, Protocollum Senatorium) 1625. VI. 2, p. 152.

196 Dávidházy, 'A soproni'; István Dávidházy, "18. és 19. századi posztósműhelyek egy városon belüli területi megoszlása," [Location of clothmakers' workshops in 18th–19th-century Sopron] in *IV. Nemzetközi Kézművesipartörténeti Szimpózium* (Veszprém: Magyar Tudományos Akadémia Veszprémi Akadémiai Bizottsága, 1994), 270–275, Jankó, Kücsán, and Szende, *Sopron*, 29, 37.

197 *Első telekkönyv. Erstes Grundbuch (1480–1553)*, ed. by Károly Mollay (Sopron: Soproni Levéltár, 1993), pp. 31. no. 208; 40, no. 268; 68, nos. 460–461; 84 no. 535; 90, no. 567; 91, no. 572. Cf. Jankó, Kücsán, and Szende, *Sopron*, 21.

198 The notion derives from the German word *mörung*. See: *Deutsches Wörterbuch von Jacob und Wilhelm Grimm*. <http://woerterbuchnetz.de/DWB> [accessed 10 January 2023]. On this, see Imre Holl, "Középkori városi élet – városi építészet," *Archaeologiai Értesítő* 116 (1989): 53–54; Harry Kühnel, "Die materielle Kultur Wiens im Mittelalter," in *Wien im Mittelalter. Katalog* (Vienna: Museum der Stadt Wien, 1975), 32. The term also appears at Bratislava: *A magyar törvényhatóságok jogszabályainak gyűjteménye / Corpus Statutorum Hungariae Municipalium*, ed. by Sándor Kolosvári and Kelemen Óvári (Budapest: Magyar Tudományos Akadémia, 1885–1904), IV/2, 172.

199 Mollay, *Első telekkönyv*, 91, no. 572. On the fish market, see: Szende, "The Sopron," 163.

200 Holl, "Sopron," [98], 25, 27. (Fig. 3). *Lackner Kristófnak, mindkét jog doktorának*, 138–139 cap. 38.

201 See, for instance Házi, *Sopron*, II/2, 191, 195, or 208. A similar situation is found in other cities: Paul Trio, "The Challenge for a Medieval Center of Industrial Growth: Ypres and the Drinking-Water Problem," in *The Nature and Function of Water, Baths, Bathing, and Hygiene from Antiquity through the Renaissance*, eds. Cynthia Kosso and Anne Scott (Leiden and Boston: Brill, 2009), 381–404; Jørgensen, "Local," Barbara Rouse, "Nuisance Neighbours and Persistent Polluters: The Urban Code of Behaviour in Late Medieval London," in *Medieval Urban Culture* (Studies in European Urban History, 43), eds. Andrew Brown, and Jan Dumolyn (Turnhout: Brepols, 2017), 75–92. https://doi.org/10.1484/M.SEUH-EB.5.114417.

202 Marián Uličný, and Peter Harčar, "Počiatky mesta Prešova na základe výsledkov najnovších archeologických výskumov," [The origins of the town of Prešov deduced from the results of the latest archaeological research] *Archaeologia historica* 35 (2010): 393–403.

203 *Eperjes szabad királyi város levéltára, 1245–1526*, 2 vols [Archive of the free royal town of Prešov], ed. Béla Iványi (Szeged: Szegedi Városi Nyomda és Könyvkiadó, 1931–1932).

204 Katalin Szende, "Gyermekek, testvérek, házastársak. Családi viszonyok és örökösödési szokások Eperjesen a középkori végrendeletek tükrében," [Children, siblings, spouses. Family ties and testamentary practices in Prešov in light of medieval last wills] *Történelmi Szemle* 46 (2004): 113–140 (with reference to the author's other works in the topic). See also: Marcela Domenová, "Prešovské testamenty do roku 1526. (Diplomaticko-obsahový rozbor)," [The last will from Prešovs until 1526 (diplomatic content analysis)] *Historica Carpatica* 33 (2002): 29–68.

205 László Szabolcs Gulyás, "Városfalépítés a középkori Eperjesen," [Town wall construction at medieval Prešov] in *Falak és választóvonalak a történelemben* [Walls and division lines in history] (A Nyíregyházi Főiskola Történettudományi és Filozófia Intézete Kiadványai. Terminus Könyvek, 1), eds. Attila Buhály, Gábor Reszler, and György Szoboszlay (Nyíregyháza: Nyíregyházi Főiskola Történettudományi és Filozófia Intézet, 2014), 127–142.

206 Béla Szalai, and Lajos Szántai, *Magyar várak, városok, falvak metszeteken, 1515–1800*, 3 vols [Hungarian castles, towns and villages on cuttings] (Budapest: Múzeum Antikvárium, 2006), vol. 2. 70–72.

207 On the routes, see: Katalin Szende, "Kraków and Buda in the Road Network of Medieval Europe," and István Draskóczy, "Highways between Buda and Kraków," in *On Common Path*, 31–37, and 38–41, respectively.

208 Erik Fügedi, "Középkori magyar városprivilégiumok," [Medieval Hungarian urban privileges] *Tanulmányok Budapest Múltjából* 14 (1961): 27–28, and more recently, see Boglárka Weisz, "Kassa kereskedelmi életének jogi háttere a középkorban," [The legal background of the trade life of Košice in the Middle Ages] in *Arcana tabularii. Tanulmányok Solymosi László tiszteletére*, 2 vols [Studies in honor of László Solymosi], eds. Attila Bárány, Gábor Dreska, and Kornél Szovák (Debrecen and Budapest: ELTE BTK, 2014), vol. 2. 899–908, and Attila Zsoldos, "The Path of Košice to the Privileges of 1347," and Boglárka Weisz, "The legal background of the trade life of Košice in the Middle Ages," in *Košice in the Coordinates of European History*, eds. Mária Hajduová, and Martin Bartoš (Košice: Kancelária Ústavného súdu Slovenskej republiky, 2013), 34–49, and 94–111, respectively.

209 MNL OL DL 41 829.

210 Kornél Divald, and Béla Iványi, "Az eperjesi Szent Miklós-templom," [The St Nicholas Church at Prešov]. *Az Országos Magyar Szépművészeti Múzeum Évkönyvei* 4 (1924–1926): 38.

211 MNL OL DF 228 604, edited in: *Eperjes*, vol. 1. 79–92 no. 211, for the inhabitants of the street by the moat, here: 90–92. The row of houses in tax lists were recurrently referred to as standing by the moat. E.g., MNL OL DF 228 989. 6. p. (1486) or MNL OL DF 229 008 (1488). See on these tax lists: Domenová, "Daňové písomnosti."

212 *Item dedimus XX, den. pro reformatione pontis cirea molendinum medians inter alia duo molendina.* MNL OL DF 228 612. See also *Eperjes*, vol. 1. 93–103 no. 220, for the relevant part, here: 97–98.

213 MNL OL DF 229 042, for its at some points erroneous editions, see: Carolus Wagner, *Diplomatarium Comitatus Sarosiensis* (Posonii and Cassoviae: Sumtibus Joannis Michaelis Landerer, 1780), 228–231 no. 17 and CD, vol. 10/7. 656–659 no. 305. Its regesta: *Eperjes*, vol. 1. 109–110 no. 237. On the case, see: *Dejiny Prešova*, vol. 1. 94, Norma Urbanová, *Prešov. Pamiatková rezervácia – Prešov* [Monument guide: Prešov] (Pamiatky mestských rezervácií, 8) (Bratislava: Tatran, 1986), 10–15 and Árpád Nógrády, "A középkor végi Magyarország mindennapi kenyere (Beregszász lakosságának gabonavásárlásai 1530-ban és a Jagelló-kori malomvámok)," [The everday bread of Hungary at the end of the Middle Ages. The grain acquisition of the inhabitants of Berehove in the 1530s and the mill customs of Jagiellonian period] *A Nyíregyházi Jósa András Múzeum Évkönyve* 42 (2000): 162. On the wars of the period, see: Pál Tóth-Szabó, *A cseh-huszita mozgalmak és uralom története Magyarországon* [The history of the Czech-Hussite movements and rule in Hungary] (Budapest: Hornyánszky ny. M. Tud. Akadémia, 1917).

214 E.g., the case of a mill at Čachtice: *Extunc prefatus dominus Georgius plebanus et sui successores plebani habeant potestatem ipsa molendina iterum cum tot rotis quibus fuerit eorum voluntas translocandi et transformandi in eodem flumine ad alium aptiorem et commodiorem locum ubi ipsis et ecclesie melius videbitur expedire, neque aliquod aliud molendinum in derogamen et preiudicium illius per nos aut successores nostros fundari poterit in futurum.* MNL OL DL 65 946, edited in: József Teleki, *A Hunyadiak kora Magyarországon*, vol. 11 [History of the age of the Hunyadis in Hungary] (Pest: Emich, 1855), 395–397 no. 501. See also: Chapter 3.1.

215 On the two burghers, see: MNL OL DF 228 604, edited in: *Eperjes*, vol. 1. 79–92 no. 211; MNL OL DF 228 669, edited in: *Eperjes*, vol. 1. 127–129 no. 283.

216 MNL OL 228 631, for its regesta: *Eperjes*, vol. 1. 110–111 no. 240; MNL OL DF 228 633, for its regesta: *Eperjes*, vol. 1. 111 no. 242; MNL OL DF 228 634, for its regesta: *Eperjes*, vol. 1. 111 no. 243.

217 MNL OL DF 229 078, for its regesta: *Eperjes*, vol. 2. 288 no. 714. See also: Nógrády, "A középkor végi," 162. On the mill incomes after 1526, see: Domenová, "K hospodáreniu mesta Prešov," 36–37.

218 On the topography of the town, see: Václav Mencl, Středověká města na Slovensku [Medieval towns in Slovakia] (Bratislava: Universum, 1938), 98, and *Dejiny Prešova*, vol. 1.

219 Discussed in: Nógrády, "A középkor végi."

220 *In plathea medii molendini penes murum civitatis.* MNL OL DF 282 529. 4. p. The relevant entry – with the omission of important words – in the town protocol

book was edited in: Béla Iványi, "Eperjes szabad királyi város jegyzőkönyve, 1424–1509," [The protocols of the free royal town of Prešov, 1424–1509] *Történelmi Tár* 10 (1909): 121–122.

221 For the latter, see e.g., MNL OL DF 282 538. fol. 112^r. See also Figures 3.9 and 3.10.

222 The town view was not included in the collection of Szalai and Szántai: Szalai and Szántai, *Magyar várak*, 70–72, even though it has already been published in 1965: *Dejiny Prešova*, vol. 1. For the original town view, see: Prešov, Štátny archív v Prešov, Magistrát Prešov, Zbierka máp a plánov (Inv. č. 2). Without referring to the repository, the map was published in *Dejiny Prešova*, vol. 1.

223 For these two, see: Domenová, "K hospodáreniu mesta Prešov," 37–38.

224 Domenová, "K hospodáreniu mesta Prešov," 38.

225 The Carmelites took over the building from the Franciscans. On the friary, see: Beatrix F. Romhányi, *Kolduló barátok, gazdálkodó szerzetesek: koldulórendi gazdálkodás a késő középkori Magyarországon* [Mendicant brothers, farming friars. Mendicant economy in late medieval Hungary] (Budapest: Martin Opitz, 2018), passim and Kund Regényi, "Az eperjesi Szentháromság karmelita konvent története," [History of the Holy Trinity Carmelite convent at Prešov] in *Tanulmányok a középkori magyar történelemről. Az I. Medievisztikai PhD-konferencia (Szeged, 1999. július 2.) előadásai* [Studies in medieval history. Proceedings of the 1st PhD conference in medieval studies, Szeged, 2 July 1999], eds. Sarolta Homonnai, Ferenc Piti, and Ildikó Tóth (Szeged: Szegedi Középkorász Műhely, 1999), 103–113. On the relationship of the Franciscan friary and the Carmelite convent: Romhányi, *Kolostorok és társaskáptalanok*, 23.

226 For Bél's description of Sáros County, see: Matthias Bel, *Notitia Hungariae novae historico geographica... Comitatuum ineditorum tomus quintus, in quo continentur... Comitatus Sarosiensis, Zempliniensis, Ungváriensis et Bereghiensis*, edited by Gergely Tóth et al. (Budapest: MTA Bölcsészettudományi Kutatóközpont Történettudományi Intézet 2018), 19–94.

227 *Moenia murus geminus, alter altero sublimior, et solidior incingit, suff ulcientibus eos propugnaculis, quorum duo praecipua, mercatorum unum, alterum vietorum non parum addunt ornamenti urbeculae, ut nihil dicamus de reliquis, quae a sartoribus, carnifi cibusque nomen habent: eae enim pariter singulari munimento moenia, et urbem ornant. Praeterea adstant portae, superior una, altera inferior, adiecta una portula a latere, quae omnia turritis tectis condecorant urbis compagem. Totam eam ambiunt fossae, alte satis depressae, quarum una, quae Tarcae vicina est, oppletur aquis ad summum.* Bel, *Notitia Hungariae*, 82.

228 MNL OL DF 282 538.

229 MNL OL DF 282 535.

230 MNL OL DF 282 538 (images: 533, 625, and 703).

231 Szende, "The Sopron Fish Market."

232 On Magyaróvár: Ágnes Aszt, "A mosonmagyaróvári Bástya utcában végzett régészeti feltárás történeti tanulságai," [Historical lessons of the archaeological excavations at Bástya Street in Mosonmagyaróvár] *Moson Megyei Műhely* no. 3 (2001): 29–32.

233 On Cluj: Elek Benkő, *Kolozsvár magyar külvárosa a középkorban* [The Hungarian suburb of Cluj in the Middle Ages] (Erdélyi Tudományos Füzetek, 248) (Cluj: Erdélyi Múzeum Egyesület, 2004).

234 An earlier version of this subchapter has been published: András Vadas, "Water-Use in the Medieval Mining Towns of the Kingdom of Hungary," in *Conflicts over Water Management and Water Rights from the End of Antiquity to Industrialisation*, eds. Michele Campopiano, and Gerrit Jasper Schenk (Vierteljahrschrift für Sozial- und Wirtschaftsgeschichte – Beihefte, 260) (Stuttgart: Franz Steiner, 2024), 71–94.

235 Boglárka Weisz, "A nemesércbányászathoz kötődő privilégiumok az Árpád- és az Anjou-korban," [Privileges connected to precious ore mining in the Árpádian and Angevin periods] *Történelmi Szemle* 50 (2008): 141–162. For a slightly modified version of this study, see: Boglárka Weisz, "A nemesércbányászathoz kötődő privilégiumok az Árpád- és az Anjou-korban," [Privileges connected to precious ore mining in the Árpádian and Angevin periods] *Bányászattörténeti Közlemények* 6 (2008): 13–43, Boglárka Weisz, "A bányaváros, mint önálló várostípus a 14. században," [Mining town as a distinct type of town in the fourteenth century] *Bányászattörténeti Közlemények* 19 (2015): 31–57, and Boglárka Weisz, "Mining Town Privileges in Angevin Hungary," *Hungarian Historical Review* 2 (2013): 288–312.

236 For this, see Weisz, "Mining Town Privileges."

237 Weisz, "A nemesércbányászathoz kötődő privilégiumok," Weisz, "A bányaváros," and most recently: Weisz, "Az alsó-magyarországi bányavárosok kiváltságai a Zsigmond-korban," [Privileges provided to the Lower Hungarian mining towns in the age of King Sigismund] *Urbs. Magyar Várostörténeti Évkönyv* 12 (2018): 21–48.

238 *Urkundenbuch zur Geschichte der Deutschen in Siebenbürgen*, 7 vols, eds. Franz Zimmermann et al. (Hermannstadt and Bucharest: Michaelis and Verlag der Rumänischen Akademie, 1892–1991), vol. 1. 396.

239 Gusztáv Wenzel, *Az alsó-magyarországi bányavárosok küzdelmei a nagylucsei Dóczyakkal, 1494–1548* [The struggles of the Lower Hungarian mining towns with the Dóczy family of Nagylucse] (Értekezések a Történeti Tudományok Köréből, VI/6) (Budapest: MTA, 1876).

240 Cf. István Batta, "Középkori bányászatunk és kohászatunk a Metercián," [Medieval Hungarian mining and metallurgy on the Metercia] *Bányászati és Kohászati Lapok. Kohászat* 121 (1988): 275–285. On the forest management of the Lower Hungarian mining towns, see: Eszter Magyar, *A feudalizmus kori erdőgazdálkodás az alsó-magyarországi bányavárosokban, 1255–1747* [Feudal-age forest management in the Lower Hungarian mining towns, 1255–1747] (Értekezések a Történeti Tudományok Köréből, Új Sorozat, 101) (Budapest: Akadémia, 1983). Because of numerous errors in the parts of the book concerning medieval forest management, the work is only relevant in its parts on the Early Modern period.

241 On the possible forms of metal processing, see: Georgius Agricola, *De re metallica libri XII* (Basileae: Hieronymus Froben & Nicolaus Episcopius, 1556). For its English translation, see: Georgius Agricola, *De re metallica*, eds. Herbert Clark Hoover, and Lou Henry Hoover (New York: [no publisher] 1950). For the existing forms and their connections to the milling industry: Adam Lucas, "Industrial Milling in the Ancient and Medieval Worlds: A Survey of the Evidence for an Industrial Revolution in Medieval Europe," *Technology and Culture* 46 (2005): 1–30, and Zsuzsa Teke, "L'energia idraulica nella siderurgia medioevale ungherese," in *Economia e energia sec. XIII–XVIII. Atti della "Trentaquattresima Settimana di Studi", 15–19 aprile 2002*, ed. Simonetta Cavaciocchi (Florence: Le Monnier, 2003), 335–341.

242 Oszkár Paulinyi, "A bányavíz leküzdése érdekében létesített első tározómedencék és váltókerekek," [The first water reservoirs and water wheels in the fight against mine water] in *Gazdag föld – szegény ország: Tanulmányok a magyarországi bányaművelés múltjából* [Rich soil – poor country. Studies in the history of Hungarian mining industry] (Gazdaság- és Társadalomtörténeti Kötetek, 3), eds. István Draskóczy, and János Buza (Budapest: Corvinus Egyetem, Budapesti Közgazdaságtudományi Egyetem Alapítvány and Miskolci Egyetem, 2005), 315–335.

243 Agricola, *De re metallica*, 313–317.

244 László Makkai, "Östliches Erbe und westliche Leihe in der ungarischen Landwirtschaft der frühfeudalen Zeit," *Agrártörténeti Szemle* 45 [Supplementum] (1974): 45. See also: Walter Endrei, "A magyar malomipar, 1550–1800," [Hungarian milling industry, 1550–1800] in *Műszaki innovációk sorsa Magyarországon: malomipar, vaskohászat, textilipar* [The fate of technological innovations in Hungary: milling industry, metallurgy, textile industry], ed. Walter Endrei (Budapest: Akadémiai, 1995), 48–54. On the question, in details, see András Vadas, *The Environmental Legacy of War on the Hungarian-Ottoman Frontier, c. 1540–1690* (Environmental Humanities in Pre-modern Cultures) (Amsterdam: Amsterdam University Press, 2023). https://doi.org/10.1515/9789048552016.

245 See for instance, the case of Baia Mare (Asszonypataka), see: Weisz, "A nemesércbányászathoz kötődő privilégiumok," 142 and 148.

246 For *hospites*, see: Erik Fügedi, "Das mittelalterliche Königreich Ungarn als Gastland," in *Die Deutsche Ostsiedlung des Mittelalters als Problem der europäischen Geschichte: Reichenau-Vorträge: 1970–1972* (Vorträge und Forschungen / Konstanzer Arbeitskreis für Mittelalterliche Geschichte, 18), ed. Walter Schlesinger (Sigmaringen: J. Thorbecke, 1975), 471–507.

247 *Item aurum, argentum et omnia alia metalla per totum comitatum de Zolum infra metas suas exquirere poterunt tam in terris quam in nemoribus, silvis, aquis, ac liberi sint infra subscriptas metas in omnibus utilitatibus suis excepta venacione et piscium capcione.* CDES, vol. 2. 340–341 no. 491.

248 *Item concessimus, quod ab omnium iudicum, baronum et specialiter comitum de Lyptou iudicio et descensu penitus sint exempti et quod nullus super eosdem descensum presumat facere violentum. Preterea aquam Reuce piscacioni congruentem reliquimus eisdem hospitibus. Ceterum concessimus, ut in silvis nostris et aquis liberam habeant facultatem querendi mineram auri, argenti et cupri ubicunque poterunt invenire. De omnibus, cum invenire poterunt nobis servient secundum, quod de aliis aurifodinis et argentifodinis regie consueverunt impedire servicium maiestati.* VMMS, vol. 1. 44–45 no. 29. On the problems of the charter and its transcriptions, see: RA, vol. 2/1. 57–64 no. 1933, and Ferdinand Ulicný, "Mestské privilégiá Nemeckej Lupce a otázka ich pravosti," [Urban privileges of Partizánska Ľupča and the question of their authenticity] *Historické štúdie* 5 (1959): 302–324.

249 *Item concessimus eisdem hospitibus, ut ipsam totamque terram Hybe, protu est et adiacet, possideant et liberam in silvis et in aquis intra metas suas eiusdem terre Hybe contemtas versus Scepes venandi et piscandi habeant facultatem, excepta aqua Vag, in qua nullus ex eis piscari poterit ullo modo.* VMMS, vol. 1. 49–50 no. 33.

250 RA, vol. 2/2–3. 385–387 no. 3464. For its edition, see: VMMS, vol. 1. 67–68 no. 66. See also: Tibor Almási, "Megjegyzések Gölnicbánya Kun László királytól

elnyert privilégiumához és megerősítéseihez," [Notes on Ladislas the Cuman's privilege to Gelnica and its confirmation] *Acta Universitatis Szegediensis de Attila József nominatae. Acta Historica* 102 (1995): 43–49 (with the edition of the different variants of the document: 46–49).

251 *Item totam terram Rosumbergh, prout est et adiacet, possidebunt et liberam in silvis et in aquis intra metas eiusdem terre contentis venandi et piscandi Iiberam habebunt facultatem, excepta aqua Wagh, in qua nullus ex eis piscare poterit ullo modo*. VMMS, vol. 1 91–92 no. 103.

252 See for instance the case of the *hospes* privilege of the settlers of Koroleve in Ugocsa County: CD, vol. 5/1. 176–178. Cf. Degré, *Magyar halászati jog*, 58. On the diplomatic analysis and criticism of the charter, see: Fügedi, "Középkori magyar," 45–46.

253 Urszula Sowina, *Water, Towns and People*, 79–82. For the privilege: *Akta Grodzkie i Ziemskie z czasów Rzeczypospolitej Polskiej z Archiwum tak zwanego bernardyńskiego we Lwowie* [Town and land deeds from the times of the Polish Republic in the so-called Bernardine Archives in Lviv], 25 vols (Lwów: Towarzystwo Naukowe, 1868–1914), vol. 3. 29–31 no. 12.

254 *Concessimus eciam, ut iidem hospites nostri intra metas eorum infrascriptas, absque preiudicio iuris alieni, ubique locorum venandi, aves capiendi, piscandi et molendina construendi liberam habeant facultatem ... Preterea aquam Reuce, piscacioni congruentem, reliquimus eisdem hospitibus piscandi*. VMMS, vol. 1. 132–133 no. 161.

255 Petrus de Rozemberk: VMMS, vol. 1. 105–106 no. 125.

256 For him and his *ispánság* at Zólyom, see: Attila Zsoldos, "Doncs mester zolyómi ispánsága," [The ispánság of Doncs at Zólyom] *Századok* 158 (2024): 1179–1197.

257 Katalin Szende, "Mennyit ér a kiváltság? Városprivilégiumok kibocsátása és rendelkezéseik betartása I. Károly alatt," [How much privileges worth? The issuing and the implementing of urban privileges during the reign of Charles I] in *Pénz, posztó, piac: Gazdaságtörténeti tanulmányok a magyar középkorról* [Money, cloth, market. Studies in the economic history of medieval Hungary], ed. Boglárka Weisz (Budapest: MTA BTK Történettudományi Intézet, 2016), 301–302 notes 70–71.

258 Degré, *Magyar halászati jog*, 58.

259 Fügedi, "Középkori magyar városprivilégiumok," 48.

260 There is no data on mills, but there is on mill stone quarry from the period: MNL OL DL 102 943. For the summary of the document, see AOklt, vol. 9. 168 no. 296.

261 Cf. AOklt, vol. 31. 457–458 no. 883. Arhivele Naţionale ale României, Serviciul Judeţean Maramureş, Fond Primăria oraşului Baia Mare, Seria 2 – Documente privilegiale, no. 1 (the document is not available at the Medieval Digital Archive of Romania (http://arhivamedievala.ro/). I am thankful to Péter Levente Szőcs for providing information on the document.

262 *Documente privind istoria României. C. Transilvania. Veacul XIV*, 12 vols (Documenta Romaniae historica) Bucureşti: Acad. Republicii Populare Române, 1953–1985), vol. 4. 673–676 no. 577. For its summary, see. AOklt, vol. 31. 457–458 no. 883.

263 Cf. Fügedi, "Középkori magyar városprivilégiumok," 47–48.

264 *Quod iam dicti ciues, et hospites nostri dictarum duarum ciuitatum nostrarum molendinum, casas, fornaces, Balnea, allodia, et alias quaslibet haereditates*

aedificari facientes, ordinandi, acea in perpetuum ritu aliarum nostrarum montanarum conseruandi habeant potestatem. CD, vol. 9/5. 96–101 no. 39. Cf. Weisz, "A nemesércbányászathoz kötődő privilégiumok," 157.

265 On urban bath-houses see: Judit Majorossy, "A test és a lélek fürdője. A középkori pozsonyi fürdőkről," [The bath of body and soual. On the medieval bathing houses of Bratislava] in *„Köztes-Európa" vonzásában: Ünnepi tanulmányok Font Márta tiszteletére* [In the attraction of "Mitteleuropa." Studies in honor of Márta Font], eds. Dániel Bagi, Tamás Fedeles, and Gergely Kiss (Pécs: Kronosz, 2012), 343–356.

266 This meaning of the word is not referred to in the Lexicon Latinitatis Medii Aevi Hungariae. See the online database: http://clph.elte.hu/llmaeh/mklsz/simplesera ch_b_hu.php.

267 MNL OL DF 250 152. Michal Matunák, *Z dejin Slobodného a Hlavného Banského Mesta Kremnice* [History of the privileges and the town of Kremnica] (Kremnica: Slobodné a hlavné banské mesto, 1928), 455–457 no. 4.

268 László Fejérpataky, *Magyarországi városok régi számadáskönyvei (Selmeczbánya, Pozsony, Beszterczebánya, Nagyszombat, Sopron, Bártfa és Körmöczbánya városok levéltáraiból)* [Old account books of Hungarian towns (from the archives of Banská Štiavnica, Bratislava, Banská Bystrica, Trnava, Sopron, Bardejov and Kremnica)] (Budapest: MTA, 1885), 85, 88, and 101. See also on that Miroslav Minárčic, "Metačná listina svätoantonského panstva z roku 1266 a jej výpovedná hodnota k dejinám banskoštiavnického baníctva," [A document from the year 1266 on the estate of Svätý Anton and its significance in the history of the mining industry at Banská Štiavnica] *Montánna história* 2 (2009): 46–69.

269 Cf. Berent Schwineköper, "'Cum aquis aquarumve decursibus' Zu den Pertinenzformeln der Herrscherurkunden bis sur Zeit Ottos I.," in *Festschrift für Helmut Beumann zum 65. Geburtstag* ed. Kurt-Ulrich Jäschke (Sigmaringen: Thorbecke, 1977), 22–56, Härtel, "Von Wasser- und Mühlenrecht," 221. See also Chapter 2.

270 On *hospes*-privileges, see: Fügedi, "Középkori magyar városprivilégiumok," 61–64, László Solymosi, "Hospeskiváltság 1275-ből," [Hospes privilege from 1275] in *Tanulmányok Veszprém megye múltjából* [Studies in the history of Veszprém County], ed. Lajos Madarász (Veszprém: Veszprém Megyei Levéltár, 1984), 17–19, and more recently: Zsoldos, "Károly és a városok," 271ff.

271 ÁÚO, vol. 8. 31–32 no. 23. For details of the charter, most recently, see: László Szabolcs Gulyás, *Városfejlődés a középkori Máramarosban* [Urban development in medieval Maramureş] (Erdélyi Tudományos Füzetek, 280) (Cluj: Erdélyi Múzeum-Egyesület, 2014), 29–39.

272 István Tringli, "Sátoraljaújhely első kiváltságai," [First privileges of Sátoraljaújhely] in *Tiszteletkör*, 251–259, István Tringli, *Sátoraljaújhely* (Magyar várostörténeti atlasz, 2) (Budapest: MTA Történettudományi Intézet, 2011), 7–9. Its edition: EFHU, vol. 3/2. 52–54 no. 43.

273 For its recent critical edition, see: Boglárka Weisz, and Attila Zsoldos, "Másolatok és kiadások. Győr 1271. évi kiváltságlevelének ma ismert szövegváltozatai," [Copies and editions. The known variants of the text of the 1271 privilege of Győr] in *Márvány, tárház, adomány. Gazdaságtörténeti tanulmányok a magyar középkorról* [Marble, storage house, donation. Studies in the economic history

of the Middle Ages], eds. Boglárka Weisz, Renáta Skorka, and István Kádas (Budapest: MTA BTK Történettudományi Intézet, 2019), 247–256.

274 EFHU, vol. 3/2. 71–72 no. 59, and 78–79 no. 63.

275 Vladimir Horváth, *Bratislavské mestské privilégium* [The urban privilege of Bratislava] (Bratislava: Archív hlavného mesta SR: Obzor, 1991). For a recent diplomatic study on the interpolated charter, see Žofia Lysá, "Privilégium pre Bratislavu z roku 1291," [The privilege of the town of Bratislava from 1291] *Historický časopis* 59 (2011): 189–214.

276 István Tringli, "Sátoraljaújhely egyházai a reformáció előtt," [The churches of Sátoraljaújhely before the Reformation] in *Erősségénél fogva várépítésre való. Tanulmányok a 70 éves Németh Péter tiszteletére* [Proper for castle building for its strength... Studies in honor of Péter Németh for his 70th birthday], eds. Juan Cabello, and Norbert C. Tóth (Nyíregyháza: Szabolcs-Szatmár-Bereg Megyei Önkormányzat Múzeumok Igazgatósága, 2011), 377–396, Tringli, *Sátoraljaújhely*, 74–75. (Gazetteer: mills), Szende, "Mills and Towns," 492–496.

277 Cf. Erik Fügedi, "Győr városának 1271. évi kiváltságlevele," [The 1271-privilege charter of the town of Győr] in *Győr. Várostörténeti tanulmányok* [Győr. Studies in the history of the town], eds. Lajos Dávid, Alfréd Lengyel, and László Z. Szabó (Győr: Győr Megyei Jogú Város Tanácsa, 1971), 109–117, and Weisz, and Zsoldos, "Másolatok és kiadások," (the latter with the literature of the privilege).

278 ÁMTF, vol. 2. 607.

279 ÁMTF, vol. 2. 598. See: Beatrix F. Romhányi, *Kolostorok és társaskáptalanok a középkori Magyarországon* Monasteries and collegiate chapters in medieval Hungary (Budapest: Pytheas, 2000), 41.

280 Lajos Juhász, "A Vas megyei Farkaserdő a XVII–XVIII. században," [Farkas Forest in Vas County in the seventeenth–eighteenth centuries] *Századok* 71 (1937) [Supplement]: 553–575 [37–69].

281 *Concessimus etiam, ut in silva Raba ligna recipiendi pro usibus suis necessaria, falcandi fenum et herbas et utendi usu aque sine exactione et impedimento aliquo liberam habeant facultatem.* EFHU, vol. 3/2. 68–71 no. 59.

282 *Hospites in eadem villa Castri Ferrei ad prepositum et capitulum predicte ecclesie Beati Michaelis pertinentes ac alii iobagiones, equites, populi, liberi et conditionarii ubique locorum constituti ad eandem spectantes ecclesiam in aquis, piscationibus fluminum Rabe et Macua, fenetis, cesu lignorum ac aliis utilitatibus quibuscunque in silva Rabe liberum et securum habeant.* EFHU, vol. 3/2. 89–79 no. 63.

283 Cf. Fügedi, "Középkori magyar városprivilégiumok," 47.

284 Cf. Katalin Szende, "Town Foundations in East Central Europe and the New World: The Use of the Grid Plan in a Comparative Perspective," in *Medieval East Central Europe in a Comparative Perspective: From Frontier Zones to Lands in Focus*, eds. Gerhard Jaritz, and Katalin Szende (New York: Routledge, 2016), 161–172.

285 On German settling and its legal background, see: Fügedi, "Das mittelalterliche Königreich Ungarn," and Katalin Szende, "Iure Theutonico? German Settlers and Legal Frameworks for Immigration to Hungary in an East-Central European Perspective," *Journal of Medieval History* 45 (2019): 360–379. https://doi.org/10.1080/03044181.2019.1612195.

286 *In eodem etiam fluvio concedimus advocatis, vel quibus ipsi contulerint, duo nostra molendina et tercium, quod fuit olim fratrum de Meccovia, nec non quartum, quod fuit monachorum de Andreowe, et si plura (sint) in eodem fluvio molendina sine aliorum tamen nocumento et prejudicio aedificare poterint, eam eis jure hereditario concedimus facultatem, ita, quod de qualibet rota tam presentium*

quam futurorum molendinorum fertonem usualis tunc argenti nobis annuatim persolvere teneantur. Its best edition: *Kodeks dyplomatyczny miasta Krakowa 1257–1506*. vol. 1 [Cartulary of the town of Kraków], ed. Franciszek Piekosiński (Kraków: Akad. Umiejętności, 1879), 1–3 no. 1. For an English translation of the charter, see Anon., "A Charter to Re-establish Kraków," in *Epistolae: Medieval Women's Latin Letters*, ed. Joan Ferrante. ([New York]: Columbia University Libraries, 2014) https://epistolae.ctl.columbia.edu/letter/26164.html

287 On the foundation-period history of Kraków, see: F[rancis] W. Carter, *Trade and Urban Development in Poland. An Economic Geography of Cracow from its Origins to 1795* (Cambridge Studies in Historical Geography, 20) (Cambridge: Cambridge University Press, 1994), 57–61. https://doi.org/10.1017/CBO9780511522222, Marcin Starzyński, *Das Mittelalterliche Krakau. der Stadtrat im Herrschaftsgefüge der polnischen Metropole* (Städteforschung, 92) (Cologne: Böhlau, 2015), 15–41, and most recently: Jacek Zinkiewicz, "Creatio. Foundation of Kraków," in *On Common Path: Budapest and Kraków in the Middle Ages. Exhibition Catalogue, Budapest, March 18 – July 24, 2016*, eds. Judit Benda et al. (Budapest: Budapest History Museum, 2016), 87–92.

288 See e.g., the example of Petrus Gallicus de Siena: Katalin Szende, "Királyi kényszer vagy közösségi akarat? Hivatali írásbeliség a magyarországi bányavárosokban a 13–14. században," [Royal pressure or common will? Pragmatic literacy in Hungarian mining towns in the 13th–14th centuries] in *Márvány, tárház, adomnány*, 519. For him, see also: Attila Zsoldos, "Sienai ötvösből szepesi alispán. Olaszok és itáliaiak a Druget-nádorok szolgálatában," [A Sienese goldsmith as alispán of Szepes: Italians in the service of the Druget palatines] *Világtörténet* 7 [39] (2017): 61.

289 Hans-Wolfgang Strätz, "Kuttenberger Bergordnung," in *Lexikon des Mittelalters*, vol. 5, eds. Robert Auty et al. (Munich: Artemis, 1991), 1594.

290 For its edition, see: *Ius regale montanorum* in *Codex juris Bohemici*, ed. Hermenegildus Jireček (Prague: Kober, 1867), 266–435. On the codification process, see: Kaspar von Sternberg, *Umrisse einer Geschichte der böhmischen Bergwerke* (Prague: Haase, 1836–1837), Adolf Zycha, *Das böhmische Bergrecht des Mittelalters*, 2 vols (Berlin: Vahlen, 1900) (with another edition of the text in the second volume), and more recently: Guido Pfeifer, *Ius regale montanorum: ein Beitrag zur spätmittelalterlichen Rezeptionsgeschichte des römischen Rechts in Mitteleuropa* (Ebelsbach: Aktiv Druck & Verlag, 2002), and Jan Libor, *Václav II. a struktury panovnické moci* (Brno: Matice moravská, 2006), 79–162. The author is thankful to Renáta Skorka for sharing these works.

291 See the database: http://vodnimlyny.cz. See also: Kateřina Vavrušková, "Po stopách mlynářů v Kutné Hoře," [In the footsteps of the millers in Kutna Hora] *Krásné Město* no. 2 (2011): 14–18. Online: www.cms-kh.cz/po-stopach-mlynaru-v-kutne-hore.

292 MNL OL DF 250 152. Edited in: Matunák, *Z dejin Slobodného*, 455–457 no. 4.

293 On him, see: Martin Stefánik, "Italian Involvement in Metal Mining in the Central Slovakian Region from the Thirteenth Century to the Reign of King Sigismund of Hungary," *I Tatti studies* 14/15 (2011–2012): 19–20. https://doi.org/10.1086/its.14_15.41781522.

294 Minárčic, "Metačná listina svätoantonského panstva," 50.

295 Referred to today as Skalka Stream, an east-west running river in the borders of Kremnica, and a tributary to the Kremnický Stream. More mills are indicated by this stream by Györffy (ÁMTF, vol. 1. 455) as well as Martin Štefánik,

"Kremnica," in *Lexikon stredovekých miest na Slovensku* [Lexicon of medieval towns in Slovakia], eds. Martin Štefánik and Ján Lukačka (Bratislava: Historický Ustav SAV, 2010), 221.

296 North-south running of the Kremnický Stream.

297 ÁMTF, vol. 1. 454–456, Štefánik, "Kremnica," 221 and 223.

298 He may be identical with a man referred to as a burgher of Hybe: Iván Borsa, *A Szent-Iványi család levéltára 1230–1525* [Cartulary of the Szentiványi family] (Magyar Országos Levéltár kiadványai, II. Forráskiadványok, 14) (Budapest: MOL, 1988), 27 no. 38.

299 Probably meaning the right to one-sixth of the water of the river.

300 As an *urburarius* of Pukanec: AOKlt, vol. 29. 296–296 no. 480. (MNL OL DF 237 039). He owned a mill by the Büksavnica Stream close to Nová Baňa, to be discussed hereafter. Later, he became chamber count of Kremnica: www.corycats.sk/mint/index.php?go=grofi_is.

301 Later called the Upper Mill (1365, 1366, 1372, 1373), which had four wheels by 1373. Cf. MNL OL DF 249 435.

302 They probably do not represent two mills, but a single mill with two wheels.

303 For Pukano – Bakabánya: Katalin Fehértói, "RATA és TILO," [RATA and TILO] *Magyar Nyelvőr* 122 (1998): 66. Lajos Kiss, *Földrajzi nevek etimológiai szótára* [Etymological dictionary of geographic names] (Budapest: Akadémiai, 1978), 76 (Bakabánya).

304 Such as the judge of Pukanec and Kadold, who was also active at Pukanec.

305 *Item molendinum circa Mazaldrum situm et infra molendinum judicis de Pukano ipsum Jeclinum contigentem, pro quinquaginta marcis eciam Regii pagamenti. Item unum confum, quod wlgariter Kolbe dicitur in eadem aqua inferius sita et ibidem Nycolaus dictus Putner habet unum contum, pro quinque marcarum pagamenti estimamus*… and *Primo unam rotam cum sexta parte aque in valle Soler pro triginta marcarum pagamenti. Item duas Rotas et chazam in superflui aqua circa molendinum Kadoldi sitam, pro quadraginta marcarum Regii pagamenti.* Matunák, *Z dejin Slobodného*, 455–457 no. 4.

306 For the *hospes*-villages and their foundations in Upper Hungary, see: Adrienne Körmendy, *Melioratio terrae: vergleichende Untersuchungen über die Siedlungsbewegung im östlichen Mitteleuropa im 13.–14. Jahrhundert* (Prace Komisji historycznej / Poznańskie Towarzystwo przyjaciół nauk, Wydział historii i nauk społecznych, 48) (Poznań: Wydawn. Poznańskiego Towarzystwa przyjaciół nauk, 1995).

307 Matunák, *Z dejin Slobodného*, 458 no. 6. See also here: 457–458 no. 5.

308 VMMS, vol. 1. 97–98 no. 110.

309 *Insuper judici praedicto et suis villanis contullimus, quod quantum in aqua sursum et deorsum poterunt aedificare, nemo eos audeat infestare.* Matunák, *Z dejin Slobodného*, 458 no. 6.

310 Matunák, *Z dejin Slobodného*, 459–461 no. 8.

311 Matunák, *Z dejin Slobodného*, 461–462 no. 9. Cf. Štefánik, "Kremnica," 218.

312 RA, vol. 1/3. 396 no. 1398, 416 no. 1354, 461 no. 1516, and 474 no. 1551. For the early history of the landownership in the area, see: Elemér Mályusz, *Turóc megye kialakulása* [The formation of Turóc County] (2nd ed. Máriabesnyő–Gödöllő: Attraktor, 2005), 99–101, and Béla Bráz, *A nagycsepcsényi és muthnai Vladár-család története és leszármazása* [The history and genealogy of the Vladár family of Veľký Čepčín and Mutna] (Turócszentmárton: Magyar Ny., 1907), passim.

313 For a map of the area, see: Ján Št. Šikura, *Miestopisné dejiny Turca* [Historical geography of Turóc]. (Bratislava: Slovenská akadémia vied a umení, 1944).
314 RA, vol. 1/3. 416 no. 1354. Its edition: ÁUO, vol. 8. 56. See also: RA, vol. 1/3. 415 no. 1351, its edition: ÁUO, 8. 55.
315 Cf. Szende, *Trust, Authority*, 66, 168, and 219.
316 MNL OL DF 249 547 (for its summary, see: AOklt, vol. 48. 287–288 no. 546).
317 On this, see: ZsO, vol. 2/1. 41 no. 328, 295 no. 2537, 359 no. 3056, 367 no. 3112, 375–376 no. 3181, 461 no. 3838, 465–466 no. 3866, vol. 2/2. 36–37 no. 5431, 230 no. 6718, and vol. 3. 217 no. 697.
318 Árpád Nógrády, "Az elakadt fejlődés," [Halted development] in *Keresztesekből lázadók: Tanulmányok 1514 Magyarországáról* [From crusaders to rebels. Studies on Hungary in 1514], eds. Norbert C. Tóth, and Tibor Neumann (Budapest: MTA BTK Történettudományi Intézet, 2015), 11–30.
319 MNL OL DL 95 433. (August 21, 1498)
320 MNL OL DL 98 227. (March 30, 1525)
321 Matunák, *Z dejin Slobodného*, 462–463 no. 10.
322 János Melich, "Három helynévről: Lehota, Vólya, Ohába," [On three placenames: Lehota, Vólya, Ohába] *Századok* 41 (1907): 321–324.
323 For an edition of the document, see: Matunák, *Z dejin Slobodného*, 463–464 no. 11.
324 MNL OL DF 249 435.
325 See further, on the interests of the burghers of Nová Baňa in the mills of Kreminca: MNL OL DF 249 443. Cf. Martin Stefánik, "Nová Baňa," in *Lexikon stredovekých*, 298.
326 Mihály Matunák, "Eredeti oklevelek a körmöczbányai püspöki malomról," [Original charters on the bishop's mill at Kremnica] *Magyar Történelmi Tár* 8 (1907): 321–324. Cf. András Kubinyi, "A budai német patriciátus társadalmi helyzete családi összeköttetései tükrében a 13. századtól a 15. század második feléig," [The social position of the German patriciate of Buda in the light of their family ties from the thirteenth to the second half of the fifteenth century] in András Kubinyi, *Tanulmányok Budapest középkori történetéről*, vol. 2. 496 (published earlier under the same title at: *Levéltári Közlemények* 42 [1971]: 203–269).
327 Erik Fügedi, "Az esztergomi érsekség gazdálkodása a XV. század végén. I–II. rész," [Farming of the archbishopric of Esztergom at the end of the fifteenth century, I–II] *Századok* 94 (1960): 102–103, and 536.
328 On the acquisition of the mill, see: Matunák, "Eredeti oklevelek," 324–326 no. 5.
329 For the most important works: Dezső Csánki, and Néda Relković, "Újbánya város levéltára és rövid története," [Short history of Nová Baňa and its archive] *Levéltári Közlemények* 2 (1924): 86–107, ÁMTF, vol. 1. 482, Kristóf Keglevich, *A garamszentbenedeki apátság története az Árpád- és az Anjou-korban, 1075–1403* [The history of the abbey of Hronský Beňadik in the Árpádian and Angevin Periods, 1075–1403] (Capitulum VIII.) (Szeged: Szegedi Tudományegyetem Törteneti Intézete, 2012), 88–90, Kristóf Keglevich, "A garamszentbenedeki bencések és Újbánya polgárai a 14. század közepén," [The Benedictines of Hronský Beňadik and the burghers of Nová Baňa in the middle of the fourteenth century] in *Tiszteletkör,* 137–144, Weisz, "A nemesércbányászathoz kötődő privilégiumok," 147–148, Weisz, "A bányaváros," 34–35, and Weisz, "Mining Town Privileges," 291–292. Slovakian researchers have also discussed the circumstances of the town's foundation: Richard Marsina, "Štúdie k Slovenskému diplomatáru. I. Druhá časť," [Studies in Slovak diplomatics. I. Second part]

Historické štúdie 18 (1973): 66–67 and more recently: Štefánik, "Nová Baňa," 294–304.

330 Weisz, "A nemesércbányászathoz kötődő privilégiumok," 147.

331 MNL OL DL 25 169. For an earlier edition: CD, vol. 8/4. 273–274. The undated document, mostly based on Fejér's *Codex diplomaticus* has been dated to 1337, erroneously. For its new edition, with a rather convincing argument on the dating, see: Kristóf Keglevich, *A garamszentbenedeki apátság 13. és 14. századi oklevelei (1225–1403)* [Thirteenth- and fourteenth-century documents of the abbey of Hronský Beňadik] (Capitulum, 9). Szeged: Szegedi Tudományegyetem Történti Intézete, 2014), 333–335 no. 5.

332 Cf. Weisz, "A nemesércbányászathoz kötődő privilégiumok," 147. *Nulla iustitia vel libertatis prerogativa, si quam ipsorum concives in Bakabanya vel hospites in novis montaniis Schewnycze vocatis ex consuetudine vel ex donatione regali haberent vel in posterum habere possent.* See: Keglevich, *A garamszentbenedeki apátság*, 334 no. 5.

333 For the best edition, see: Keglevich, *A garamszentbenedeki apátság*, 335. no 6.

334 Keglevich, *A garamszentbenedeki apátság*, 336 no. 7.

335 Keglevich, *A garamszentbenedeki apátság*, 336–337 no. 8.

336 Cf. Keglevich, "A garamszentbenedeki bencések és Újbánya polgárai," 142–143.

337 Szende, "Királyi kényszer," 522, and 535–536 (Appendix).

338 Paul Jacobus du Plessis, "An Annotated Translation of Bartolus' Tractatus de Fluminibus seu Tyberiadis (Book 1)," (MA-Th., Potchefstroomse Universiteit, 1999), 35. (Corrected after the translation of du Plessis). For the best edition of the prologue, see Osvaldo Cavallar, "River of Law: Bartolus's Tiberiadis (De alluvione)," in *A Renaissance of Conflicts: Visions and Revisions of Law and Society in Italy and Spain*, eds. John A. Marino, and Thomas Kuehn (Toronto: Centre for Reformation and Renaissance Studies, 2004), 84–116. For the printed editions of the work, see Bartolus de Saxoferrato, *Tractatus de fluminibus seu Tyberiadis* (Bononiae: Johannes Roscius, 1576), and Bartolus de Saxoferrato, *Tractatus de fluminibus seu Tyberiadis*, ed. Guido Astuti (Turin: Bottega d'Erasmo, 1964) (with the reprint of the 1576 edition of the text).

339 An earlier version of this subchapter has been published: András Vadas, "Border by the River – But Where is the River? Hydrological Changes and Borders in Medieval Hungary," *Hungarian Historical Review* 8 (2019): 336–360.

340 For the political context of the writing, see Helmut Walther, "Wasser in Stadt und Contado. Perugias Sorge um Wasser und der Flußtraktat Tyberiadis des Perusiner Juristen Bartolus von Sassoferrato," in *Mensch und Natur im Mittelalter*, 2 vols (Miscellanea Mediaevalia, 21), eds. Albert Zimmermann, and Andreas Speer (Berlin: de Gruyter, 1991), vol. 2. 889–890. See also: Cavallar, "River of Law."

341 Cf. Cavallar, "River of Law."

342 Although the problem of emerging islands also appears in Hungarian legal documents. E.g., AO, vol. 1. 94–95 no. 87, AO, vol. 4. 10–12 no. 10.

343 His disciples however took some Bartolus' ideas further, and considered many aspects of water law. Cf. Osvaldo Cavallar, "The Wheels of Watermills and the Wheel of Fortune A consilium of Donatus Ricchi de Aldighieris." *Rechtsgeschichte* 13 (2008): 80–116. http://doi.org/10.12946/rg13/080-116

344 Joseph L. Sax, "The Accretion/Avulsion Puzzle: Its Past Revealed, Its Future Proposed," *Tulane Environmental Law Journal* 23 (2010): 305–367.

345 Beatry Augustyn, "Evolution of the dune Ecosystem in Flanders during the Middle Ages: Anthropogenic Factors Versus Sea Level Change Theory,"

online document: www.vliz.be/imisdocs/publications/100070.pdf, and Beatry Augustyn, "De evolutie van het duinecosysteem in Vlaanderen in de Middeleeuwen: antropogene factoren versus zeespiegelrijzingstheorie," [Evolution of the dune ecosystem in Flanders during the Middle Ages: anthropogenic factors versus sea level change theory] *Historisch-Geografisch Tijdschrift* 13, no. 1 (1995): 9–19.

346 For Muratori's reading of the 589 floods see Lodovico Antonio Muratori, *Annali d'Italia, dal principio dell'era volgare fino all'anno MDCCL*, vol. 3/2 (Rome: Stamperia degli eredi Barbiellini, 1752), 339. On this, see Paolo Squatriti, "The Floods of 589 and Climate Change at the Beginning of the Middle Ages: An Italian Microhistory," *Speculum* 85 (2010): 801. https://doi.org/10.1017/S0038713410002290. More recently: Alessandro Fontana, Matteo Frassine, and Livio Ronchi, "Geomorphological and Geoarchaeological Evidence of the Medieval Deluge in the Tagliamento River (NE Italy)," in *Palaeohydrology. Geography of the Physical Environment*, eds. Jürgen Herget and Alessandro Fontana (Cham: Springer, 2020), 97–116. https://doi.org/10.1007/978-3-030-23315-0_5.

347 *Eo tempore fuit aquae diluvium in finibus Veneciarum et Liguriae seu ceteris regionibu Italiae, quale post Noe tempore creditur non fuisse. Factae sunt lavinae possessionum seu villarum hominumque pariter et animantium magnus interitus. Destructa sunt itinera, dissipatae viae, tantum tuncque Atesis fluvius excrevit, ut circa basilicam Beati Zenonis martyris, quae extra Veronensis urbis muros sita est, usque ad superiores fenestras aqua pertingeret* [...] *Urbis quoque eiusdem Veronensis muri ex parte aliqua eadem sunt inundatione subruti.* Paulus Diaconus, *Historia Langobardorum*, III. 23. For its edition, see Paulus [Paul the Deacon], *Historia langobardorum* in *MGH Scriptores Rerum Germanicarum*, vol. 48, ed. Georg Waitz (Hannover: Hahn, 1878), 127–128.

348 To this day, the most detailed treatise of the problem is: Squatriti, "The Floods of 589," 799–826 (with a thorough criticism of the earlier secondary literature).

349 Maarten G. Kleinhans, Henk J. T. Weerts, and Kim M. Cohen, "Avulsion in Action: Reconstruction and Modelling Sedimentation Pace and Upstream Flood Water Levels Following a Medieval Tidal-River Diversion Catastrophe (Biesbosch, The Netherlands, 1421–1750 AD)," *Geomorphology* 118 (2010): 65–79. https://doi.org/10.1016/j.geomorph.2009.12.009. See also: Torbjörn E. Törnqvist, "Middle and Late Holocene avulsion history of the River Rhine (Rhine-Meuse delta, Netherlands)," *Geology* 22 (1994): 711–714. https://doi.org/10.1130/0091-7613(1994)022<0711:MALHAH>2.3.CO;2, Tim Soens, "The Origins of the Western Scheldt. Environmental Transformation, Storm Surges and Human Agency in the Flemish Coastal Plain (1250–1600)," in *Landscapes or Seascapes? The History of the Coastal Environment in the North Sea Area Reconsidered* (Comparative Rural History of the North Sea Area 13), eds. Erik Thoen et al. (Turnhout: Brepols, 2013), 287–312. https://doi.org/10.1484/M.CORN.1.101557, and Winfried Trusen, " 'Insula in flumine nata.' Ein kanonischer Zivilprozeß aus den Jahren 1357 bis 1363/64 um eine Insel im Mittelrhein," *Zeitschrift der Savigny-Stiftung für Rechtsgeschichte: Kanonistische Abteilung* 68 (1982): 294–338. http://doi.org/10.7767/zrgka.1982.68.1.294.

350 Mireille Provansal, Georges Pichard, and Edward J. Anthony, "Geomorphic Changes in the Rhône Delta During the LIA: Input from the Analysis of Ancient Maps," in *Sediment Fluxes in Coastal Areas* (Coastal Research Library, 10), eds. Mohamed Maanan, and Marc Robin (Dordrecht: Springer Netherlands, 2015), 47–72. https://doi.org/10.1007/978-94-017-9260-8_3 and Jean-Michel

Carozza et al., "Lower Mediterranean Plain Accelerated Evolution during the Little Ice Age: Geoarchaeological Insight in the Tech Basin (Roussillon, Gulf of Lion, Western Mediterranean)," *Quaternary International* 266 (2011): 94–104. https://doi.org/10.1016/j.quaint.2011.06.049.

351 *Landscapes or Seascapes?*

352 Christoph Sonnlechner, Severin Hohensinner, and Gertrud Haidvogl, "Floods, Fights and a Fluid River: The Viennese Danube in the Sixteenth Century," *Water History* 5 (2013):173–194. https://doi.org/10.1007/s12685-013-0077-z, Severin Hohensinner et al. "Two Steps Back, One Step Forward: Reconstructing the Dynamic Danube Riverscape Under Human Influence in Vienna," *Water History* 5 (2013): 121–143. https://doi.org/10.1007/s12685-013-0076-0, and Severin Hohensinner et al., "Changes in Water and Land: The Reconstructed Viennese Riverscape from 1500 to the Present," *Water History* 5 (2013): 145–172. https://doi.org/10.1007/s12685-013-0074-2.

353 E.g., Peter Pišút, "Príspevok historických máp k rekonštrukcii vývoja koryta Dunaja na uhorsko-rakúskej hranici (16.–19. storočie)," [Contribution of historical maps to the reconstruction of the Danube channel evolution near the old Hungarian-Austrian Border (16th–19th centuries)] in *Historické mapy* [Historical maps], ed. Ján Pravda (Bratislava: Kartografická spoločnosť SR– Geografický ústav SAV, 2005), 167–181 (for the Danube), Gábor Timár, Pál Sümegi, and Ferenc Horváth. "Late Quaternary Dynamics of the Tisza River: Evidence of Climatic and Tectonic Controls," *Tectonophysics* 410 (2005): 97–110. https://doi.org/10.1016/j.tecto.2005.06.010 (for the Tisza), András Vadas, *Egy határfolyó környezettörténete. Háború és vízgazdálkodás a kora újkori Rába-völgyben* [The environmental history of a frontier river. War and water-management in the Early Modern Rába River valley] (Budapest: Bölcsészettudományi Kutatóközpont, 2021) (for the Rába), Bence Péterfi, "Debates Concerning the Regulation of Border Rivers in the Late Middle Ages: The Case of the Mura River," *Hungarian Historical Review* 8 (2019): 313–335, and Renáta Skorka, "On Two Sides of the Border. The Hungarian–Austrian Border Treaty of 1372," *Hungarian Historical Review* 8 (2019): 290–312 (on the Mura), and *"Per sylvam et per lacus nimios" – The Medieval and Ottoman Period in Southern Transdanubia, Southwest Hungary: the Contribution of the Natural Sciences*, eds. Gyöngyi Kovács, and Csilla Zatykó (Budapest: MTA BTK Régészeti Intézet, 2016), esp. the contribution by István Viczián, "Geomorphological Research in and around Berzence, on the Border of the Drava Valley and Inner Somogy Microregion, Hungary," in *"Per sylvam,"* 75–91 (on the Dráva).

354 On the source type, see Péter Szabó, "Sources for the Historian of Medieval Woodland," in *People and Nature in Historical Perspective* (CEU Medievalia, 5), eds. József Laszlovszky, and Péter Szabó (Budapest: CEU Press – Archaeolingua, 2003), 268–271. On the proportion of perambulation documents in the surviving documentary evidence, see: Péter Szabó, "Medieval Trees and Modern Ecology: How to Handle Written Sources." *Medium Aevum Quotidianum* 46 (2002): 12–17.

355 E.g., Sax, "The Accretion."

356 MNL OL DL 68 795 (10 April 1324). Cf. Attila Zsoldos, "Hűséges oligarchák," [Faithful oligarchs] in *A történettudomány szolgálatában. Tanulmányok a 70 éves Gecsényi Lajos tiszteletére* [In the service of historical research. Studies in

honor of Lajos Gecsényi on his 70th birthday], eds. Magdolna Baráth, and Antal Molnár (Budapest and Győr: Magyar Országos Levéltár Győr-Moson-Sopron Megye Győri Levéltára, 2012), 347.

357 On the Drugets, see Attila Zsoldos, *A Druget-tartomány története, 1315–1342* [The history of the Druget lands, 1315–1342] (Budapest: MTA Bölcsészettudományi Kutatóközpont Történettudományi Intézet, 2017).

358 On the noble judges of Sáros County, see István Kádas, "Sárosi 'reform' Miklós fia Miklós ispánsága idején (1374–1380)?," ['Reform' in Sáros County during the ispánság of Miklós son of Miklós (1374–1380)?] in *Micae mediaevales V. Fiatal történészek dolgozatai a középkori Magyarországról és Európáról* [Studies of young historians on medieval Hungary and Europe] (ELTE BTK Történelemtudományok Doktori Iskola Tanulmányok – Konferenciák, 9), eds. Laura Fábián et al. (Budapest: ELTE BTK Történelemtudományok Doktori Iskola, 2016), 127–144. For Tivadar see here: 135.

359 MNL OL DL 68 895. (For a summary, see AOklt, vol. 33. 255 no. 505), MNL OL DL 68 894. (For a summary, see: AOklt, vol. 33. 335–336 no. 684), and MNL OL DL 98 895 (for a summary, see: AOklt, vol. 33. 364 no. 745).

360 On similar cases, and the role of floods in that, see Kiss, *Floods and Long-Term.*

361 MNL OL DL 68 916 and MNL OL DL 68 917.

362 MNL OL DL 68 950. (For a summary, see ZsO, vol. 2/1. 641 no. 5091). For an edition of the document, see the appendix (1. no. 3).

363 *ITem quia plurimarum civitatum, villarum ac possessionum multorumque opidorum & prędiorum metć ac cursus metales fluviorum & aquarum fluvialium distinctione cinguntur pariter & clauduntur per huiusmodi autem fluvii inundationem atque vehementiam magnę sćpenumero terrarum, pratorum & silvarum particulć scinduntur & ad partem alterius vicinć civitatis aut oppidi vel possessionis segregantur & accedunt. Impetus enim & vehementia fluminis de suo solito cursu alveoque & meatu in alium novum meatum sćpenumero vergi dilabique solet. Unde nonnulli putaverunt & opinati sunt eiuscemodi terras, prata vel silvas quć scilicet ex ipsius aquę cursus meatusque seu alvei transmutatione in partem alterius vicinć civitatis oppidique vel possessionis ceciderint & segregatć fuerint ad eandem civitatem aut oppidum vel possessionem postea spectare pertinereque debere allegantes & astruentes metas suas per cursum meatumque & alveum fluminis dirigi. Quod tenendum non est. [§1] Nam hoc modo plurimć fraudes committi, & aquć seu flumina per occulta canalia ac fossata interdum satis exigua vel aggerum positionem & repleturam in eum quem quis mallet locum & alveum deflui terrćqua & silvć aut prata alterius facilime per hoc usurpari possent. [§2] SIC ITAque contraria opinio intelligendaest…* – Werbőczy, *Tripartitum*, Book 1. Ch. 87. 168–170.

364 Rezső Dell'Adami, *Az anyagi magyar magánjog codifikátiója. I. A nemzeti eredet problémája. Külön lenyomat a «Magyar Themis»-ből* [Codification of the Hungarian common law I. The problem of the national origin. Separatum from the Magyar Themis] (Budapest: Eggenberger, 1877).

365 Degré, *Magyar halászati jog a középkorban* and Tringli, "A magyar szokásjog."

366 See e.g., Johannes Engels, "Der verklagte Flußgott Mäander." *Historia: Zeitschrift für Alte Geschichte* 51, no. 2 (2002): 204 and Sax, "The Accretion."

367 *2. Svat so dat water afschevet deme lande, dat hevet die verloren des dat lant is. Brict it aver enen nien agang, dar mede ne verlüset he sines landes nicht. § 3. Svelk werder sik ok irhevet binnen enem vliete, svelkeme stade he nar is, to*

dem stade hort die werder; is he vormiddes, he hort to beiden staden. Dat selve dat die agang, of he verdroget. – *Sachsenspiegel Landrecht.* in *MGH Fontes iuris Germanici antiqui, Nova series Fontes iuris N.S.*, 1.1., ed. Karl August von Eckhardt (2nd ed. Göttingen: Musterschmidt, 1955), II. 56.

368 *Digesta* 41.1.7.2. For its English translation, see: *The Digest of Justinian*, eds. Alan Watson, Theodor Mommsen, and Paul Krueger (Philadelphia, PA: University of Pennsylvania Press, 1985).

369 MNL OL DF 249 510. (For a summary, see AOklt, vol. 22. 12–13 no. 5).

370 MNL OL DL 102 905 (September 4, 1339) For the edition of an incomplete version of the document (MNL OL DL 58 505): AO, vol. 3. 597–598 no. 394. For a summary, see AOklt, vol. 23. 250–251 no. 529. See in the context of floods Kiss, *Floods and Long-Term*, 245–246. The document is fully edited in the appendix (1. no. 1).

371 MNL OL DL 11 742 (January 2, 1340). For a summary, see AOklt, vol. 24. 9 no. 2.

372 MNL OL S 73. no. 102. (Pál Szattmári, *Borsod megyebeli Ónod m. város és határának szabályozás előtti térképe*, 1852 [The map of the town of Ónod and its borders before water regulations, 1852]). Available online: https://maps.hungaricana.hu/hu/MOLTerkeptar/11395/.

373 *Autem dominum Nicolaum episcopum Agriensem pro indebita earumdem particularum terrarum litigiosarum requisicione in emenda estimacionis earumdem, scilicet in predictis tredecim marcis contra eundem Stephanum filium Pauli commisimus sentencialiter aggravari.* MNL OL DL 3932. (September 14, 1347), Edited (with parts left out): AO, vol. 5, 118–120 no. 55; For a summary, see AOklt, vol. 31. 445–447 no. 862. See on the document, Kiss, *Floods and Long-Term*, 259–260.

374 MNL OL S 73. no. 102. Online: https://maps.hungaricana.hu/hu/MOLTerkeptar/11395/.

375 Tringli, "A magyar szokásjog," and Chapter 3.1 in the present work.

376 MNL OL DL 98 381 (for its summary, see ZsO, vol. 8. 251–252 no. 859). See Kiss, *Floods and Long-Term*, 293.

377 *Digest.* 43. 12., 13. 1-13., 15. For an English translation, see *The Digest of Justinian.*

378 E.g., MNL OL DL 52 420, 91 893 (For a summary, see ZsO, vol. 2, 186 no. 61), MNL OL DF 207 457 (for its summary, see ZsO, vol. 1, 15, no. 138; edited in: Gábor Dreska, "A pannonhalmi konvent hiteleshelyének működése a Zsigmond-korban: a pannonhalmi konvent hiteleshelyének 1387 és 1437 között készült kiadványai a Pannonhalmi Bencés Főapátsági Levéltárban," [The functioning of the place of authentication of the abbey of Pannonhalma in the Sigismundian period: the documents issued by the place of authentication of Pannonhalma between 1387 and 1437 in the Archive of the Benedictine Archabbey of Pannonhalma] *Levéltári Közlemények* 68 [1997]: 13), DL 53 871 (For a summary, see ZsO, vol. 5, 191 no. 571, edited: Tivadar Ortvay and Frigyes Pesty, *Oklevelek Temesvármegye és Temesvár város történetéhez. I. 1183–1430* [Documents to the history of Temes County and the town of Timişoara] [Temesvármegye és Temesvár város története, 4] [Bratislava: Eder István Könyvnyomdája, 1896], 511–512), 53 984. (For a summary, see ZsO, vol. 6, 380 no. 1377, edited: Ortvay and Pesty, *Oklevelek Temesvármegye és Temesvárváros történetéhez*, 543), DL 53 990 (ZsO, vol. 6. 405 no. 1499), 66

938, 16 498, 39 456 (for a summary, see Antal Fekete Nagy, "A Petróczy levéltár középkori oklevelei: első közlemény," [Medieval documents of the Petróczy Archive. Part I] *Levéltári közlemények* 8 [1930]: 261–262 no. 202), DL 36 393 (here: 87–88, no 2. For a summary, see Zsigmond Jakó, *A kolozsmonostori konvent jegyzőkönyvei, 1289–1556*, 2 vols [The protocol of the abbey of Kolozsmonostor] (Magyar Országos Levéltár kiadványai, II, Forráskiadványok, 17) Budapest: Magyar Országos Levéltár, 1990), vol. 1. 717–718 no. 2022), DL 17 372 (Ferenc Olexik, "Középkori levéltártörténeti adatok," [Medieval archival history data] *Levéltári Közlemények* 13 [1935]: 270–271 no. 9), DL 65 632, 83 932, 95 726, 106 744. (K. Németh, "Vizek és vízgazdálkodás, I.," 7 and idem, "Vizek és vízgazdálkodás, II.," 9. [erroneously dating the document to 1505]), DL 29 981, and 63 037.

379 E.g., MNL OL DL 30 554.

380 Anon., "Vízimalmok," in *Magyar néprajz*, vol. 3. 169–187. Available online: http://mek.niif.hu/02100/02152/html/03/23.html and Sándor Takáts, *Művelődéstörténeti tanulmányok a XVI–XVII. századból* [Studies in the intellectual history of the 16th–17th centuries], ed. Kálmán Benda (Budapest: Gondolat, 1961), 177, and 350.

381 MNL OL DL 92 239. (For a summary, see ZsO, vol. 2/1, 446 no. 3726). See the edition of the document in the appendix (1. no. 2).

382 OSZK K Cod. Lat. 197. 2r. Edited in: Petri Lossai, *Notationes et Delineationes 1498. Az Országos Széchényi Könyvtár Cod. Lat. 197 jelű kódexe* [Notationes et Delineationes 1498. The Cod. Lat. 197 of the Széchényi National Library], eds. Zoltán Poronyi, and Alajos Fleck (Pécs: Pécsi Geodéziai és Térképészeti Vállalat, 1969). For Lossai and his treatise, see: Levente Takács, "A magyar földmérés(i irodalom) kezdetei és az ókori hagyomány," [The beginnings of geodesy (geodetic literature) in Hungary and the Ancient tradition] in *Hunyadi Mátyás és kora* (Memoria Hungariae, 12 = Studia Folkloristica et Ethnographica, 75), eds. Attila Bárány et al. (Debrecen: MTA-DE Lendület „Magyarország a középkori Európában" kutatócsoport, 2019), 24–33, and Bálint Lakatos, "Lossai Péter," [Peter Lossai] in *Magyar művelődéstörténeti lexikon* [Lexicon of cultural history in Hungary], 14 vols, ed. Péter Kőszeghy (Budapest: Balassi, 2003–2014), vol. 7. 84–85. On the codex, see: Anna Boreczky, "Mikor és hol készült Lossai Péter kódexe?," [When and where was Peter Lossai's codex produced?] *Magyar Könyvszemle* 113 (1997): 249–265.

383 For legal practice that was not in accordance with the *Tripartitum*, see e.g., Lajos Szádeczky, *Székely oklevéltár V. kötet (1296–1603.)* [Székely cartulary, vol. 5] (Cluj: Ajtai K. Albert Könyvnyomdája, 1896), 61–63 no. 935 and 63–65 no. 936 (Cf. Degré, *Magyar halászati jog*, 138 and Tringli, "A magyar szokásjog," 262. [1547]). See also Péter Tóth, *Vas vármegye közgyűlési jegyzőkönyveinek regesztái I. 1595–1600* [Summaries of the minutes of the gatherings of Vas County, I] (Vas megyei levéltári füzetek, 2) (Szombathely: Vas Megyei Levéltár, [no date]), 243 no. 719 (November 23, 1600)

384 *In facto territoriorum, per exundationem; vel sensim factam aut fiendam alluviem aquarum, ab uno territorio avulsorum, et alteri adjectorum; observetur tit. 87. partis 1. §. 1608–1657. évi törvényczikkek / Corpus Juris Hungarici 1608–1657*, ed. Dezső Márkus (Budapest: Franklin, 1900), 632.

385 Skorka, "On Two Sides of the Border," and Péterfi, "Debates."

386 *S egiebirantis az Dobrai főldeket azide valo hatarhoz szakasztotta az raba visza de attal ugian nem idegenithetik el Dobrától.* (The Rába did attach the lands in the borders of Dobra to our borders, but it does not mean that they could be alienated from Dobra). The letter of Magdolna Batthyány to Ádám Batthyány, MNL OL P 1314 no. 5104. (December 30, 1655).

387 Cf. the recent works of Péterfi, "Debates," and Skorka, "On Two Sides of the Border." See also: Károly Vörös, "Ein Grenzkonflikt zwischen Steiermark und Ungarn im Jahre 1742," *Zeitschrift des Historischen Vereines für Steiermark* 82 (1991): 195–197.

388 E.g., OSZK TK no. 1397. (del And. Locatelli, *Planum incipit ad terrena Petánc, et Lafeld, de sinitque ad i[nclytum] co[mi]t[a]tum Zaladiensem seu terenum Deklesin, linea sic dicta Theresiana hinc inde interrupta* [Vienna: (1790?)]), MNL OL S 11 no. 635/1. (*Planum VII. Mappa exhibens limites inter Cottum. Castriferrei et Circulum Marburgensem seu a Lapide 34 usque 55 juxta Basim Linae Theresianae* [1829]).

389 Cf. Gunter Wesener, *Einflüsse und Geltung des römisch-gemeinen Rechts in den altösterreichischen Ländern in der Neuzeit (16. bis 18. Jahrhundert)* (Vienna and Cologne: Böhlau, 1989).

390 E.g., Kyra Lyublyanovics, *New Home, New Herds: Cuman Integration and Animal Husbandry in Medieval Hungary from an Archaeozoological Perspective* (Archaeolingua. Central European Archaeological Heritage series, 10) (Oxford: Archaeopress, 2017).

391 Nonetheless, related problems certainly did appear in mining areas in different parts of Central Europe. See: Kiss, *Floods and Long-Term*, 104–114.

392 Hans-Rudolf Bork et al., *Landschaftsentwicklung in Mitteleuropa: Wirkungen des Menschen auf Landschaften* (Gotha and Stuttgart: Klett-Perthes, 1998), 159–214.

393 On the population of the mountain areas, see: Körmendy, *Melioratio terrae*.

394 Cf. Tamás Vajda, "Meghatározható-e a Dél-Alföld késő középkori lélekszáma?," [Is it possible to give an estimate of the late medieval population of the Southern Great Hungarian Plain] in *Műhelyszemináriumi dolgozatok I* [Research seminar papers, 1], eds. Éva Révész, and Szilvia Kovács (Szeged: SZTE BTK Történelemtudományi Doktori Iskola, 2013), 145–180, Beatrix F. Romhányi, "Szempontok a Kárpát-medence térszervezésének változásaihoz (5–14. század)," [Considerations on the changes of the spatial organization of the Carpathian Basin (5th–14th centuries)] in *Hatalmi központok az Avar Kaganátusban – Power Centres of the Avar Khaganate*, eds. Csilla Balogh, József Szentpéteri, and Erika Wicker (Kecskemét: Katona József Múzeum – MTA BTK MŐT, 2019), 399–420, and Beatrix F. Romhányi, "A középkori magyar plébániák és a 14. századi pápai tizedjegyzék," [The medieval parishes in Hungary and the 14th-century papal tithe records] *Történelmi Szemle* 61 (2019): 339–360.

395 MNL OL P 507 Nádasdy család nádasdladányi levéltára [Nádasladány archive of the Nádasdy family] Levelezés [Letters] II. 341. 1770.

4

THE MEN BEHIND THE WHEEL... BUILDING ON WATER IN THE MIDDLE AGES

As discussed in the introductory chapter of this book, water regulation works can be traced back to the establishment of the Hungarian state in Central Europe. In this context, Károly Takács's research, which highlighted extensive water-regulation projects during the Árpádian period, is of great importance. If his assertions are accepted, ditch systems, water reservoirs, and fisheries were constructed and maintained, at the latest, from the eleventh or twelfth centuries in various lowland areas of the Carpathian Basin.[1] He suggested that, although in different forms, dug channel systems existed not only in the parts of the Little Hungarian Plain that he studied in detail but also in the Transtisza region, the lowland areas east of the Tisza. While examples from a much later period indicate the existence of channel systems that differed in nature from the dug channels described by Takács, in many other parts of the Carpathian Basin, from the eleventh century onward, as discussed in Chapter 2, water mills also began to operate alongside these channel systems.[2] However, the planning and operation of these channels, mills, and other elements of water-related infrastructure have not yet been thoroughly addressed. In this chapter, primarily utilizing late medieval and, in some cases, Early Modern source material, the individuals involved in water construction are discussed. While scattered references mention individuals who participated in similar works during the early centuries of Hungarian statehood, it is only from the fourteenth century onward that a relatively large number of documents detail professions directly related to water construction. The main questions explored in the following pages are which specific knowledge was required for these professions and which particular skills these individuals possessed. How did they attempt to resolve or prevent disputes that could potentially arise from their interventions in waterways?

DOI: 10.4324/9781003596905-4

Were there planning and construction steps intended to mediate potential problems?

In doing so, representatives of two professions are discussed primarily. First, millers, the operators of water mills, are in the focus, as they had several duties beyond supervising the grinding of grains and other raw materials. Secondly, it examines a previously underrepresented group: ditch and pond diggers (and pond masters). The roles of practitioners in these professions are initially highlighted by a specific source: a section of the town of Sopron's account book from the year 1540. Then, other evidence, including legal sources, account books, terriers, and letters, is used to illuminate the various activities in which they were involved while practicing their professions. All these sources contribute to our understanding of the professions and the planning of water use in different contexts. Furthermore, the sources emphasize the high level of integration among the various professions in water construction during the medieval and Early Modern periods.

4.1 The Renovation of the Moat of Sopron in 1540: A Unique Source

Water management and the water mills in the town of Sopron are discussed in Chapter 3.2.3.2. The goal here is not to examine the town's mills in general, but rather to focus on their connection to the reconstruction of the urban moat. The aforementioned Foregate Mill (*Előkapu malom*) sheds light on who was involved in, and how, a sixteenth-century water construction project was carried out. Sopron's moat was covered in Chapter 3.2.3.2, which notes that it was built from the 1340s onward; two centuries later, according to the Sopron account book, it required significant renovation. The sluice at the Foregate may have been in place since the fourteenth century, as it was likely the key infrastructural element in maintaining a consistent water level and current in the moat. The ongoing flow of water in the moat, as noted earlier, played a crucial role in preventing silting and in operating the mill connected to the gate.

This latter building was probably not operational by the 1540s, and the rebuilding of part of the moat was a prerequisite for the use of the lost water mill. The work was carried out in the 1540s, when, according to the Sopron account book for that year, it was one of the town's most significant expenses, certainly the largest among its non-regular expenditures.[3] The source provides insights into labor organization on the one hand, and the involvement of experts in water construction projects on the other. First, the building work itself will be detailed, followed by a brief discussion of work organization. In this subchapter, the professionals involved in the various phases of work are presented. Additionally, the source offers insights into the wages of various professionals, highlighting their prestige. This is followed by an analysis of sources from late medieval and early modern Hungary concerning practitioners of the aforementioned professions: millers, ditch diggers, and pond masters.

As noted previously, the rebuilding work at the Foregate took place in 1540, when the related payments were recorded in Sopron's account book. A ten-folio-long list of expenditures is preserved in the account book,[4] and, for some reason, immediately afterward, two folios recorded expenses for the moat for the previous year, 1539.[5] The latter, based on the various items, likely documents a regular annual task performed to keep the moat clean and maintain a consistent flow of water. This work was supervised by a pond master (*teichtmaister*) named Sigismund. The account on these folios recorded the payments he and his subordinates received. The total sum is not significant, and nothing indicates a connection between the major rebuilding and this minor work, aside from the fact that they were recorded consecutively in the account book. The total cost of the 1540 work was twelve times the expenditures of 1539. Based on this and the individuals listed, the work performed in 1539 was intended for routine maintenance rather than rebuilding.

The major rebuilding work preserved on ten folios of the account, however, reveals numerous details about the projects, their organization, and, in particular, the miller's role in the construction process. The document lists expenses in three parts, organized by the main stages of the work. The records detail the reconstruction of the moat. Work commenced at the end of February in 1540. During this phase, neither millers nor pond masters were involved, as basic groundwork had to be completed at that time. Instead, mainly stonecutters and their workers were actively engaged. The leading stonecutter, a certain Bartholomew, was first paid on February 21st (Saturday),[6] indicating that he began work at the beginning of that week, that is, February 15th. Following the stonecutter, the pond digger was the next master involved in the work. A man named Martin (*Martl*) Pehem was referred to in the German text as *deichtknecht*. His task was to dig and remove earth from the *zwinger*.[7] His name, *Pehem*, is particularly interesting as it suggests his Czech (Bohemian) origin. This may be significant, as discussed in the book's introduction, since fishponds were notably important in the Czech lands during the late medieval economy.[8] The building project at Sopron began relatively early in the year, already in late February. Due to the occasionally frozen soil, these works were typically undertaken in the later months of the year. However, as indicated by numerous sources from Central Europe, the weather was exceptionally mild in February that year.[9]

The stonecutter and his assistant continued their work in the following weeks. For one day, Bartholomew, the stonecutter, was not preparing the larger blocks for the moat bed but was instead breaking smaller stones, which were likely also required in significant quantities during the rebuilding.

Although the paleography of the word in the account book is not entirely clear, it seems that the cutter supervised the breaking of tombstones from the Jewish cemetery (*iud[en]hof* [?]) of the town.[10] This was probably the easiest and cheapest way to obtain ballast at the time, and since the Jewish community was forced to leave the city following the Battle of Mohács (1526), there was no objection to the act.[11]

The cemetery in question stood next to Saint Michael's Gate in the suburb, and the above line in the account book is not the only evidence of its use as a quarry. The earliest source indicates this dates back to 1539. At that time, as shown by Michael Töltl's account book, stone was transported in four trips from the cemetery to the construction works of the nearby Saint Michael's Church between the 20th and 30th of March.[12] In light of this, it is very likely that a year later, the remaining stones from the cemetery were used as ballast for the rebuilding of the section of the moat around the Foregate. The breaking of the stones for ballast continued at the beginning of March, as did the digging of the ditch. During these weeks, the account records payments to the two masters and their assistants in alternating order.

Payment for a new master appears during the week of the 13th of March. From that time on, not only was the digging done in the moat, but work to incorporate the stones prepared by the cutters began. The stonemason (*maurer*) started construction of the mill race for the water mill at the Foregate.[13] While most similar mill races were made of timber, it is certainly worth noting that a more expensive and durable material was chosen here. A few days later, on the 19th of March, a new master (or rather assistant, a *merterknecht*) appears in the account, whose duty was to provide the mortar for the building work.[14] He likely worked under the direction of the stonemason on incorporating the major blocks into the bed of the mill race by the moat. Although he is referred to as an assistant (*knecht*), sources from the fifteenth century onward continue to refer to individuals working specifically on producing mortar for construction projects.[15]

The masons and diggers worked continuously in March and April. Since the beginning of May, a new person, the town's master builder (*statmaister*), has been involved in the works.[16] This position first appeared in Sopron in the fifteenth century. They are mentioned for the first time in 1427, and then again in 1432 in the accounts of the mayor, Thomas Schadendorfer, as well as in the records of the chamberlain of the town, Ulrich Abenstorffer, in 1466.[17] Unlike most people involved in the building work, he received a daily payment in 1540, which reflects the prestige of his position. In addition to the daily payment, he and his assistants sometimes received gratuity and a bathing fee.[18] The parts of the work connected to the moat itself were completed in mid-June, as this is the latest period for which expenses are recorded in the accounts. While the stonecutters and masons were probably not involved in planning the water construction but were instead instructed to carry out well-defined tasks, the master builder played a key role in identifying the consequences and interests of the work.

Alongside the rebuilding of the moat and the construction of the mill race, work also began on reconstructing the mill building itself. The miller, Urban Kümpf, supervised this phase of the project.[19] The initial steps of the reconstruction started in February, with the construction of the moat. In the first phase, aside from unspecified payoffs, there are recorded payments for transporting timber to the mill. One of these payments highlights that the miller's first assistant was not from Sopron but from Ágfalva (Agendorf), one of the tenant peasant villages for which Sopron was the landlord.[20] After February, timber continued to be transported to the construction site. A well-known burgher, Francis Siebenbürger, received payment because his forest was used to harvest the timber needed.[21] Siebenbürger came from one of Sopron's most prominent families and was recognized for his trade connections that extended well beyond Hungary's borders; he also served as the town's mayor.[22]

Fewer details are available about the specific work performed at the mill. However, the miller's weekly payments recorded in the accounts indicate ongoing construction work, underscoring his crucial role in planning and supervision. He likely had to repeatedly check whether the new bed and the race leading to the mill met the mill's and the moat's requirements. In addition to the miller, two assistants (*muelknecht*) were involved in the works during spring, with the first payment to them recorded at Pentecost (May 27).[23]

In the third part of the account, the costs of materials and shipping are listed. In addition to detailing various expenses, the accounts also reference the use of these materials. However, this section of the source includes significant additions regarding the locations of the different works undertaken by the moat. Based on one of the expenses, work was also done at the fishpond or fish keepers. This site was in the southwestern part of the town and, as discussed in Chapter 3.2.3.2, was connected to the water system of the moat.[24] This suggests that the works extended beyond the immediate vicinity of the Foregate, located at the northern tip of the moat. It is possible that the fish keeper and the channel linking the small pond to the moat were covered over during that time.

This would not have been unusual in medieval Hungary, as similar constructions were noted at the above-discussed Felhévíz, where multiple races leading to mills were likely covered over. For the works performed in 1540 at Sopron, fir and other timber species were used. It is probable that in some areas, water flowed through a timber joint, which would be common in such mills. It is somewhat surprising that, as previously mentioned, parts of the race might have been built of stone. The account in this third section lists numerous additional metal and wooden materials; however, their usage is not discussed in detail. Some of the timber was undoubtedly used to cover the mill building. The acquisition of the millstone is a noteworthy aspect of the account and somewhat indicative of the roles and significance of millers.[25] The millstone, a crucial part of a mill, was procured from as far away as Vienna.[26] This may appear somewhat surprising compared to similar acquisitions. There is relatively limited data on the Hungarian market for millstones. Still, by the seventeenth century, and likely earlier, Győr had become an important center for the trade of this item. In most cases, the millstones for the mills in neighboring Vas County were purchased there, and during the late medieval and Early Modern periods, another nearby Hungarian town, Pápa, is also mentioned in some instances as a significant center of millstone trade.[27] In 1540, despite customs (notably the *tricesima*, or thirtieth, to be paid at the border between Hungary and the Austrian duchies), they appear to have opted for Vienna to obtain the millstone.[28] This choice may be explained by the occasional poor communication toward the Fertő and Hanság areas, where Győr is located; however, this was typically due to wet weather. As noted earlier, 1540 was one of the driest and warmest years of the last millennium.[29]

Even if the nature of the source type prevents the account from providing insight into many aspects of the building work, the materials and the masters mentioned can help us understand how and by whom water construction projects were carried out. Before discussing in detail the role of the different masters in the building work, the scope of the work is examined.

4.1.1 The Expenditures on the Work

The account discussed here likely represents a comprehensive rebuilding effort, from the removal of the initial land layer to the completion of renovations to the mill, its race, and a section of the moat at Sopron. Since expenses are recorded in the account with meticulous detail, it is possible to determine the scope of the building project and calculate the proportions of various costs, particularly those related to wages and building materials. This facilitates an understanding of the importance of managing water resources effectively in this specific settlement. While this may not be directly linked to disputes, the costs incurred for one relatively small (likely one-wheeled) mill and its surroundings could shed light on the efforts of different parties to secure their rights to use water resources.

During the rebuilding, under the threat of the Ottoman Empire, military considerations, such as the proper functioning of the moat, were likely crucial. However, a significant portion of the expenditures was directly linked to the management of the town's water resources. First, it is worth noting the total cost of the work, which amounted to 297 pound denars, or 170 golden florins. In the absence of comparable regional data, it is not easy to assess the scale of the work. What can be done, however, is to compare the costs of the work with Sopron's budget during the years surrounding the rebuilding efforts. Given that the town's account books from the 1530s and 1540s have survived, we have a clear source for this purpose. Very little research has been conducted on the account books from this period. Except for an article in the 1960s that analyzed the town's account book from 1535/1536, there has been no study of the account books for the early sixteenth century.[30] In the financial year of 1535/1536, the town's total income was approximately 2992 pound denars, while expenses totaled 2979 pound denars. To achieve a more comprehensive understanding, the accounts from 1530 to 1543 (*see* Figure 4.1) have been examined for this book. Throughout this period, income and expenses fluctuated between 2000 and 3000 pound denars, with significant variability over the fifteen years studied. The data, of course, must be understood within the context of political changes. The turbulent military and political circumstances following the Battle of Mohács, along with the conflicts between the two contenders for the Hungarian throne, Ferdinand I and John I Szapolyai, significantly affected the town and its economy. In the financial year of 1539/1540, total expenditure rose to 2697 pound denars. The approximately 300-pound denar cost of rebuilding Sopron's moat represents precisely eleven percent of all the expenditure for that year. This is undoubtedly a substantial amount and, as noted, the town's most significant non-regular expense in that year.

It may be worth briefly noting the proportions of material costs and workers' wages. Generally, one can assume that the latter comprised the majority of the building costs.[31] The construction work at Sopron was no

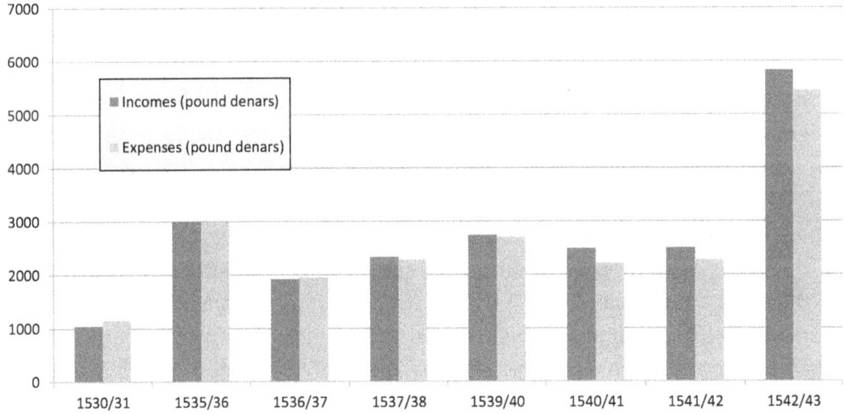

FIGURE 4.1 The incomes and expenses of the town of Sopron between 1530 and 1543 in light of its account books.

Sources: MNL GYMSML SL IV. 1009. a. 1530/1531 117. p, 1535/1536 112. p, 1536/1537 143. p, 1537/1538 120. p, 1539/1540 50, 132. p, 1540/1541 46, 121. p, 1541/1542 38, 90. p, 1542/1543 55, 141. p.

exception; this is unsurprising, given the nature of the work, as significant groundwork occurred in the initial phase without any material installation. Wages accounted for eighty two percent of costs, with the remaining eighteen percent allocated to building materials and shipping. The cost proportion is somewhat more balanced in the case of house construction, which is unsurprising.[32]

The reconstruction of the moat required nearly 1500 working days. There were weeks when about a dozen workers were engaged in the digging and assembling different materials in the bed of the moat and the mill race. When the mill's construction and groundworks were occurring simultaneously, this number undoubtedly exceeded fifteen workers per day. In light of this, a relatively large-scale building project unfolds. The following subchapter discusses labor organization and the involvement of different masters.

4.1.2 The Labor Organization

One of the problems closely connected to the project's costs is the organization of wage laborers. A brief overview of how the different workers were organized in their tasks and compensated is provided here. By doing so, one may gain insight into the status of the various laborers. Jenő Szűcs, one of the most innovative Hungarian social and economic historians of the twentieth century, conducted a detailed study of the organization of work during the major rebuilding and extension of the Bratislava Castle in 1434. Based on six months of surviving accounts, he highlighted the importance of studying professions that lacked guild organization in medieval Hungary. The building industry, as well as millers, undoubtedly belonged to this group.[33] Despite the works ordered by King Sigismund of Luxembourg at Bratislava, one of his principal seats in Hungary, close to the Holy Roman Empire, the works there varied in scale from those at Sopron; this is by far the best comparative material.[34] Even if the scale of the work at Sopron was more limited, many different professions were referenced in the accounts. The accounts not only help identify the various wages professionals received for their labor but also are essential for understanding how the work was organized under the direction of different building masters. In most cases, the accounts from 1540 describe hired hands who were paid daily under a master, who typically received weekly payments. However, for the masters, as previously noted and linked to the master builder of Sopron, one-time special payments were also made. The various individuals involved had significantly different wages. It is therefore possible to comprehend the different levels of status and, likely, the requisite expertise at various stages of the construction work.

At the works analyzed by Szűcs, the building master Conrad of Erling earned seven times the wage of the hired hands and twice as much as the other masters (carpenter, smith).[35] In the case of Sopron, the difference is also significant; the miller's wage was five times that of the hired hands. The lowest daily wage was that of women, at sixteen denars, a point to be revisited hereinafter, while the highest was the miller's wage (77 denars). Other professions fell within the 30–44 denar range, followed by assistants to various masters, and finally hired hands and women (*see* Table 4.1).

The daily wage in similar accounts largely depended on the season. In medieval Hungary, wages were categorized into two seasons: winter and summer. The account registers indicate that work began on February 22, marking the start of the summer season for wage payments. This period had significantly more daylight and therefore more working hours than the winter half-year.[36] Differences in compensation for the same types of tasks can be attributed to experience and proficiency.[37]

The presence of female laborers in an account such as this warrants attention. While Western Europe has relatively abundant data on women's participation in building works, there is an almost complete lack of comparable data from the medieval Kingdom of Hungary.[38] According to the account book, their wages were significantly lower, at sixty percent of what was paid to men. This percentage closely resembles that observed for female workers in English towns such as Durham, Newcastle, Lincoln, and York.[39] There is one notable exception: when building the castle of Builth in Wales in 1278, women were paid only slightly less—approximately fifteen percent—than men.[40] In England, it was not until the seventeenth century that women received approximately equal wages for a day's work in the building industry.

As mentioned, due to the lack of data from Hungary, there is no basis for offering similar comparative material. However, it is important to note that in Sopron, female participation in similar works was likely common. They appear not only in this account but also in minor rebuilding projects in 1541, where the account book again confirms their presence as hired hands.[41]

TABLE 4.1 Wages according to the 1539/1540 account book of Sopron

Labor	*Wage calculated for a day (in denars)*
Miller (*mülner*)	77.14
Miller assistant (*mühlknecht*)	40 and 44
Pond master (*deichtmaister*)	40
Stonemason (*maurer*)	40
Stonecutter (*steinmetz*)	40 (1539)
Pond/ditch master assistant (*deichtknecht*; *maisterknecht*)	28, 30, 32 (1539)
Mortar maker assistant (*merterknecht*)	4 and 28
Stone breaker (*steinbrecher*)	24
Hired hand (*raicher*)	16 and 24
Female worker (*weib*)	16

Women were paid daily, whereas the miller received a fixed weekly wage. In every case, the masters of the crafts received the highest wages, followed by their assistants, who earned two to four denars less daily. In addition to daily and weekly payments, single payments are associated with distinct work phases. The largest of these sums is ten pound denars (2400 denars), which the miller received for installing the millstone after traveling to Vienna for this purpose. This clearly indicates the importance of this work stage and the crucial role of millers in the construction.

As this brief analysis shows, there is a significant difference in the prestige of various professions. Some professions are held in higher esteem than others, as evidenced by the salaries of different masters. The miller's high salary reflects their crucial role in water construction projects. In the following sections, these roles will be highlighted using various types of written evidence. In addition to the miller, the pond master also earned higher wages than other masters. Their roles are examined in the same way as those of millers. While millers and their duties have been emphasized in many studies, the tasks of pond masters and pond diggers have yet to be studied in depth. Given their high wages, their roles were likely also crucial in the planning and execution of water construction.

4.2 Millers

Medieval and Early Modern millers were likely just as much carpenters by modern standards as they were leaders in grinding grain and other products. It is no surprise that the construction of the mill at Sopron in 1540 was done under the guidance of the miller, Urban Kümpf.[42] This is further underscored by the testimony of Menocchio, one of the most famous millers of all time. The miller of Friuli was immortalized in Carlo Ginzburg's classic work, *The Cheese and the Worms*. When Menocchio was asked about his profession during his inquisition trial, he responded: "miller, carpenter, sawyer, mason, and other things..."[43] It appears that representatives of the milling and carpentry professions often complemented one another, although this sometimes led to tension among their practitioners. Perhaps it is no coincidence that the otherwise negatively portrayed, drunken miller tells the story of the repeatedly cuckolded carpenter in Chaucer's *Canterbury Tales*.[44] The close association of the two professions is well illustrated by the fact that in medieval towns, the living spaces of millers and carpenters often coincided. For example, as demonstrated in Bratislava, both carpenters and millers lived in the suburbs by the Danube.[45]

Even if sources sometimes mention tensions between millers and carpenters, one should not assume that the two were distinctly separated in medieval Hungary.[46] Legal documents reveal the merging of the two Latin terms—*molendinator* and *carpentarius*—in several instances, and millers are even referred to as "Hungarian carpenters."[47] Early Modern terriers (*urbaria*) from Hungary, as well as the income list of the Gyula manor from 1525, reference millers whose obligations to their landlord included carpentry work.[48] In some cases, millers were assigned specific carpentry tasks to benefit their community. For example, in Sopron, according to data from 1595, millers oversaw the construction of the gallows.[49] The less distinct expertise of millers and carpenters is further emphasized by the fact that when the first miller guilds were established in Hungary during the sixteenth century, some included carpenters among their members.

Although mills were familiar places for socializing during the Middle Ages and the Early Modern period, millers were not the most popular members of society. The seigniorial ban on milling imposed high costs on tenant peasants,[50] and several sources suggest that millers intentionally harmed the interests of those who came to their mills to grind grain.[51] The unpopularity of millers is primarily linked to their most apparent role: overseeing the milling process in their mills.

As previously implied in earlier sections of this chapter, this only constituted a small part of the millers' work. The actual daily tasks at the mills were mainly performed by their assistants. However, beyond this rather obvious role, at least three additional functions, some of which have already been highlighted in previous chapters, can be linked to millers. First, as shown in Sopron's account books, was their role in planning the construction and overseeing the mill building. This task, unlike the others discussed here, was not exclusive to millers. In certain instances, carpenters or masons could also be involved as builders or rebuilders of mills.[52] Nevertheless, in most cases, millers managed the mill buildings, particularly their maintenance. Referring to the 1540 account of Sopron, some phases necessitated the expertise of the miller himself. The most crucial of these was the acquisition and supervision of the millstone installation. These tasks were almost always carried out by the miller responsible for the particular building.

As part of an earlier book by the present author, a considerable amount of data on the mills in the Early Modern manor of Körmend in Western Hungary, then owned by the aforementioned Batthyány family, was collected.[53] The rebuilding efforts of the two largest water mills in the manorial complex, those of Körmend and the neighboring Szecsőd (part of present-day Molnaszecsőd), demonstrate the millers' roles in maintenance work. These mills, like most others, were frequently damaged by floods, particularly ice floods. In cases of minor flood damage, it was always the miller's duty to oversee the rebuilding. However, when the losses were greater, others were also involved in the efforts. For instance, the head of the estate complex's administration (*provisor*) of Körmend, Stephen Nemsem, once traveled to the nearby Egervár, which also belonged to the same landlords at that time, to observe the construction work on the underwater parts of the water mill there. Thus, he appears not to have overseen the work but rather to have been sent by his landlord to learn and apply the knowledge acquired at the mills in Körmend and Szecsőd.[54]

The most frequently damaged parts of mill complexes were their dams. If both the logs that formed the upper part of a dam and its foundations were destroyed, the *provisores*, along with the millers, sometimes sought the assistance of carpenters. This indicates that although the millers were the primary experts in constructing a mill, the carpenters had greater expertise in reconstructing a dam from its foundations. This is well illustrated by the letters sent by the aforementioned Stephen Nemsem. In this, the *provisor* informed the landlord, Adam Batthyány, that he had requested a carpenter from Batthyány's further estate, Strem, to rebuild the mill's foundation. The German carpenter prepared the necessary timber and then returned home. However, the miller was unsure which materials the carpenter intended for specific functions.[55] The letter notes that the carpenter was German and the miller was Hungarian, suggesting that distinct building traditions may have shaped their respective expertise. Nevertheless, despite the few examples cited in the sources, it was likely the millers who primarily served as building masters for the mills they managed.

Beginning in the thirteenth century, sources indicate a different role with which millers were associated. In water-related legal disputes, they served as experts to assess whether water-use norms were followed.[56] It has already been noted regarding the lawsuits at Felhévíz that millers served as witnesses and experts; however, the earliest evidence of their role as advisors dates to much earlier than the late medieval period. In 1282, two tenant peasants in the village of Patacs (within the borders of present-day Kamenný Most) decided to construct new mills near an existing mill operated by a certain Simon. Simon filed a complaint regarding the construction of the new mills, alleging that they would hinder the operation of his mill. The cathedral chapter of Esztergom, the landlord of the village, sent the dean of the chapter, Serafin, to investigate the matter.[57] With the involvement of millers, Serafin needed to determine whether Simon's claim was valid. Serafin, along with the "expert millers and miller masters" (*cum peritis molendinariis et molendinorum magistris*), likely influenced by the fact that the chapter would benefit financially from the new mills (probably indeed one mill with two wheels), did not support Simon's position. Ultimately, the chapter allowed the construction of the mills on the condition that the dam of the new mill would not alter the average water level of the River Hron.[58]

Several documents attest to millers' expertise in water-related matters during the Angevin period. To mention but one, in 1325, a lawsuit arose between the Austin friars and the Paulines in Újhely over the mills of the two institutions on the River Ronyva.[59] Since the church institutions were under the authority of the Diocese of Eger, Bishop Csanád Telegdi (later archbishop of Esztergom) was entrusted with investigating the case. He appointed experts in milling (*in arte molendinaria peritos et expertos*) to assist with the decision.[60]

Numerous other cases from the fifteenth and sixteenth centuries also survive, among which one previously studied in scholarship is worth mentioning.[61] In the mid-fifteenth century, a lengthy lawsuit unfolded between a lesser noble, Nicholas of Szalonta, and the Pauline monks of nearby Lád. According to the monks, Nicholas constructed the dam of his new mill at Szalonta so high that it prevented their downstream mill at Keresztúr from operating. The Paulines claimed that not only did the mill obstruct the functioning of the mill at Keresztúr, but it also caused water backup that flooded their lands. Half a dozen documents concerning the case survive, most of which vividly illustrate the role of millers in such investigations.[62] The noted close association between millers and carpenters is well reflected in the Ács (*Alch*) forename of one of the appointed millers, John of Bőcs, which is the Hungarian term for carpenter.[63]

This expertise in milling later took on new forms and eventually gave rise to a new role documented in Early Modern and Modern sources: the mill judge (*molendinae dominus*).[64] In his book on the legal literacy of market towns in late medieval Hungary, Bálint Lakatos demonstrated the existence of such experts as early as the turn of the fifteenth century. He highlighted a document from Vințu de Jos, dated 1510, in which the local Hungarian townspeople demanded equality in the settlement's self-governance, which, according to the document, would have included the election of a second mill judge.[65] Their responsibilities, based on Early Modern parallels, may have included overseeing mills within their settlement's boundaries and assisting in resolving disputes concerning water use for milling.

In some cases, millers were entrusted with rather specific roles. From the mid-sixteenth century onward, the Rába became one of the most important natural obstacles between the Ottoman Empire and the Kingdom of Hungary, with Vienna in its immediate hinterland. Sources indicate that millers along the River Rába in Western Hungary played an important role in organizing the frontier defense system on the Hungarian side of the Ottoman–Hungarian frontier.[66] Because millers' tasks required considerable expertise in regulating the water level, they knew how to impound the river; consequently, their expertise was inestimable.[67]

The first millers' guilds in Hungary emerged only in the Early Modern period; therefore, it is likely that, prior to this and probably even afterward, the primary place where millers learned their skills was the mills themselves.[68] Before being entrusted with a mill, millers likely spent many years as assistants to experienced millers. A source from Körmend illustrates how long it might take for someone to become a miller. In a letter dated 31 March 1650, the above-mentioned Stephen Nemsem wrote a complaint to his landlord, Adam

Batthyány, expressing his grave doubts about the new miller who had been sent to Körmend:

> My lord, if I may write to you about the German miller whom Marko Simoncsics sent to the local mill, I wonder whether he knows milling or not. My lord, this mill is not suitable for him to be a miller at, as the Rába is not like the Lafnitz [a minor left side tributary of the Rába]. Here, a miller trained at a mill by the Rába for 30 or 35 years is needed, as he knows how to catch water. I showed him the damage the Rába caused here, and when I asked how the water should be regulated, he simply suggested filling it [probably the dam] again. Even if we solidify it once more, we cannot back up the water; it just flows beneath it [the structure of the dam]. He is not cut out to be a miller here, as he lacks any mastery, and even if he did, he doesn't speak Hungarian, so the people wouldn't be able to understand him.[69]

The letter, aside from being one of the most vivid sources on the problem of milling at the Körmend manor, also introduces aspects of millers' position and milling that are rarely mentioned. The source confirms that milling was anything but routine, especially at a mill located by a major, strategically significant river such as the Rába. According to the source, unlike Adam Batthyány's leading official, Marc Simoncsics, the *provisor* of Körmend, Nemsem believed that river milling was a complex task. This was likely partly due to the Rába experiencing numerous disastrous floods in the previous period.[70] Learning to operate a mill, like many other professions, likely began at a very young age. Even if thirty or thirty-five years of training may seem somewhat exaggerated, it likely took decades for someone to progress from being an assistant to becoming the head of a mill. This has similarly been demonstrated in several professions, including the construction industry.[71] Milling, due to the specific characteristics of different rivers and landscapes, as mentioned in the letter, was probably one of the most complex professions that was not organized within a guild system in medieval Hungary, yet remained a highly esteemed profession, as evidenced by the wages offered by the town of Sopron.

4.3 Ditch and Pond Diggers and Pond Masters

Even if large-scale regulations that modified entire water systems in Europe were typical undertakings of modern times, as shown in the previous chapters, water was backed up, new channels were dug, and in several other ways, water was manipulated to provide societies with water, water power, irrigation water, and so on. The planning of these water construction projects, however, remains almost invisible, particularly in the early medieval period, despite sources frequently mentioning ditch and pond digging in Europe. Some of these written records suggest that, from that time onward, such construction efforts were regarded as requiring specialized expertise.[72] Although only once, even the *Domesday Book* refers to a ditch digger, a certain Herberd *fossator*.[73] In England, various sources mention individuals who served as masters of ditch digging, responsible for planning major water-regulation projects. To mention but one, in 1277, King Edward I decided to surround the castle of Rhuddlan in Wales with seawater, both to enhance its protection and to provide the castle with access to the sea. For a time, this project was supervised by a certain William of Boston, who personally hired diggers to complete the work. The scale of the project is evident in the fact that, on average, 66 diggers were involved, working six days a week for at least three years. This totals around 20,000 working days. Although similar-sized projects from the early medieval period onward exist, in most cases, the sources remain silent about the organization of work and the individuals who planned and oversaw them.[74] Based on the complexity of this work, research indicates that by the thirteenth century, leading comparable projects required specific expertise and had become a distinct profession.[75]

Of the many possible examples, two are mentioned here to demonstrate that the digging of ditches and ponds required careful planning and substantial knowledge. In the thirteenth century, the town of Milan undertook a major water-regulation and canalization project. Access to Milan by water was virtually impossible, which hindered the town's development. Therefore, the town council decided to connect the city to the Po water system via a navigable canal. The best approach was to create a link to the River Ticino, which flowed slightly west of Milan from the north toward the Po. The city commissioned the most renowned architect of the time, Giacomo Arribotti, to plan and execute what is known as the Naviglio Grande.[76] In this instance, the task involved much more than merely supervising the diggers. It demanded specific architectural planning, engineering expertise, and knowledge from several other fields.

Another example from England, from the same period, is worth mentioning. The project is much smaller than the one in Milan; nonetheless, a detailed account of the work makes it valuable for the present investigation. In the late thirteenth century, the Cistercian abbey founded by King Edward I at Vale Royal in England commissioned a fishpond. The construction expenses were precisely recorded, and the account survives in the monastery's archive.[77] The document, as the one discussed above from Sopron, records the wages of various laborers and distinguishes between the pond-digger master and the diggers. While the master, William le Fox earned eighteen denars per day, the diggers received only eight or ten denars.[78] Although the project was relatively small and the pond-digger master was likely also involved as a manual laborer still earned twice the others' wages due to his master status.

In the Central European context, similar experts appear somewhat later, even though the aforementioned, supposedly Árpádian-age channel system also required complex planning and execution. From the Late Middle Ages onward, as a growing body of evidence reveals the construction of channels and, most importantly, fishponds, there is an increasing need for similar masters.[79]

As discussed above, in Bohemia from the fourteenth century onward, hundreds of fishponds were in operation, and probably somewhat fewer, but the Kingdom of Hungary and Poland also utilized their streams and rivers to create fishponds. During the reign of the Luxembourg dynasty, particularly under Emperor Charles IV and Wenceslas IV, nearly a hundred major fishponds were constructed with royal support. By the end of the Middle Ages, almost 10,000 ponds, both small and large, were in use in the Kingdom of Bohemia.[80] The construction of these ponds, especially the larger ones, was overseen by pond masters. Research in Bohemia concluded that, due to the difficulty of digging, the daily wage of diggers by the early sixteenth century was four times that of the average agricultural laborer. This research attributed the higher salaries to job-related challenges, but it likely has much more to do with the specific knowledge required.[81]

It is also from the late medieval period onward that sources in Hungary repeatedly record ditch and pond diggers, as well as pond masters. This, of course, does not mean that the profession came into existence at this time; as noted, it may have existed since the early Árpádian period. Nonetheless, written evidence attests to these professions only from the fifteenth century onward; this is true of hundreds of occupations.[82] By then, they were regarded as a group of professionals with greater expertise in water infrastructure construction and were primarily hired for such projects. They appear in the account books of towns as well as in other forms of evidence.[83] In recent years, a significant body of research has examined family names in late medieval sources to gain at least some insight into the professions present in free royal or market towns that lacked guild organization (or for which sources on their guilds are lost). He regarded pond masters as officeholders and did not regard their work as a profession.[84] It is important to emphasize that, apart from sporadic data in urban account books, most references to pond masters and pond diggers are family names. However, based on the surviving sources, their tasks are relatively well defined. Of course, being a pond master was a vastly different profession from being a pond digger, with the former being a much more responsible task.

Based on evidence regarding pond masters, they must have been responsible for planning fish-harvesting dates and scheduling pond sluicing.[85] This task required years of practice, and, as with the millers' work, pond masters also needed site-specific knowledge because hydrological conditions varied significantly from one pond to another. In some cases, pond masters were indeed regarded as practitioners of an independent profession. This is further illustrated by an example from Sopron, specifically a fishpond in the town's eastern suburb near the Ikva Stream. The earliest record of the pond dates to 1486 when a burgher of Sopron established a chapel dedicated to Saint Sebastian near the pond. Half a century later, in 1525, the Sopron town council decided that the pond should be utilized again, indicating its previous use and subsequent abandonment. The council assigned the task to an expert, a pond master, who, however, remains unnamed in the source. This still implies that, even if not in every case, pond mastery was viewed as a profession.[86]

Most references to pond masters, as noted above, survive as family names in various sources. The first reference dates to 1418, when a source mentions a tenant peasant named Lawrence Tómester (meaning "pond master" in Hungarian) and a man named Phillip who shared the same family name, according to a source from 1431.[87] From the second half of the fifteenth century, the name continues to appear in sources, and several pieces of evidence suggest that the name of the profession (or office) became a standard family name. This is evident in the case of George Tómester, who worked as a customs officer.[88] This assumption is also supported by another example from the early sixteenth century, when three tenant peasants bore the name Tómester on the same estate within a short period. Given the size of the estate, it is doubtful that all three would have been able to earn their living as pond masters.[89]

It is difficult to say much about the social status of pond masters, as in most cases, the name is the only available evidence. Both burghers and tenants are mentioned as pond masters. However, at least one person may truly be worthy of reference here: the archdeacon of the cathedral chapter of Győr, Paul Pápay. As part of his office, Pápay was also responsible for the chapter's ponds, according to the sources, albeit with some errors.[90] More relevant to the present analysis is that he was highly educated and literate. However, in this instance, pond mastership certainly represented an office, not a profession for Pápay. Since the chapter of Győr held a significant share in the fishing business in the vicinity of this important town, the chapter usually designated someone from its members to undertake this task.[91] In this case, it is most likely that the designated person collaborated with the actual pond masters across the chapter's various estates.

Data on pond diggers are available later than on pond masters. The earliest date, 1466, also originates from Sopron. Payments are recorded for ditch diggers who worked on clearing the previously discussed town moat.[92] As with the data on pond masters, most of the fifteenth-century information on pond and ditch diggers derives from personal names.[93] However, one source from the early sixteenth century is certainly worth considering. The 1524 terrier of the bishopric of Veszprém exempts a tenant in the market town of

Sümeg, a certain Peter Tóásó (meaning "pond digger" in Hungarian), from any seigniorial dues in exchange for taking care of the bishopric's ponds there.[94] The relatively high number of references to pond diggers from the sixteenth century leaves little doubt that their task was regarded as a specific profession. The previously analyzed building account also supports this, as the significantly higher daily wage paid to the hands is difficult to explain otherwise. By the Early Modern period, the role of ditch diggers was clearly distinguished from that of simple building workers. Sources suggest that workers' expertise in water construction was carefully considered during the hiring process. The already mentioned Batthyány manorial complex was one of the best-organized in the Kingdom of Hungary, and many of its estates had fishponds. From the mid-sixteenth century onward, the family's manors were governed from the manor of Güssing in present-day Austria. In a letter dated 1545 sent from Güssing by Francis Batthyány, ban of Croatia and *alispán* (*vicecomes*) of Vas County at the time, he informs his nephew, Christoph Batthyány of having sent a man with expertise in pond digging *(fodiator piscinarum*) to assist in building new ponds.[95]

As part of the economic reforms a century later, Adam Batthyány introduced numerous new fishponds on his various manors. In the instructions given to the leading official of the manor of Güssing, Jeremiah Jobbágy, Batthyány cautioned his new officer to carefully select the individuals to dig the ponds on the estate he was overseeing: "Try your best in hiring good and apt pond diggers and hire them according to the conventions of hospites. If they find potential or lost ponds, renew them and introduce fish into them."[96] Though pond diggers by training were most experienced in water construction, they were sometimes also employed at other construction sites. For instance, in 1634, while building a new manor house at Gattendorf, pond diggers were hired to excavate the foundations of the new building, according to the testimony of the building accounts.[97]

Finally, it is important to understand the relationship between the pond masters and the pond diggers. Based on an analysis of the Sopron account book, the two professions were distinct, as they received markedly different wages. A source from the town, however, sheds a somewhat different light on the question. Sources on the so-called Tómalom ("Lake Mill" in Hungarian) and the associated fishpond mention a Sopron burgher named John Schöberel. In 1495, he is referred to as a pond digger (*teichtknecht*), while the pond master was a certain George (*teichtmaister*). Three years later, however, the pond master was none other than Schöberel. It is likely that after years of learning at the pond of the Tómalom as an assistant and a pond digger, he was appointed as master.[98] Although pond masters were considered masters, their work was still not among the most valued in the construction industry. The carpenter at the building works of the Tómalom in Sopron received a wage that was 40 percent higher than Schöberel's.[99]

In light of this, it is worth revisiting the question of whether pond masters and pond diggers can be regarded as practitioners of a distinct profession. The answer is both yes and no, as both positive and negative examples can be drawn from medieval and Early Modern source material. In some instances, pond masters were specifically designated to supervise certain ponds. In contrast, in other cases, they, and likely even more frequently, pond diggers, were hired for various types of construction work.

4.4 Conclusions

This short chapter provided insights into the social groups involved in the water construction projects discussed in Chapter 3. First, a unique source—the accounts of the reconstruction of Sopron's moat from 1540—was analyzed. The varied professions mentioned in the source, along with the number of water-related occupations, makes this source particularly valuable for the present analysis. The role of millers in water construction is especially emphasized in the source, as is a previously overlooked group of individuals: pond masters and pond diggers. These professions are then discussed in detail, utilizing a wide variety of still extant account books, terriers, and letters.

As shown by these sources, millers had responsibilities that extended beyond supervising milling, which remained their most crucial task. They also participated in more complex activities, such as mill construction, and, at times, served as experts in water-related legal disputes. They possessed considerable technical and theoretical expertise—maybe even had legal knowledge that from the sixteenth century onward was increasingly available in vernacular Hungarian.[100] The role of pond masters was likely less complex but still required considerable expertise and extensive practical experience. Several other professions contributed to water construction in the Middle Ages, and many more individuals were involved in related tasks. This chapter demonstrated that, although these professions in the Middle Ages were not among the most esteemed, they were, in fact, complex and required significant expertise.

Notes

1 Károly Takács, "Medieval Hydraulic Systems in Hungary: Written Sources, Archaeology and Interpretation," in *People and Nature in Historical Perspective* (CEU Medievalia, 5), eds. József Laszlovszky, and Péter Szabó (Budapest: Archaeolingua – CEU Press, 2003), 289–312.

2 Bertalan Andrásfalvy, *Duna mente népének ártéri gazdálkodása Tolna és Baranya megyében az ármentesítés befejezéséig* [Flood plain economy of the peoples of the Duna valley in Tolna and Baranya Counties until the completion of the water regulation works] (Tanulmányok Tolna megye történetéből, 7) (Szekszárd: Tolna Megyei Tanács, 1975) and the debate that unfolded on the flood plain economy of premodern Hungary such as *Korall* no. 1 (2000).

3 MNL GYMSML SL IV. 1009. a. vol. 6. 1540/1541. 76–82. p. (also accessible as a microfiche copy: MNL OL X 591 – MNL OL Box 5202). For the source, see also: Katalin Szende, "Mills and Towns: Textual Evidence and Cartographic Conjectures from Hungarian Towns in the Pre-Industrial Period," in *Extra muros. Vorstädtische Räume in Spätmittelalter und Früher Neuzeit. Espaces suburbains au bas Moyen Âge et à l'époque moderne* (Städteforschung, A 91), eds. Guy Thewes, and Martin Uhrmacher (Cologne and Vienna: Böhlau, 2019), 485–516. https://doi.org/10.7788/9783412515164.485. Published in extenso in Appendix 1. no. 5 in this book.

4 MNL GyMSML SL IV. 1009. a. vol. 6. 1540/1541. 76–85. p.

5 MNL GyMSML SL IV. 1009. a. vol. 6. 1540/1541. 86–87. p.

6 MNL GyMSML SL IV. 1009. a. vol. 6. 1540/1541. 76. p.

7 MNL GyMSML SL IV. 1009. a. vol. 6. 1540/1541. 76. p.

8 Jiří Andreska, "Development of Fish-Pond Culture in Bohemia," in *The Fishing Culture of the World*, 2 vols, ed. Béla Gunda (Budapest: Akadémiai Kiadó, 1984), vol. 1. 77–83, Frantisek Graus, *Dějiny venkovského lidu v Čechách v době předhusitské*, 2 vols [History of the rural population in Bohemia in the pre-Hussite period] (Prague: Státnínakladatelstvi politické literatury, 1953–1957), vol. 2, 32–35 and 483–486. See also: Richard C. Hoffmann, "Economic Development and Aquatic Ecosystems in Medieval Europe." *The American Historical Review* 101 (1996): 660. https://doi.org/10.1086/ahr/101.3.631.

9 Petr Dobrovolný et al., "Monthly, Seasonal and Annual temperature Reconstructions for Central Europe Derived from Documentary Evidence and instrumental Records since AD 1500," *Climatic Change* 101 (2010): 69–107. https://doi.org/10.1007/s10584-009-9724-x. For the data from 1540, see: https://www.ncei.noaa.gov/pub/data/paleo/historical/europe/dobrovolny2010temperature-mar-noaa.txt. The whole year was probably one of the warmest and driest in the last millennium: Oliver Wetter et al., "The Year-Long Unprecedented European Heat and Drought of 1540 – A Worst Case," *Climatic Change* 125 (2014): 349–363. https://doi.org/10.1007/s10584-014-1184-2.

10 MNL GyMSML SL IV. 1009. a. vol. 6. 1540/1541. 76. p.

11 Cf. Katalin Szende, "Scapegoats or Competitors? The Expulsion of Jews from Hungarian Towns in the Aftermath of the Battle of Mohács, 1526," in *Expulsion and Diaspora Formation: Religious and Ethnic Identities in Flux from Antiquity to the Seventeenth Century* (Religion and Law in Medieval Christian and Muslim Societies 5), ed. John Tolan (Turnhout: Brepols, 2015), 64–68. https://doi.org/10.1484/M.RELMIN-EB.5.109160. For the re-use of tomb stones in general, see Carsten Wilke, "Medieval Hebrew Inscriptions: Towards a European Database," *Jewish Studies at the Central European University* 7 (2013): 147–172.

12 For the edition of this part of the account book, see: *Magyar-Zsidó oklevéltár*, 18 vols [Hungarian–Jewish cartulary], eds. Ármin Friss et al. (Budaest: Magyar Izraeliták Orzságos Képviselete, 1901–1980), vol. 5/2. 408–409 no. 15. For the fate of the Jewish cemetery and its stones: Sándor Scheiber, "Középkori zsidó sírkő-töredékek Sopronban," [Medieval Jewish tombstone-fragments at Sopron] *Soproni Szemle* 14 (1960): 357–361.

13 MNL GyMSML SL IV. 1009. a. vol. 6. 1540/1541. 77. p.

14 MNL GyMSML SL IV. 1009. a. vol. 6. 1540/1541. 77. p. For mortars and other professions, see the illustrations of the *Hausbuch der Mendelschen*

Zwölfbrüderstiftung from the same period as the Sopron account book analyzed here. For descriptions and illustrations of the different professions, see the online database: www.nuernberger-hausbuecher.de/.

15 As *cementarius*, see for instance: Árpád Mikó, "Rész és egész. A magyarországi reneszánsz kőfaragóműhelyek és kutatásuk," [Part and the whole. Hungarian Renaissance stone-carvingworkshops and research into them] in *Koppány Tibor hetvenedik születésnapjára* [For the 70th birthday of Tibor Koppány] (Művészettörténet – műemlékvédelem 10), eds. István Bardoly, and Csaba László (Budapest: Országos Müemlékvédelmi Hivatal, 1998), 218 (1491) and as a name István Draskóczy, "Az erdélyi sókamarák ispánjai, 1529–1535: az erdélyi sóbányák sorsa a Szapolyai korszakban," [Counts of the Salt Chambers of Transylvania, from 1529 to 1535: the fate of the Transylvanian salt mines in the Szapolyai era] *Levéltári Közlemények* 75, no. 1 (2004): 30 (1532). For similar from Sopron: Házi, *Sopron*, vol. 2/5. 98 (1498), and 438 (1524/25).

16 MNL GyMSML SL IV. 1009. a. vol. 6. 1540/1541. 79. p.

17 Házi, *Sopron*, vol. 2/2. 395, vol. 2/3. 3, and vol. 2/4. 262.

18 MNL GyMSML SL IV. 1009. a. vol. 6. 1540/1541. 79. p. For the functions of the latter on the example of Bratislava, see: Judit Majorossy: "Church in Town. Urban Religious Life in Late Medieval Hungary," (PhD-diss., Central European University, 2006), 236–240, and eadem, "A test és a lélek fürdője. A középkori pozsonyi fürdőkről," [The bath of body and soul. On the medieval bathing houses of Bratislava] in *„Köztes-Európa" vonzásában: Ünnepi tanulmányok Font Márta tiszteletére* [In the attraction of "Mitteleuropa" Studies in honor of Márta Font], eds. Dániel Bagi, Tamás Fedeles, and Gergely Kiss (Pécs: Kronosz, 2012), 354.

19 MNL GyMSML SL IV. 1009. a. vol. 6. 1540/1541. 80. p.

20 MNL GyMSML SL IV. 1009. a. vol. 6. 1540/1541. 80. p.

21 MNL GyMSML SL IV. 1009. a. vol. 6. 1540/1541. 80. p.

22 On him in details, see: Károly Goda, "A soproni városvezető réteg a 15–16. században. A polgármesteri és városbírói tisztség összehasonlító igazgatási és társadalomtörténete," [The leading elite of Sopron in the 15th–16th centuries. The administrative and social history of the positions of majors and town judges] (PhD-diss., Eötvös Loránd Tudományegyetem, 2011), 193–194.

23 MNL GyMSML SL IV. 1009. a. vol. 6. 1540/1541. 82. p.

24 MNL GyMSML SL IV. 1009. a. vol. 6. 1540/1541. 83. p. For the fish keepers: István Dávidházy, "A soproni posztókallók," [The fulling mills of Sopron] *Soproni Szemle* 43 (1989): Fig. 5. This is used in: Ferenc Jankó, József Kücsán, and Katalin Szende, *Sopron* (Magyar Várostörténeti Atlasz, 1) (Sopron: Győr-Moson-Sopron Megyei Levéltár Soproni Levéltára, 2010), 27. Fig. 10 and 82 (Fishponds). For the ownership of the fish keepers, see: Károly Mollay, "A Tómalom középkori előzményei (Fejezet a soproni határ történetéből)," [Medieval predecessors of the Tómalom (An episode from the history of the borders of Sopron)] *Sopron Szemle* 46 (1992): 154.

25 MNL GyMSML SL IV. 1009. a. vol. 6. 1540/1541. 83–84. p.

26 MNL GyMSML SL IV. 1009. a. vol. 6. 1540/1541. 85. p.

27 On millstone mining in Hungary: Gergely Csiffáry, "A gyöngyössolymosi malomkőbányászat története," [History of mill stone quarrying at Gyöngyössolymos] in *Tanulmányok a kézműipar történetéből* [Studies in the history of craftsmanship], eds. Gergely Csiffáry, and Klára Dóka (Veszprém: Veszprémi

Akadémiai Bizottság, 1999), 161–189. For the latter problem, see MNL OL P 1314 no. 7606 (September 6, 1637, Ladislas Bozay to Adam Batthyány) no. 33 860 (August 5, 1646 [this refers to the fair]), and no. 33 647 (September 28, 1646, both by Stephen Nemsem to Adam Batthyány).

28 MNL OL P 1314 no. 43 747 (February 9, 1611, Andrew Somogy to Adam Batthyány). For the role of Pápa as a trade center, see: András Kubinyi, "A középkori Pápa," [Medieval Pápa] in *Tanulmányok Pápa város történetéből. A kezdetektől 1970-ig* [Studies in the history of the town of Pápa. From the beginnings to 1970], ed. András Kubinyi (Pápa: Pápa Város Önkormányzata, 1994), 75–124.

29 Cf. the directions of lime trade: Mollay, "A Tómalom," 152–153 (map).

30 István Baraczka, "Az 1535/36. évi Sopron városi kamarai számadáskönyv néhány tanulsága," [Some lessons of the account book of Sopron from 1535/1536] *Soproni Szemle* 23 (1969): 207–215.

31 Cf. e.g., Christopher Dyer, *Everyday Life in Medieval England* (London: Hambledon, 1994), 133–165. See also: Jenő Szűcs, "A középkori építészet munkaszervezetének kérdéséhez," [To the question of the labor organization in medieval building industry] *Budapest Régiségei* 18 (1958): 320.

32 See e.g., István Hermann, "Egy Pápa belvárosi ház felújítása az 1670-es években. (forrásközlés)," [Rebuilding of a house in the downtown of Pápa in the 1670s (source edition)] *Soproni Szemle* 70 (2016): 226 (here the ratio of costs of the material and the wages was 56 percent to 44 percent).

33 Szűcs, "A középkori építészet."

34 See his itinerary on the frequency of his visits at Bratislava: Pál Engel and Norbert C. Tóth, *Királyok és királynék itineráriumai (1382–1438), Zsigmond király és császár (1382–1437)* [Itinerary of kings and queens, 1382–1438; King and Emperor Sigismund (1382–1437)] (Segédletek a középkori magyar történelem tanulmányozásához, 1) (Budapest: Magyar Tudományos Akadémia Támogatott Kutatóhelyek Irodája, 2005).

35 Szűcs, "A középkori építészet," 328.

36 Szűcs, "A középkori építészet," 319–320. See for the winter and summer wages at Sopron: Dezső Dányi, and Vera Zimányi, *Soproni árak és bérek a középkortól 1750-ig* [Prices and wages from Sopron from the Middle Ages to 1750] (Budapest: Akadémiai, 1989), 448–507. By the nineteenth century the summer half year was between St George's (24 April) and St Michael's (29 September) day for which period significantly higher wages were paid. Cf. Ottó Domonkos, *Ár- és bérlimitációk Sopron városban és Sopron megyében, XVI–XIX. század* [Price and wage limitations at Sopron and in Sopron County, 16th–19th centuries] (Előmunkálatok a magyarság néprajzához, 8) (Budapest: MTA Néprajzi Kutató Csoport, 1980), 111. In Austria, three seasonal wages were identified by research. Different wages were paid to hired hands in summer, winter and autumn-spring: Elisabeth Gruber, *"Raittung und außgab zum gepew." Kommunale Rechnungspraxis im oberösterreichischen Freistadt Edition und Kommentar der Stadtgrabenrechnung (1389–1392)* (Quelleneditionen des Instituts für Österreichische Geschichtsforschung, 14) (Vienna: Böhlau, 2015), 73.

37 For the different day wages of masons, see e.g., Házi, *Sopron*, vol. 2/5. 285.

38 See for instance for the female laborers at Hall in Tyrol: Klaus Brandstätter, "Schwerarbeitende Frauen im Mittelalter: Bauarbeiterinnen in Hall in Tirol,"

in *Margarete „Maultasch": zur Lebenswelt einer Landesfürstin und anderen Tiroler Frauen des Mittelalters; Vorträge der wissenschaftlichen Tagung im Südtiroler Landesmuseum für Kultur- und Landesgeschichte Schloss Tirol, Schloss Tirol, 3. bis 4. November 2006*, ed. Jutta Hörmann-Thurn und Taxis (Innsbruck: Wagner, 2007), 277–281. For similar at Paris, see: Ilona Sz. Jónás, "Női munka a 13. század végi Párizsban," [Female work in late-13th-century Paris] in eadem, *Sokszínű középkor. Természet, társadalom, kultúra a középkori Európában* [Colorful Middle Ages. Nature, society, culture in medieval Europe], eds. Balázs Nagy, and Veronika Novák (Budapest: ELTE Eötvös Kiadó, 2010), 341. With several examples from different parts of Europe: Shelley E. Roff, "'Appropriate to Her Sex?' Women's Participation on the Construction Site in Medieval and Early Modern Europe," in *Women and Wealth in Late Medieval Europe* (The New Middle Ages), ed. Theresa Earenfight (New York: Palgrave, 2010), 109–134. https://doi.org/10.1057/9780230106017_7. For female work in towns in Hungary, see: Katalin Szende, "A női munka a középkori gazdaságban," [Female work in medieval economy] *Aetas* 11, no. 1–2 (1995): 179–193 and Judit Benda, "A csendes háttér. A nők munkalehetőségei a középkori Budán," [The silent background. Possibilities of female work in medieval Buda] *Korall* no. 60 (2015): 11–40.

39 Donald Woodward, *Men at Work: Labourers and Building Craftsmen in the Towns of Northern England, 1450–1750* (Cambridge: Cambridge University Press, 1995), 112. See also: Roff, "Appropriate to Her Sex," 116.

40 *The History of the King's Works. The Middle Ages*, 2 vols, ed. Howard M. Colvin (London: H. M. Stationery Office, 1963), vol. 1. 297.

41 MNL GYMSML SL IV. 1009. a. vol. 7, 1540/1541, 43–44. p (can be researched as a microfiche: MNL OL Box 5203).

42 For millers in general, with mostly Early Modern examples: Sándor Takáts, "Műveltségtörténeti közlemények: a magyar molnár," [Studies in intellectual history: the Hungarian miller] *Századok* 41 (1907): 52–56, Sándor Takáts, "Műveltségtörténeti közlemények: a magyar malom I–II," [Studies in intellectual history: the Hungarian mill] *Századok* 41 (1907): 143–160 and 236–249.

43 Carlo Ginzburg, *The Cheese and the Worms. The Cosmos of a Sixteenth-Century Miller* (Baltimore, MD: Johns Hopkins University Press, 1992), 1.

44 Geoffrey Chaucer, *The Canterbury Tales* in *The Complete Poetry and Prose of Geoffrey Chaucer*, ed. John H. Fisher (New York: Holt, Rinehart and Winston, 1977), 57–68. For the reputation of millers: Paul Freedman, "The Miller," in *Historians on Chaucer: The General Prologue to the Canterbury Tales*, eds. Stephen H. Rigby, and Alastair J. Minnis (Oxford: Oxford University Press, 2014), 368–385. https://doi.org/10.1093/acprof:oso/9780199689545.003.0021

45 Judit Majorossy, "A foglalkozás topográfiája. A társadalmi tértől a személyes térig: a társadalmi mobilitás térbeli elemei a 15. századi Pozsonyban," [Topography of employment. From the social space to the personal space: elements of the spatial mobility in 15th-century Bratislava] *Korall* no. 45 (2011): 102–135.

46 It probably was somewhat more clearly distinguished in medieval England. See: Richard Holt, *The Mills of Medieval England* (Oxford and New York: Basil Blackwell, 1988), 101.

47 *Magistros carpentarios videlicet molendinorum* or *molendinatoribus alias hungaris carpentariis. A nagy-károlyi gróf Károlyi család oklevéltára*, 5 vols [Cartulary

of the Károly family of Carei], ed. Kálmán Géresi (Budapest: Nyomatott Kocsi Sándornál, 1882–1897), vol. 1. 362–367 no. 227. Cf. Tamás Vajda, "Vízimalmok építésével előidézett, mesterséges áradások a középkori Magyarországon," in *Középkortörténeti tanulmányok 8. A VIII. Medievisztikai PhD-konferencia (Szeged, 2013. június 17–19.) előadásai* [Studies in medieval history, 8. The 8th PhD conference in medieval studies (Szeged, 17–19 June 2013)], eds. Márta Tóber, and Ágnes Maléth (Szeged: Szegedi Középkorász Műhely, 2015), 348, and Mihály Détshy, "Munkások és mesterek az egri vár építkezésein 1493 és 1596 között. 1. közlemény," [Workers and masters at the building works of the castle of Eger between 1493 and 1596, I] *Az Egri Múzeum Évkönyve* 1 (1963): 190.

48 Endre Veress, *Gyula város oklevéltára. 1313–1800* [Cartulary of the town of Gyula] (Budapest: Stephaneum, 1938), 89, and in terriers, see e.g.: MNL OL E 156 (Magyar Kamara Archívuma. Urbaria et Conscriptiones) a. Fasc. 089. No. 007. 4. p. (1565); MNL OL E 156 a. Fasc. 065. no. 090. 4, 9. p. (early seventeenth century); MNL OL E 156 a. Fasc. 037 No 057 31. p. (February, 1639); MNL OL E 156 a. Fasc. 036. No 016. 10, 12. p. (August 1, 1670); MNL OL E 156 Fasc 36. No 009 4. p. (April 6–9, 1718); MNL OL E 156 a. Fasc. 005. No 029c. 108. p. (July 6, 1720). For further examples, see: Takáts, "A magyar molnár."

49 Elemér Winkler, *A soproni céhek története a XV–XIX. században* [Life of the guilds of Sopron in the 15th–19th centuries] (Sopron: [Tóth Kny.], 1921), 11.

50 For England: Adam Lucas, *Ecclesiastical Lordship, Seigneurial. Power and the Commercialization of Milling in Medieval England* (Farnham and Burlington, VT: Ashgate, 2014), 269–306. https://doi.org/10.4324/9781315578620, for Hungary: Tamás Vajda, "Földesúri jogok és kötelezettségek a középkori vízimalmok körül," [Seigniorial right and duties connected to medieval waster mills] in *Víz és társadalom Magyarországon a középkortól a XX. századig* [Water and society in the Hungary from the Middle Ages to the 20th century], ed. Gergely Krisztián Horváth (Budapest: Balassi, 2014), 79–110.

51 With mostly English examples, see Richard Bennett, and John Elton, *History of Corn Milling*, 4 vols (London: Simpkin, 1898–1904), vol. 3. 106–176. The same is reflected in the 1812 price limitations of Sopron County, in which numerous fines connected to fraud with grain is mentioned: Domonkos, *Ár- és bérlimitációk*, 118–119. See also: Holt, *The Mills of Medieval England*, 104–105.

52 For the former, see e.g., the Czech carpenters paid by the king to build to mills (or a mill with two wheels) at Belgrade: *Registrum proventuum regni Hungariae: A Magyar Királyság kincstartójának számadáskönyve (1494–1495)*, ed. Tibor Neumann (Budapest: MTA BTK Történettudományi Intézet, 2019), 1111–1112, 1852, and 1948, or MNL OL E 41 1578. no. 317. For the latter, see e.g., Prímási Levéltár Esztergom, Acta Radicalia, Classis X.24. File 18. 49. p. (August 8, 1650, contract concluded with Balthasar Boniger, *murarius*).

53 For a detailed analysis of the sources: András Vadas, *Egy határfolyó környezettörténete. Háború és vízgazdálkodás a kora újkori Rába-völgyben* [The environmental history of a frontier river. War and water-management in the Early Modern Rába River valley] (Budapest: Bölcsészettudományi Kutatóközpont, 2021) and András Vadas, *The Environmental Legacy of War on the Hungarian-Ottoman Frontier, c. 1540–1690* (Environmental Humanities in Pre-modern Cultures) (Amsterdam: Amsterdam University Press, 2023). https://doi.org/10.1515/9789048552016.

54 MNL OL P 1314 no. 33 782 (June 25, 1650, Stephen Nemsem to Adam Batthyány).
55 MNL OL P 1314 no. 33 741 (August 12, 1649, Stephen Nemsem to Adam Batthyány).
56 István Tringli, "A magyar szokásjog a malomépítésről," [Customary law on construction of mills] in *Tanulmányok a középkorról* [Studies in medieval history] (Analecta Medievalia, 1), ed. Tibor Neumann ([Budapest–Piliscsaba]: Pázmány Péter Katolikus Egyetem, 2001), 251–268 and Vajda, "Vízimalmok építésével előidézett."
57 For Serafin, see Norbert C. Tóth, *Az esztergomi székes- és társaskáptalanok archontológiája 1100–1543* [Archontology of the cathedral and collegiate chapters of Esztergom, 1100–1543] (Subsidia ad historiam medii aevi Hungariae inquirendam, 9) (Budapest: Magyar Tudományos Akadémia Magyar Medievisztikai Kutatócsoport, 2019), 131, 200.
58 MES, vol. 2. 152–153 no. 126. For the case: Tamás Vajda, "Okleveles adatok Árpád-kori vízimalmainkról," [Documentary data on Árpádian-age water mills] in *Medievisztikai tanulmányok. A IV. Medievisztikai PhD-konferencia (Szeged, 2005. június 9–10.) előadásai* [Studies in medieval history. Papers read at the 4th conference of medievalists, Szeged, 9–10 June 2005], eds. Szabolcs Marton, and Éva Teiszler (Szeged: Szegedi Középkorász Műhely, 2005), 202–203, Tamás Vajda, "Árpád- és Anjou-kori vízimalmaink tájalakító hatása," [Water mills as driving forces of landscape change in the Árpádian and Angevin Periods] in *Micae mediaevales II. Fiatal történészek dolgozatai a középkori Magyarországról és Európáról* [Studies of young historians on medieval Hungary and Europe], eds. Péter Jakab et al. (Budapest: ELTE BTK Történelemtudományok Doktori Iskola, 2012), 65, and Vajda, "Vízimalmok építésével előidézett," 359.
59 stván Tringli, *Sátoraljaújhely* (Magyar várostörténeti atlasz, 2) (Budapest: MTA Történettudományi Intézet, 2011), 14 and Szende, "Mills and Towns," 492–496. See also Chapter 3.3.2.
60 MNL OL DL 2310 (June 10, 1325), edited in: AO, vol. 2. 201–203 no. 182. Cf. Vajda, "Hazai vízimalmaink," 198–199, and for the mills at the town, see: Tringli, *Sátoraljaújhely* (Gazeteer: Mills) and Szende, "Mills and Towns," 492–496.
61 E.g., MNL OL DL 54 994, 18 275, 70 714, etc.
62 MNL OL DL 14 969, 15 098, 15 133, 16 417, 16 288. The documents were summarized in: Zsuzsanna Bándi, "Északkelet-magyarországi pálos kolostorok oklevelei." [Charters of the Pauline abbeys of northeastern Hungary] *Borsodi Levéltári Évkönyv* 5 (1985): nos. 86, 94, 95, 103, 105, and 106. The case is briefly mentioned in Éva Gyulai, "Miskolc középkori topográfiája," [Medieval topography of Miskolc] in *Miskolc története I. A kezdetektől 1526-ig* [History of Miskolc, I. From the beginnings to 1526], ed. István Dobrossy (Miskolc: Borsod-Abaúj-Zemplén Megyei Levéltár, 1996), 215, and Tringli, "A magyar szokásjog," 256–257.
63 Éva Gyulai, "Termelés és kereskedelem a középkori Miskolcon," [Production and trade at medieval Miskolc] in *Miskolc története*, 312–313.
64 László Lukács, "Sokféle bírák. A faluközösség önigazgatásának „népi" tisztségviselői," [Many kinds of judges. The "folk" office holders of the self-governance of the village communities] in *Halmok és havasok. Tanulmányok a hatvan esztendős Bárth János tiszteletére* [Mounds and mountains. Studies in

honor of the 60th birthday of János Bárth], eds. Dániel Bárth, and János Laczkó (Kecskemét: Bács-Kiskun Megyei Önkormányzat Múzeumi Szervezete, 2004), 405–407.

65 MNL OL DF 246 004 (January 2, 1510). Edited in: Gustav Gündisch, "Winz und die Sächsische Nationsuniversität," in *Emlékkönyv Kelemen Lajos születésének nyolcvanadik évfordulójára* [Memorial book in honor of the 80th birthday of Lajos Kelemen] (A Bolyai Tudományegyetem kiadványai, I-1), eds. András Bodor András et al. (Bucharest and Cluj: Tudományos Könyvkiadó, 1957), 323–324. Cf. Bálint Lakatos, "Mezővárosi és falusi önkormányzati testületek Magyarországon a késő középkorban," [Councils in late medieval Hungarian market towns and villages] *Századok* 148 (2014): 524, and Bálint Lakatos, *Mezővárosi oklevelek: Települési önkormányzat és írásbeliség a késő középkori Magyarországon, 1301–1526* [Market town charters. Urban autonomy and literacy in late medieval Hungary] (Budapest: Bölcsészettudományi Kutatóközpont Történettudományi Intézet, 2019), 76 and 225 (no. 3).

66 Péter Dominkovits, "Folyóvizek és a XVII. századi vármegyei közigazgatás, bíráskodás (A nyugat-dunántúli törvényhatóságok példáján)," [Waterways and 17th-century county administration, jurisdiction (on the example of Western Transdanubian courts of laws] in *Víz és társadalom*, 166 and 168.

67 See in details, Vadas, *The Environmental Legacy of War*.

68 The earliest known millers' guilds: 1550: Sassin (part of present-day Šaštín–Stráže), 1570: Pápa, 1576: Vas County, 1577: Bistrița, 1581: Sibiu. Cf. *A magyarországi céhes kézművesipar forrásanyagának katasztere*, 2 vols [Cadastre of the sources of Hungarian guilds], eds. István Éri, Lajos Nagy, and Péter Nagybákay (Budapest: [no publisher], 1975–1976), vol. 2. (Millers guilds), and *Magyar néprajz*, vol. 3/2. 44–45 (map 5), and 199–207.

69 *Kegmes Uram, Nagnak azt irhatom, Hogy Simonczicz Marko egy nemet molnart kűldőt az itt valo malomban, hogy megh probaliuk ha erti az molnarsagot avagy nem. Azert Kegmes uram ez nem arra valo hogy az itt valo malomban molnar lenni mert az Raba nem olyan minth az Lapincz, ide oly molnar kől az ki 30. s 35 eztendeigh az Raban valo malomban tanult, s, az viznek az eő cziniat tudgya. Az minemő szakasztast az Raba itten tett, mutattam neki, hogy mikepen köllene vizet el fogni, csak azt mondgya megh köl tölteni. elege töltiük de ugian nem tudgiuk az vizet el vennj, mind altta fol el az viz s alol szagattia, ez ataliaban nem ide valo molnar, mert sem mestersege semmi ninczen, mas az ha mestersege volnais magiarul nem tud s itt az emberek nem tudnak valaszolni.* MNL OL P 1314 no. 33 778.

70 On floods, see: András Vadas, "The 'waters leave their beds frequently': A Western-Hungarian Town and the Flooding of the Rába/Raab River in the Seventeenth Century," *Water History* 5 (2013): 267–286. https://doi.org/10.1007/s12685-013-0087-x.

71 Philippe Bernardi, "Apprentissage et transmission du savoir dans les métiers du bâtiment à Aix-en-Provence à la fin du Moyen Age (1400–1550)," in *Actes du 1er colloque du CRISIMA. Apprentissages, éducation, initiation, éducation au Moyen Age* (Les cahiers du CRISIMA, 1) (Montpellier: [Univ. Paul-Valéry CRISIMA], 1993), 69–79.

72 Paolo Squatriti, "Digging Ditches in Early Medieval Europe," *Past and Present* 176 (2002): 11–65. https://doi.org/10.1093/past/176.1.11.

73 K[atharine] S. B. Keats-Rohan, *Domesday People: a Prosopography of Persons Occurring in English Documents, 1066–1166. I. Domesday Book* (Woodbridge: Boydell, 1999), 251.

74 See the sources related to the building of the *Fossa Carolina*. Cf. Squatriti, "Digging Ditches."

75 Richard Holt, "Medieval England's Water-Related Technologies" in *Working with Water in Medieval Europe: Technology and Resource-Use* (Technology and Change in History), ed. Paolo Squatriti (Leiden, Boston, and Cologne: Brill, 2000), 80. https://doi.org/10.1163/9789047400110_006, and more recently: Keith D. Lilley, Christopher D. Lloyd, and Steven Trick, "Designs and Designers of Medieval 'New Towns' in Wales," *Antiquity* 81 (2007): 279–293. https://doi.org/10.1017/S0003598X0009517X, and James Bond, "Canal Construction in the Early Middle Ages: An Introductory Review," in *Waterways and Canal-Building in Medieval England*, ed. John Blair (Oxford: Oxford University Press, 2007), 201. https://doi.org/10.1093/acprof:osobl/9780198723134.003.0008. Another ditch digger master is mentioned in 1278: *The History of the King's Works*, vol. 1. 297.

76 For the building of the ditch: P[ietro] Ghinzoni, "Maestro Giacomo Arribotti e il Naviglio Grande reso navigabile," *Archivio Storico Lombardo* 20 (1893): 200–203, with the charter summarizing the works at 202–203. See further: Roberta Magnusson, and Paolo Squatriti, "The Technologies of Water in Medieval Italy," in *Working with Water in Medieval Europe* (Technology and Change in History, 3), ed. Paolo Squatriti (Leiden, Boston, and Cologne: Brill, 2000), 229–230. https://doi.org/10.1163/9789047400110_010, Patrick Boucheron, "Water and Power in Milan, c. 1200–1500," *Urban History* 28, no. 2 (2001): 180–193. https://doi.org/10.1017/S0963926801002024, for Arribotti see 182–183. For the later usage of the Naviglio Grande, Giuliana Fantoni, *L'acqua a Milano. Uso e gestione del basso medioevo* (1385–1535) (Bologna: Cappelli, 1990), 27–60.

77 For the case, see: Roberta J. Magnusson, *Water Technology in the Middle Ages. Cities, Monasteries, and Waterworks after the Roman Empire* (Baltimore, MD: The Johns Hopkins University Press, 2001), 57.

78 *The Ledger Book of Vale Royal Abbey*, ed. John Brownbill (Manchester: [no publisher], 1914), 192–231. Online document: British History Online: http://www.british-history.ac.uk/lancs-ches-record-soc/vol68/pp192-231.

79 Richard Hoffmann, "Medieval Fishing," in *Working with Water*, 331–393. https://doi.org/10.1163/9789047400110_012 and Richard Hoffmann, "Economic Development."

80 Andreška, "Development of Fish-Pond Culture," 77–80, and Hoffmann, "Medieval Fishing," 382–383.

81 Andreška, "Development of Fish-Pond Culture," 81.

82 Cf. László Szabolcs Gulyás, "Középkori mezővárosi foglalkozásneveink forrásértékéről," [On the source value of profession names from medieval Hungarian market towns] *Századok* 142 (2008): 437–462.

83 Fejérpataky, *Magyarországi városok régi számadáskönyvei*, 214, 218–219, 240, etc.

84 László Szabolcs Gulyás, "A középkori magyar városfejlődés migráció- és ipartörténeti vonatkozásai a történeti személynévtan tükrében II. Személynévadás és kézműipar," [Migration and industrial historical aspects of medieval Hungarian urban development in light of historical onomastics, II. Personal naming and crafsmanship] *Névtani Értesítő* 37 (2015): 139.

85 Cf. Hoffmann, "Economic Development," 667–668.

86 *Ain ersame gemain begert, das ain ersamer rat den teucht bei Sand Sebastian sol machen lassen vnd mit teuchtmaister hanndlen, was gestallt man den aufs*

nutzlichest, sol auffueren. Házi, vol. 2/2. 198. Cf. Károly Mollay, "Kőszeg 1532. évi ostroma és Sopron," [The siege of Kőszeg in 1532 and Sopron] *Soproni Szemle* 37 (1983): 208 note 28. For the location of the St Sebastian Chapel: Jankó, Kücsán, and Szende, *Sopron*, 68.

87 Zichy, vol. 6. 516, 518 (1418), vol. 8. 450 (1431), vol. 8. 473. (1432), etc. For further examples, see: Miklós Kázmér, *Régi magyar családnevek szótára, XIV–XVII. század* [Dictionary of old Hungarian family names, 14th–17th centuries] (Budapest: Magyar Nyelvtudományi Társaság, 1993), 1073.

88 *Gregorius filius Johannis dicti Tomester tributarius de Zeremlyen*. Zichy, vol. 8. 604.

89 *A Perényi család levéltára 1222–1526* [Cartulary of the Perényi family, 1222 to 1526] (Magyar Országos Levéltár kiadványai, II. Forráskiadványok, 44), ed. István Tringli (Budapest: Magyar Országos Levéltár, 2008), 354 (1509) and 376 (1513).

90 Vince Bedy, *A győri székeskáptalan története* [The history of the chapter of Győr] (Győregyházmegye múltjából, 3) (Győr: Győregyházmegyei Alap Nyomdája, 1938), 359.

91 Bedy, *A győri székeskáptalan*, 256.

92 *Item vmb I fueder holcz zu sprewczen vnd grrüsten den teichknechten, als sy den graben gerawmbt haben, XXXII den.* and *Item den teichknechten von dem statgraben beÿm Sÿnnig Thoman ze rawmen XVIII sol. VI den. Item drein knechten, die den teichmaistern den ersten tag ze rüssten geholffen haben, XXXVI den.* Házi, *Sopron*, vol. 2/4. 268.

93 Kázmér, *Régi magyar családnevek*, 1068–1069.

94 *Petrus Thoaso integram, exempta ab orani solutione, quia procurat cum piscinis domini reverendissimi etc.* László Kredics, and László Solymosi, *A veszprémi püspökség 1524. évi urbáriuma. Urbarium episcopatus Vesprimiensis anno MDXXIV* (Fontes Minores ad Historiam Hungariae Spectantes – Új Történelmi tár, 4) (Budapest: Akadémiai, 1993), 94.

95 MNL OL P 1314 no. 3609. (May 1, 1545, Francis Batthyány to Christoph Batthyány).

96 *Azonnis légy, hogy valahonét valami jó értelmes tóásókat fogadj, azokkal hospes szerént konvenciót tégy, és ha oly alkalmatos és puszta halásztók találtatnak avagy vannak azokat meg szépetesd és meg halasítsad* (Make sure to hire good, and smart pond diggers, and provide them with the rights of *hospites*. And if you find suitable and abandoned dish ponds or ones that still exist, have them restored and stock fish in them). MNL OL P 1322 (A herceg Batthyány család levéltára. Instrukciók) no. 131. Quoted in: Borbála Benda, "Táplálkozási változások a főúri udvarok és uradalmak étkezésében a 17. században," [Changes in the diets of aristocratic courts and manors in the 17th century] in *Környezetváltozás, termelés, fogyasztás: a történeti ökológia kérdései* [Environmental change, production and consumption: the questions of historical ecology] (Annales Tataienses, 5), eds. János Fatuska, Éva Mária Fülöp, and László Gyüszi, Jr. (Tata: Mecénás Kulturális Közalapítvány, 2006), 85 note 13, edited in: Elek Virágh, "A csicsvavári udvarbíráknak, Eszterházy és Homonnay részrül adott utasítása," [The instructions to court judges of the castle of Čičva given by the Eszterházy and the Homonnay families] *Magyar Gazdaságtörténelmi Szemle* 4 (1897): 341. For the water management works of the Batthyány family, see: Vadas, *Egy határfolyó környezettörténete*.

97 *Az tó Asso legeninek ki az pincze fundamentumot segitet ásnj* (The pond diggers should help to dig out the foundations of the cellar.). Tibor Koppány, "A gátai számadáskönyvek. Adatok egy 17. század eleji dunántúli uradalom építkezéseihez," [The account books of Gattendorf. Data to the building works of an early-17th-century manor in the Transdanubia] *Műemlékvédelmi Szemle* 10, no. 1–2 (2000): 150 and 153–154.

98 For the building of the Tómalom, and the pond diggers and masters there: Mollay, "A Tómalom," 155–156.

99 Házi, vol. 2/5. 42, 61, 89 and 91. See also: Mollay, "A Tómalom," 155 and Mollay, "Egy évtized soproni külkereskedelmének társadalomtörténeti háttere (1483–1489)," [The social historical background of a decade in the foreign trade of Sopron (1483–1489)] *Soproni Szemle* 49 (1995): 311.

100 István Tringli, "A magyar szokásjog és a protestáns természetjog magyarul," [Hungarian customary law and Protestant natural law in Hungarian] *Történelmi Szemle* 64 (2022): 188–196.

5 CONCLUSIONS AND OUTLOOK

The book's chapters demonstrate that a complex water management system existed in the Kingdom of Hungary for at least a millennium. The diverse needs of stakeholders led to intricate norms and customs governing the use and management of water. Despite the absence of detailed written regulations, many elements of water rights formed throughout the Middle Ages. In the absence of specific laws on water construction, the resolution of various legal disputes relied either on property rights, partly influenced by Roman legal principles, or on prevailing norms and customs. A recurring question addressed in the book concerns the consistency of these customs.

In the introduction to this book, the question was raised regarding the extent to which water rights were viewed as distinct from rights to the exploitation of other natural resources. The case studies presented demonstrate that landowners and the law indeed treated water-use customs as distinct. The inherent qualities of water led to numerous legal cases in medieval Hungary and, of course, in other parts of Europe. Any interventions to waterways, whether through obstacles or outlets for mill races, irrigation channels, or similar features, could have significantly affected adjacent lands upstream and downstream, making them hotspots for resource conflicts. Consequently, from the early period of the Kingdom of Hungary onwards, not independent of Roman law, waterways were never regarded as private property. This concept began to take shape during the Árpádian period, coinciding with the proliferation of water mills and other water-related infrastructure, including channels, dikes, fishponds, and bridges, that emerged in the region.

The book demonstrated that water mills were the most important infrastructural elements of built-up waterscapes in medieval Hungary, likely appearing sometime in the first half of the eleventh century. Their spread led

DOI: 10.4324/9781003596905-5

to increasing conflicts from the second half of the Árpádian period onward. However, by the end of the Árpádian reign, the frequency of mills along some of the smaller rivers and streams may have reached a point where newly erected ones threatened the functioning of others. This threat may have been twofold. On one hand, they could physically hinder the effectiveness of other mills by reducing the amount of necessary water or slowing down the current required to run the other water wheels. On the other hand, it may have been in the interests of other landlords to have more mills, as, despite the seigniorial ban on milling, people might have ground their grain in mills other than those of the landlords if that proved cheaper.[1]

As a result, disputes arose, and norms were established, even though these norms had likely been accepted since the early Árpádian period. However, these principles primarily existed as oral traditions passed down by landowners from one generation to the next. In the latter half of the Árpádian period, as previous research has suggested, a Latin tradition began to emerge in the country's legal system, encouraging the documentation of various disputes and their resolutions.[2]

From the fourteenth century onward, an increasing number of sources attest to disputes over water use. This is likely due in part to a general increase in literacy, the growing necessity of securing one's rights in writing, and the higher number of constructed infrastructures in waterways mentioned earlier. Thus, the book's main chapter addressed the various conflicts over water use from the fourteenth to the sixteenth century. The increase in literacy strongly indicates that well-defined norms already existed concerning the most critical legal issue: the construction of obstacles in waterways to impound the flow. By then, legal experts possessed considerable knowledge and experience in providing guidance on the construction of mills, dams, bridges, sluices, and other works. The legal principles that followed the established norms were quite clear; an earlier mill had precedence over those built later, and no one had the right to cause losses to another landowner's property by constructing obstacles in water flows. These principles were not unique to the Kingdom of Hungary. The problem likely existed elsewhere, as in other parts of Western and East Central Europe, where smaller rivers and streams were also the most significant waterways for milling. The norms were remarkably similar throughout Europe. However, some attempts were made to establish precise regulations, such as in Mecklenburg, Florence, and some areas of France, where either the distance between mills was fixed, or, in some cases, the maximum height of dams to be built was specified. Nevertheless, the main principle appears to be the same as in Hungary.[3] It has also been argued that, although norms existed in the Kingdom of Hungary to address less frequent issues, such as changes in property rights resulting from shifts in riverbed courses, this knowledge was not widely shared among legal experts or among the parties involved in legal disputes.

While clarifying certain points regarding the norms of water use in medieval Hungary, the book also provided insights into the economic exploitation of waterways across various environments. The archival and edited primary evidence points to numerous water-related problems in towns—specifically, free royal towns, mining towns, or cities of other legal standing. This can be readily explained by the need to accommodate a larger population in a relatively confined space compared with the countryside. Urban industries, such as grain grinding, the need to catch and keep fish near markets, the construction of moats to protect towns, the development of harbors, navigation, wastewater drainage, and the provision of drinking water, led to numerous resource conflicts in urban settings. The use of urban waterways increased environmental pressures on urban areas during the premodern period, as illustrated by diverse urban waterscapes across Central Europe.[4] These environments proved to be the most significant study areas for this book. The analysis revealed that elaborate hydraulic systems existed to use urban waters as efficiently as possible in the Late Middle Ages. The most important norms were established, even though neither laws nor many privilege charters awarded to various urban communities explicitly mentioned water rights and privileges regarding waterways. The problems in urban environments were somewhat distinct from those involving smaller rivers in rural areas, where most registered disputes concerned landlords with interests in similar activities, primarily milling or fishing. In many instances, the conflicts in urban landscapes centered on the differing interests of ferrymen, fishers, millers, and others. These enduring conflicts attest to the challenges of addressing such disputes.

The manipulation of the waterscapes led to numerous intended and unintended consequences. The silting up of riverbeds, the lack of water to run water mills, flooding other people's lands and mills, and the disannexation of lands all appear in the legal evidence. To operate these systems, intricate arrangements and infrastructure, as well as both physical and human capital, were necessary. Millers and ditch diggers played a significant role in these arrangements. Millers were not only involved in milling but also served as key personnel in water construction and regulation during the premodern period.

The book focused on the late medieval period, but in some cases, certain norms and arrangements appear to have longer trajectories. Some of the seventeenth- and eighteenth-century lawsuits cited in the book sought to demonstrate the enduring nature of these norms in Hungary. Most of the issues studied in the context of the Late Middle Ages remained largely unchanged until the nineteenth century. However, with the development of the power grid, many water mills were quickly abandoned. Very few premodern structures remained by the beginning of the twentieth century. Fewer constructions led to less competition for water resources, but this does not mean they completely disappeared. Conflicts took different forms, which are briefly outlined in the epilogue to this concluding chapter.

5.1 The Politics of Water: An Epilogue

Conflicting interests in water use are not just a premodern phenomenon. As coal use spread throughout Europe and the power grid developed, water mills—which, as discussed, are key to transforming riverscapes, especially along smaller water bodies—were quickly abandoned. Although mills powered by water wheels became obsolete in the nineteenth and early twentieth centuries, and freshwater fishing declined significantly relative to the premodern era, conflicts over the use of waterways persisted. New actors and interests emerged in modern times regarding the exploitation of rivers and riverbanks—the rights of landowners along these water bodies created just as many conflicts as in the previous millennium.

Most factors influencing legal conflicts regarding water usage did not change significantly between the Middle Ages and the Modern period. These factors include natural processes such as hydromorphological changes. Struggles over the ownership of alluvial islands have persisted into modern times, as exemplified by the Danube islands between Bulgaria and Romania.[5] Access to water and riversides has remained just as important as in the premodern period. While access to drinking water, fish catches, and milling potential were most critical in the premodern era, a new use—recreation—has gained increasing importance over the past century and a half, followed by an even more recent concern: environmental protection. The Gabčíkovo–Nagymaros Dams mentioned in the book's introduction clearly illustrate both factors.

This hydrological construction proved to be one of the most controversial projects in the history of both Hungary and Czechoslovakia. The project had already been initiated during the interwar period. At that time, the two primary motivations were to improve navigation along the Danube and to exploit waterpower to produce inexpensive electricity. However, partly due to financial constraints and partly for political reasons, the project never entered the execution phase before World War II. In the 1950s, the project once again gained traction, and with the Warsaw Treaty and the establishment of Comecon, it became a symbolic endeavor. In 1963, Czechoslovakia and Hungary signed a treaty to initiate the planning efforts. From the project's inception, controversial opinions about the dams were expressed. Increased navigation, affordable energy, flood protection, and even water levels were traditional arguments for constructing similar dams. A new argument in this case was the potential enhancement of recreational opportunities. This idea was emphasized in contemporary media to popularize the concept of such dams.[6] Meanwhile, strong skepticism arose from various perspectives from the outset. Beyond technical concerns, by the 1980s, environmental protection had come to dominate discussions, displacing other sources of criticism. Ultimately, only a portion of the project was completed, resulting in prolonged political and legal disputes between the newly formed Slovakia and Hungary, with the protection of habitats in the floodplains of the Szigetköz area (in northwestern Hungary, on the border with Slovakia) becoming the most critical issue.[7]

This is not the only instance in which nature and habitat protection have become a significant source of conflict. In recent years, a flood control dike system has been planned for the northern part of Buda, along the so-called Római-part (meaning "Roman Bank").[8] The controversial dike construction that threatens one of the last undeveloped riverbanks in Budapest was intended to protect some houses that had been illegally built in the river floodplain. The homeowners, of course, are interested in protecting their properties, while others are eager to preserve the riverbanks, partly for environmental reasons and partly because this area is one of the most popular riverside recreational spaces in the Hungarian capital. The list of modern water-resource conflicts could be expanded indefinitely. Still, the purpose of recalling these two recent cases is to illustrate that issues concerning water in premodern Hungary remain salient. The conflicts persist, and their resolution does not seem any more straightforward than it was in the Middle Ages.

Notes

1 Tamás Vajda, "Földesúri jogok és kötelezettségek a középkori vízimalmok körül," [Seigniorial right and duties connected to medieval water mills] in *Víz és társadalom Magyarországon a középkortól a XX. századig* [Water and society in the Hungary from the Middle Ages to the 20th century], ed. Gergely Krisztián Horváth (Budapest: Balassi, 2014), 85.

2 István Tringli, "A magyar szokásjog a malomépítésről," [Customary law on construction of mills] in *Tanulmányok a középkorról* [Studies in medieval history] (Analecta Medievalia, 1), ed. Tibor Neumann ([Budapest–Piliscsaba]: Pázmány Péter Katolikus Egyetem, 2001), 261–265.

3 Cf. J. Getzler, *A History of Water Rights at Common Law* (Oxford: Oxford University Press, 2004), 338. https://doi.org/10.1093/acprof:oso/9780199207602.001.0001.

4 Cf. Urszula Sowina, *Water, Towns and People. Polish Lands against a European Background until the Mid-16th Century* (Polish Studies – Transdisciplinary Perspectives, 17) (Frankfurt am Main: Peter Lang, 2016) or *Wasser Stadt Wien – Eine Umweltgeschichte*, ed. ZUG – Zentrum für Umweltgeschichte, Universität für Bodenkultur Wien (Vienna: Universität für Bodenkultur Wien, 2019).

5 Stefan Dorondel, Stelu Serban, and Daniel Cain, "The Play of Islands: Emerging Borders and Danube Dynamics in Modern Southeast Europe (1830–1900)," *Environment and History* 25 (2019): 521–547. https://doi.org/10.3197/096734018X15254461646413 (with further modern examples).

6 See e.g., an interview published in the journal of the Hungarian National Defense Sports Association (Magyar Honvédelmi Sportszövetség): *Look at what has been built on the Váh. See how the backed up water transformed the climate. All I could say, that is a fairy garden. (...) If we can do it at least at a comparable quality to that our northern and western neighbors did, I suggest, let's build as fast as possible.* Emil Kiss, "Mit jelent a 885 megawatt?" [What does 885 Megawatts mean?] *Lobogó* June 14, 1979.

7 See Judit Galambos, "An International Environmental Conflict on the Danube: the Gabcicovo – Nagymaros Dams," in *Environment and Democratic Transition in Central and Eastern Europe. Policy and Politics in Central and Eastern Europe* (Technology, Risk and Society, 7), eds. Anna Vári, and Pál Tamás (Dordrecht–Boston–London: Springer, 1993), 146–156, and John Fitzmaurice, *Damming the Danube: Gabčikovo and Post-Communist Politics in Europe* (Boulder, CO and Oxford: Westview, 1996).

8 Aron Buzogány, Szabina Kerényi, and Gergely Olt, "Back to the Grassroots? The Shrinking Space of Environmental Activism in Illiberal Hungary," *Environmental Politics* 31 (2022): 1267–1288. http://doi.org/10.1080/09644016.2022.2113607.

APPENDICES

Appendix A Documents

No. 1.
MNL OL DL 102 905. (transcription: MNL OL DL 58 505). Edited, with some gaps from the transcription in: AO, vol. 3. 597–598. no. 394., summary: AOklt, vol. 2. 250 no. 529.
Date: October 11, 1339

The palatine, Nicholas Druget, orders the reinstatement of the lands between Čoltovo and Lekeňa (today part of Bohúňovo) to the sons of Nicholas Forgács.

Villermus Drugeth palatinus et iudex cumanorum sincere sibi dilectis comitibus a quatuor judicibus nobilium de comitatu Gumuriensi comitatu sue dilectionis plenitudinem cum salutem bene vobis constare credimus, quod quia Paulus filius Galli de Choltou in nostra congregatione generali universitati nobilium de comitatu Gumuriensi feria tercia ante festum beati Egidii abbatis in villa Gumur una nobiscum celebrata de medio universorum consurgendo, Nicolaum et Andream filios Nicolai dicti Fargach quandam particulam terre Choltou predicte occupasse dicebat, prohibendo eosdem ab occupatione eiusdem, et litere honorabilis capituli ecclesie Agriensis per utramque partem ibidem exhibite quendam rivulum Hoblukapataka vocatum pro meta assignatum inter possessionem eorumdem filiorum Nicolai Lekunye vocata et inter possessionem eiusdem Pauli Cholthou predictam incipiendo a metis filiorum Scheruh usque ad fluvium Sayo vocatum separantem lucide exprimebant, prefatus vero Andreas filius Nicolai prefatum rivulum Hablucapataca vocatum de suo antiquo meatu per eundem Paulum in alium cursum dimissum et eandem particulam terre intra suas metas scilicet intra verum et antiquum meatum eiusdem rivuli esse allegabat et Dominicus filius Thyba homo noster unacum Philippo filio Pangracii de Iwanussy uno ex vobis quatuor iudicibus nobilium ad videndam et sciendum inter partes super premissis veritatem per nos transmissi, redeundo retulerunt requisiti, quod predictus rivulus Hablukapataca descendens de medio quorumdam moncium in declivio quarundam terrarum arabilium per inundationes aquarum de suo antiquo meatu per quandam supplecionem suam meati in alium cursum non perhumanam operationem sed per proprium cursum declinasset ipsaque particula terre littigiosa inter antiquum meatum seu cursum eiusdem rivuli a parte possessionis eorumdem filiorum Nicolai Lekunye vocatam remaneret et usque nunc iidem filii Nicolai usi extitissent, prout a vicinis et commetaneis suis rescire potuissent sed nos una nobiscum a nobilibus dicti comitatus taliter decrevimus quod idem Paulus filius Galli cum iudicio trium marcarum eandem particulam terre eisdem filiis Nicolai prout eorum ius resignare teneantur eosdem idem Paulus prefatum hominem nostrum et vestrum inquisitores super premissis veritatis ad suum gravamen instanter mitti postularat, et insuper in estimatione eiusdem particule terre litigiose convictus habeatur contra filios Nicolai memorata quo circa petimus vos et nichilominus precipimus vobis firmiter per presentes mandantes quatenus presentibus unum vel duos ex vobis transmittatis quo vel quibus presentibus Gallus filius Thyba de Rekethye vel Georgius filius Johannis de Malah altero absente homo noster in octavis festi Sancti Michaelis archangeli [October 6] nunc venturis ad faciem predicte particule terre littigiose accedendo eandem presentibus vicinis commetaneis suis legitime convocatis reambulando respectu omnium utilitatum eiusdem estiment in valore suo regni consuetudine requirente et statuat eisdem filiis Nicolai dicti Fargach si non fuerit per aliquos alios contradictum contradictores autem si qui alii apparuerint ad nostram citet presenciam ad terminum competentem etc. post hec totius facti seriem si expediens fuerit nomina citatorum et terminum assignatum ad octavum diem predictarum octavarum in villam Wysul rescribtatis. Datum quinto die congregationis predicte in loco memorato, anno domini millesimo trecenstimo tricesimo nono.

[At the back, with different handwriting:]
Sincere sibi dilectis comitibus et quatuor judicibus nobilium de comitatu Gumuriensi pro Andrea filio Nicolai dicti Forgach super statucione cuiusdam particule terre in octavis sancti Michaelis archangeli nunc venturis fienda memorie octavas die ad judicem.

No. 2.
MNL OL DL 92 239; summary: ZsO, vol. 2/1. 446 no. 3726.
Date: March 17, 1405, Buda

King Sigismund orders the chapter of Vasvár to investigate the claim made by John and Nicholas, sons of Pető of Gerse.

Sigismundus dei gratia rex Hungariae, Dalmatiae, Croatiae etc. marchioque Brandenburgensis etc. Sacri Romani imperii vicarius generalis et regni Bohemiae gubernator fidelibus suis capitulo ecclesiae Castriferrei salutem et gratiam dicunt nobis magistri Johannes et Thomas filii Petheu de Gerse ut quia ipsi pro molendinis ipsorum quendam fluvium Sarvize vocatum inter possessiones Gerse predictam et Sarmelleke ipsorum ac possessionem Telekus nobilium de eadem metaliter currere et separatae de antiquo suo alveo seu meatu per fossatum in superficie terre ipsus possessionis Sarmelleke recepisset dictusque antiqus meatus seu alveus dicti fluvii per continuas inundationes aquarum et infusiones arenarum adimpleri coepisset ex eo nobiles de dicta Telekus dictum fluvium Sarvize in ipso fossato currere et metaliter separare et distinguere asserere cogerentur per quod factum magnam particulam ipsorum signanter pratorum occupare intenderant in preiudicium ipsorum valde magnum super quo fidelitati vestre firmiter precipiendo mandamus quatenus vestrum mittatis hominem pro testimonio fidedignum quo presente Johannes filius Andrea de Telekus praedictam vel Johannes de Sarfeu aut Brictius de eadem seu Paulus de Karathfeld allis absentibus homo noster ab omnibus quibus incumbit meram de premissis experiatur[1] veritatem quam tandem nobis fideliter rescribatis. Datum Budae feria tertia proxima post festum Beati Gregorii. Anno Domini millesimo quadringesimo quinto.

No. 3.
MNL OL DL 68 950; summary: ZsO, vol. 2/1. 641. no. 5091.
Date: November 5, 1406, Košice

The alispán (vicecomes) of Sáros County proved that the claim of John, son of John, son of Rikalf, and John, son of Coloman, son of Rikalf, was valid, and that the lands driven by the Poprad to Orlov had initially belonged to them.

Nos Nicolaus de Gara regni Hungarie palatinus et judex Cumanorum memorie commendamus quod in congregationem nostram generali universitati nobilium Abawjvariensis et de Sarus comitatuum feria secunda proxima ante festum beatorum Symonis et Iude apostolorum prope civitatem Cassa celebrata Johannes filius Johannis filii Rykolphi ac alter Johannes filius alterius filii dicti Rykoplhi de Tarkw de medio aliorum exurgendo proposuerunt eo modo quod particulam terre quam antiquus meatus fluvii Pabrad a possessione eorum Lybytynye vocata excidisse et ad possessionem regalem Orlw vocatam annexisse semper et ab antiquo ad dictam possessionem eorum Lybytynye pertinentes et usa extitisse (...) debet pertinere. Ut hoc vicecomiti nostro judicibus nobilium juratisque assessoribus dicti comitatus de Sarus constaret evidenter petentes nos cum instancia ut iidem vicecomes necnon iudices nobilium iuratique assessores qualiter de premissis scirent per nos requisiti faterentur veritatem cumque nos eosdem vicecomitem nostrum judices nobilium ac juratos assessores super premissis requisitos habuissemus iidem ad fidem eorum deo debitam fidelitatemque domino nostro regi et suo sacro dyademati conservandam pro dicenda veritate et iustitia observandam (...) si dicta particula terre modo prehabito per dictum antiquum meatum ipsius fluvii Papbrad a dicta possessione Libitynye excisa et ad dictam possessionem Orlow annexa semper ab antiquo ad dictum possessione Libitynye pertinens et usa extitisset et nunc deberet pertinere unanimi et concordi testificarunt affirmarunt in cuius assertionis testimonium presentes litteras nostras patentes eisdem utriusque Johanni duximus concedendas communi iustitia exigente. Datum duodecimo die congregationis nostre predicte in loco memorato. Anno domini millesimo quadringentesimo sexto.

No. 4.
MNL OL DL 15 540.
Date: January 9, 1461

The convent of Zalavár confirms that Ladislas Therek of Pat and Nicholas and Ladias, sons of Matthew of Mátésfalva, agree to share their rights to construct mills and fishponds along the River Sárosd, which separates their estates.

Nos conventus monasterii beati Adriani martyris de Zala, memorie commendamus quod egregius Ladislaus Therek de Path, ab una parte vero ex altera, Nicolaus et Ladislaus filii Mathei de Mathesfalwa coram nobis personaliter constituti per eosdem communiter confessum extitit pariter et relatum in hunc modum quod si qua pars ipsorum in fluvio Sarusd inter predictas possessiones Path et Mathesfalwa vocatas decurrenti in Comitatu Simighiensi existenti molendinum seu piscinam ordinaret et appromtuaret aut ordinari et appromtuari faceret necnon uni et alteri eorum per inundationes aquarum exinde dampna evenirent, extunc nulla pars earundem, super unam et alteram eorum reacquirere ac uni et alteri eorum contradicere ac in piscinis unius et alteri ipsorum modo simili nulla pars contra partem alteram piscari facere valeat atque possit hoc etiam per ipsas partes specialiter declarato et specificato, ut si ipsi Nicolaus et Ladislaus, aut ipsorum heredes et posteritates universi predictam possessionem eorum Mathesfalwa, aliquando temporis in eventu perpetuum vendere aut in concambium alieni vel aliquibus donare tradere et contulere [sic!] seu propigneri obligare conarentur extunc nulli alteri nisi praefato egregio Ladislao Therek aut suis heredibus et posteritatibus dare et conferri valeant atque possint e converso vero prefatus Ladislaus Therek aut sui heredes et posteritates universi possessionem suam Mendzenth vocatam in predicto Comitatu Symigiensi existentem modo simili aliquando temporibus [sic!] in processu perpetuo vendere aut in concambium aliorum vel aliquibus donare trdere et contulere seu proprigneri obligare niterentur extunc nemini alteri, nisi solum modo prenominatis Nicolao et Ladislao aut ipsorum heredibus et posteritatibus dare et conferri valeant atque possit ad quod se partes praedicte sponte obligarunt coram nobis vigore et testimonio praesentium mediante. Datum feria sexta proxima post festum Epiphaniarum, domini anno eiusdem Millesimo quadringentesimo sexagesimo primo.

No. 5.

MNL GYMSML SL IV. 1009. a. vol. 6. 1540/1541. 76–82. p.[2]

The account of rebuilding the moat and Foregate Mill at Sopron in 1540.

[76. *p.*]

Hernach volgt, so ich ausgeben zu z dem paw der muel im graben beim Vorderthor auf stainmetzen, maurer, deichtknecht, zimmerleut und umb holczwerg, stain und ander noturft darzue etc., anno 40.

Item dem Berthl stainmetzen und seinen gesellen gelont am samstag nach Invocavit [*21 February*] 6 tag stainhauen, je 1 tag 20 cr., facit 1 t. d.

Item adi gelont dem Mertl Peham deichtknecht und seinen gesellen 7 tag zu 7 cr. 4 tag in 6 cr. im zwingl die erd auszustechn und ausfuren, facit

1 tal. 1 sol. 22 den.

Samstag nach Reminiscere [*28 February*] den stainmetzen gelont stain hauen 12 tag zu 10 cr. facit

2 tal. den.

Adi ainem gelont, so inen im Judenhof geholfen stain brechen 1 tag

24 den.

Samstag nach Oculi [*6 March*] dem Frantzen und den andern stainmetzen stain prechen und hauen gelont 20 tag zu 10 cr., facit

3 tal. 2 sol. 20 den.

Adi dem deichtknecht 3 ½ tag zu 7 cr. seinen gesellen [5]3 tag zu 6 cr. erd auszuschieben und den pastein zu schütten etc., facit

5 tal. 5. sol. 20 den.

Lateralis 13 tal. 2 sol. 26 den.

[77. *p.*]

Item samstag nach Letare [*13 March*] dem Frantzen und seinen gesellen gelont 12 tag an dem muelgraben zu maurern per 10 cr. ainem raicher, 5 tag zu 6 cr., facit

2 tal. 4 sol. den.

Adi dem deichtknecht gelont 5 tag zu 7 cr. seinen gesellen, 20 tag zu 6 cr., facit

2. tal. 4. sol. 20. den.

Item freitag nach Iudica [*19 March*] dem Frantzen gelont 19 tag zu 10 cr. ainem merterknecht, 5 tag zu 6 cr. ainem weib, so geraicht 4 tag zu 4 cr., facit

3 tal. 7 sol. 14 den.

Item adi dem deichtknecht 5 tag zu 7 cr., seinen gesellen 15 tag zu 6 cr., facit

2 tal. 20 den.

Item am osterabent [*28 March*] dem Frantzen mauern gelont 20 tag zu 10 cr,. dem ~~raicher~~ merterknecht[3] 5 tag zu 7 cr., den weibern so geraicht 15 ½ tag zu 4 cr., facit

4 tal. 7 sol. 28 den.[4]

Adi dem deichtknecht ~~12 tag zu 7 cr.~~ 6 tag per 1 sol., seinen gesellen 12 tag zu 7 cr., facit

2 tal. 1 sol. 6 den.

Am samstag darnach [*3 April*] gelont mauern 22 tag zu 10 cr., dem merterknecht 4 tag zu 7 cr., den weibern 8 tag zu 4 cr., facit

4 tal. 5 sol. 8. den.[5]

Dem deichtknecht 4 tag per 1 sol., seinen knechten 14 tag zu 7 cr., 8 tag zu 4 cr., facit

2 tal. 5 sol. 10 den.

Lateralis 25 tal. 4 sol. 18 den.[6]

[*78. p.*]
Item samstag nach Quasimodo [*10 April*] dem maurern gelont 34 tag zu 10 cr., dem merterknecht 6 tag per 7 cr., raich 8 taglon zu 4 cr., facit

6 tal. 7 sol. 6 den.

Item di dem deichtknecht 5 ½ tag per 1 sol., den knechten 21 tag zu 7 cr., 16 ½ tag zu 4 cr. facit

4 tal. 1 sol. 23 den.[7]

Samstag nach Misercordia [*17 April*] den maurern gelont 27 tag zu 10 cr., dem merter knecht 6 tag zu 7 cr., raich 12 tag zu 4 cr., facit 6 tal. den.

Adi dem deichtknecht 5 tag per 2 sol., 48 taglon zu 7 cr., 16 tag zu 4 cr., facit

7 tal. 2 sol. 22 den.[8]

Item freitag vor Georgii [*23 April*] den maurern gelont 23 tag zu 10 cr., ainem merterknecht und aim raicher jedlichen 5 tag zu 7 cr., mer 5 tag raich zu 4 cr., facit

5 tal. 2 sol. 20 den.

Adi dem deichtknecht 5 tag per 1 sol., den knechten 25 tag zu 7 cr., zwaien weibern 10 tag zu 4 cr., facit 4 tal. 2 sol. 20 den.

Freitag nach Georgii [*30 April*] dem maister Frantzen zalt mauern 19 tag werg zu 10 cr., dem raicher und merterknechten 8 tag werg zu 7 cr., 3 tag per 4 cr., facit

4 tal. 1 sol. 10 den.[9]

Lateralis 38 tal. 1 sol. 11 den.

[*79. p.*]
Ime dem statmaister und seinen gesellen von den gewellen umb schloss wein und zu ainem padgelt geben 2 sol.

Item adi dem Merten deichtknecht 5 taglon per 1 sol., seinen gesellen 29 tag zu 7 cr., den weibern 9 tag zu 4 cr., facit

4 tal. 4 sol. 24 den.[10]

Item samstag vor Exaudi [*8 May*] dem Merten gelont 4 tag zu 1 sol., seinen gesellen 20 tag per 7 cr., facit

2 tal. 6 sol. 20 den.

Samstag vor pfingsten [*15 May*] ime gelont 6 tag zu 1 sol., 29 tag zu 7 cr., facit

3 tal. 7 sol. 22 den.[11]

Freitag darnach [*21 May*] ime gelont 3 tag zu 1 sol., seinen gesellen 13 tag zu 7 cr., den damm gar zu machen, facit 1 tal. 2 sol. 4 den.

Freitag vor Viti [*11 June*] 4 halb taglon gelont im graben bei der müel zu schutten und gleichen, facit 1 sol. 26 den.

Item dem Mertel deichtmaister zu pesserung geben, das er die pastein, die dämm geschütt und öbristen ist gewesen, die knecht erhalten, facit

4 sol. den.

Lateralis 14 tal. 2 sol. 6 den.

[*80. p.*]

Vermerkt, was auf den maister Urban Kümpf mülner, so die müel paut und seine gesellen gangen, anno 40

Samstag nach Invocavit [*14 February*] hab ich ime 1 wochen lon geben, darumb er zu der arbait wochenlich zu haben bestelt worden und mit weniger nemen, facit

2 tal. 2 sol. den.

Seinen gesellen 37 tag zu 10 cr., ainem von Agendorf[12] 2 tag 12 cr., dem Philippen 1 taglon 8 cr. holz zu schlagen zu der müell, facit

8 tal. 6 sol. den.[13]

Item samstag nach Reminiscere [*28 February*] ime dem maister Urban sein wochensolt geben 18 sol., seinen gesellen 9 tag zu 10 cr., das zimmerholtz aufzuschneiden, facit

3 tal. 6 sol. den.

Samstag nach Oculi [*6 March*] ime seinen wochensolt geben 18 sol., seinen gesellen 19 taglon, in her Frantz Sübenbürger waldholtz zu schlagen und zimern je 1 tag per 10 cr., facit

5 tal. 3 sol. 10 den.

Samstag nach Letare [*13 March*] seinen gesellen gelont, dann er ain Pfinstag darvor hinweg geritten, 36 tag zu 10 cr., facit 6 tal. den.

Freitag nach Iudica [*19 March*] ime dem maister Urban, dieweil er nit anhanng gewest, dise wochen nit bezaltn seinen gesellen 29 taglon bezalt per 10 cr., facit

4 tal. 6 sol. 20 den.

Lateralis 31 tal. den.

[*81. p.*]

Item am osterabent [*28 March*] ime dem maister die halb wochen gelont ~~vo~~ nach Letare, da er hinweg ist, und diese wochen thuet 1 ½ wochenlon, seinen gesellen 20 tag zu 10 cr., facit

8 tal. 10 den.[14]

Item samstag nach dem ostertag [*3 April*] ime sein wochensolt 18 sol., seinen gesellen 38 taglon zu 11 cr., dan er inen so viel verhaissen, facit

9 tal. 1 sol. 22 den.

Item samstag nach Quasimodo [*10 April*] sein wochenlon 18 sol., seinen gesellen 82 tag zu 11 cr., facit

17 tal. 2 sol. 8 den.

Samstag nach Misericordia [*17 April*] ime sein wochensolt 18 sol., seinen gesellen 74 ½ taglon zu 11 cr., facit

15 tal. 7 sol. 8 den.

Item freitag vor Georgii [*23 April*] ime sein wochensolt geben 18 sol., seinen gesellen gelont 58 tag zu 11 cr., facit

12 tal. 3 sol. 2 den.[15]

Item freitag nach Georgii [*30 April*] ime sein wochensolt geben 18 sol., seinen gesellen 69 tag zu 11 cr., facit

14 tal. 7 sol.6 den.

Lateralis 77 tal. 5 sol. 26 den.

[*82. p.*]

Samstag vor Exaudi [*8 May*] ime sein wochensolt geben 18 sol., seinen gesellen 62 ½ taglon zu 11 cr., facit 11 tal. 3 sol. 20 den.

Samstag vor Phingsten [*15 May*] ime sein wochensolt 18 den., den gesellen 84 taglon zu 11 cr. thuet, facit 17 tal. 5 sol. 6 den.

Item adi seinen gesellen zu ainem padgelt, da sie gar fertig worden

4 sol. den.

Item in der phingstwochen [*16 May–22 May*] zwaien muelknechten gelont 12 taglon zu 11 cr., die müelstain zusammenzufüren und die müel gar zu fertigen, facit

2 tal. 1 sol. 18 den.

Item ime den maister Urban zu ainer pesserung und abfertigung auss beuelch meiner herren, dieweil er das paw erlich und wol verricht, geschenken, damit er auch was hinfüran doran zu wenden oder pessern zu thuen bewilligt und thuet. 10 tal. den.

Lateralis 41 tal. 6 sol. 14 den.

[*83. p.*]

Umb holtzwerg, eisenzeug und ander noturft zu der muel etc.

Item freitag vor Viti [*11 June*] zalt ich 18 laden schwartz grossen, die wasserrinnen auss der insetz zu bedecken und ander noturft, je ainen per 16 cr., thuet

4 tal. 6 sol. 12 den.

Item samstag darnach [*19 June*] dem waldpaurn der die rinnen pracht, den rest auf die 1 tal. 44 cr., so er vormals daran emphangen und ich verrait geben

1 tal. 5 sol. 4 den.

Ime umb 6 lang feichten laden zu 10 cr., ainen umb 5 rotferhen[16] (?) laden zu 4 cr., geben

1 tal. 2 sol. 20 den.

Item mer kauft ich damals 8 dick rotferhen (?) laden zu 12 elen, je 1 per 10 cr., facit

1 tal. 2 sol. 20 den.

Samstag nach Colomanni [*16 October*] kauft ich mer 15 schwartz grossin laden, je ainen per 8 cr., facit

2 tal.

Item dem maister Urban fur die zerung auf 3 raisen, so er hieher gethan von wegen der müel und wassergraben zu besichten und das paw aufzunemen 2 tal. und, so er zu zwaien maln hie verzert, selbander 4 sol. den. thuet 2 tal. 4 sol. den.

Lateralis 13 tal. 4 sol. 26 den.

[*84. p.*]

Item umb 50 flider laden, so maister Urban kauft, geben	5 tal. den.
Item kamp rad laden 4, je 1 per 6 cr., facit	3 sol. 6 den.
Umb 8 pieg (?) aufs rad	1 tal. den.
Umb 10 melstettl laden zu 6 cr. geben	1 tal.
Umb ain ler paumen grindel	4 sol. den.
Umb ain par ärm	2 sol. 20 den.
Umb 5 fueder gemain laden, je ains per 4 sol. den., facit	2 tal. 4 sol.
Item imb 300 nagel zum fliederrad	1 tal. den.
Umb 7 dick laden, so er nachmals kauft	7 sol. den.

Item da die wägen nach den zeug gefaren und der mair mit dem statzug hat der wasserbauch und er verzert und zu maut gebn 5 sol. 6 den.

Montag nach Palmarum [*22 March*] umb 4 sol. latten zum muelhauss zu decken, geben

1 tal. 3 sol. den.

Dem Clement Irher umb 3 h. (?) leim umb 1 h. 5 cr. erben 2 sol. den.

Lateralis 14 tal. 7 sol. 2 den.

[*85. p.*]
Item den fuerleuten, so maister Urban bestellt von dem überigen holzwerg, so sie auf 4 wägen gefürt, je von aim wagen 12 sol. den. und umb 1 fueder, so sie auch bracht 4 sol. 10 den. zu maut 4 cr., facit

6 tal. 4 sol. 26 den.

Item am montag Misericordia [*12 April*] kaufft ich zu Wienn ain par stain auf der müel, darumb geben 10 tal., dem jungen Simänl und seinen gesellen den fuer leuten darvon haimzuführen 1 tal. den., facit 11 tal. den.

Item am samstag nach pfingsten [*22 May*] mit maister Urban abgerait all ander zeug und noturft, so ime noch umbezalt ist beliben, 1 cr., 35 h. (?) eisen zeug, je 1 h. per 10 den., 40 kämp, je 1 per 1 den., 10 spindel, je aine per 1 den., 29 flider laden, je 1 per 7 cr., zwen ärm schwartz forhens(?) holz, so noch verhanden, 1 per 4 cr., thuet alles

9 tal. 2 sol. 24 den.

Lateralis 26 tal. 7 sol. 20 den.

Notes

1 The scribe accidentally repeated the word.
2 For the principles of the edition, see: Katalin Szende, "A magyarországi középkori német nyelvű források kiadásának elvei és gyakorlata," [Principles and practice of editing medieval German-language sources in Hungary] *Fons* 7 (2000): 49–62.
3 Added posteriorly to the margin of the line.
4 Miscount.
5 Miscount.
6 Miscount.
7 Miscount.
8 Miscount.
9 Miscount.
10 Miscount.
11 Miscount.
12 The village of Ágfalva close to Sopron.
13 Miscount.
14 Miscount.
15 Miscount.
16 If the word is indeed *rotferhen*, it is baltic redwood (*pinus sylvestris sylvestris*), see Matthias Lexer, *Mittelhochdeutsches Handwörterbuch*, vol. 3: (Leipzig Hirzel: 1878), 469.

Appendix B List of Charters of Privilege studied in Chapter 3.3

	Date	*Privileged settlement*	*Edition / Archival registry number*
1	Before 1241	Rodna	-
2	(1243) – 1255	Banská Štiavnica	Ilpo Tapani Piirainen, *Das Stadt- und Bergrecht von Banská Štiavnica/ Schemnitz: Untersuchungen zum Frühneuhochdeutschen in der Slowakei* (Oulu: Universität Oulu, 1986), 31.
3	1244	Krupina	CDES, vol. 2. 113–114 no. 168.
4	1255	Banská Bystrica	CDES, vol. 2. 340–341 no. 491.
5	1261	Dej	ÁÚO, vol. 8. 9–11 no. 6.
6	1265	Hybe	VMMS, vol. 1. 48–50 no. 33. (B)
7	1263 /1276 (?)	Partizánska Ľupča	VMMS, vol. 1. 44–45 no. 29.
8	1268	Rimavská Baňa	ÁUO, 8. 212–213 no. 143.
9	1287	Gelnica	VMMS, vol. 1. 67–68 no. 66 and Almási, "Megjegyzések Gölnicbánya," 46–49.
10	1291	Ocna Dejului	ÁÚO, vol. 5. 20–22 no. 13.
11	1291	Turda / Cojocna / Sic	*Urkundenbuch zur Geschichte der Deutschen*, vol. 1. 181–182 no. 249.
12	1318, 1340	Ružomberok	VMMS, vol. 1. 91–92 no. 103 and VMMS, vol. 1 132–133 no. 161.
13	1325	Baia de Arieş	*Urkundenbuch zur Geschichte der Deutschen*, vol. 1. 395–396 no. 437.
14	1327 (confirmed in 1338)	Smolník	VMMS, vol. 1. 128–129 no. 155.
15	1328	Kremnica	VMMS, vol. 1. 115–116 no. 136.
16	Before 1329 (?) (confirmed in the fifteenth century)	Pukanec	CD, vol. 8/5. 425–426.
17	Before 1329	Baia Sprie	Gusztáv Wenzel, *Magyarország bányászatának kritikai története* [Critical history of mining in Hungary] (Budapest: M. Tud. Akadémia Könyvkiadó Hivatala, 1880), 408–414.
18	1341	Telkibánya	Wenzel, *Magyarország bányászatának*, 346–349.
19	1347 (before 1329) (confirmation)	Baia Mare	CD, vol. 9/1. 498.
20	1349	Zlatá Idka	VMMS, vol. 1. 163–164 no. 205.
21	1357	Zlatna	MNL OL DL 36 543.

	Date	*Privileged settlement*	*Edition / Archival registry number*
22	1376	Baia Mare / Baia Sprie	*Documente privind istoria României*, 673–676 no. 577.
23	1380	Brezno	CD, vol. 9/5. 390–391.
24	1382	Ľubietová	CD, vol. 9/5. 577.
25	1453	Banská Belá	MNL OL DF 235 588.

Appendix C List of Cases Studied in Chapter 3.4

	Date	*Archival registry number*	*Summary / Reference*	*Edition*
1	April 6, 1305	MNL OL DL 1687.	AOklt, vol. 1. 347–348 no. 719.	AO, vol. 1. 94–95 no. 87.
2	January 6, 1338	MNL OL DF 249510.	AOklt, vol. 22. 12–13 no. 5.	–
3	September 4, 1339	MNL OL DL 102 905.	AOklt, vol. 23. 250–251 no. 529 and Kiss, *Floods*, 347 and 628.	AO, vol. 3. 597–598 no. 394 and App. no. 1.
4	October 6, 1339	MNL OL DF 248 651.	AOklt, vol. 23. 284–285 no. 614. See: Kiss, *Floods*, 628.	MES, vol. 3. 351–352.
5	January 2, 1340	MNOL OL DL 11 742.	AOklt, vol. 24. 9 no. 2.	–
6	March 19, 1340	MNL OL DL 58 506.	AOklt, vol. 24. 72–73 no. 154.	AO, vol. 4. 10–12 no. 10.
7	February 22 to March 7, 1347	MNL OL DL 3932.	AOklt, vol. 31. 93–94 no. 131 and 445–447 no. 862. See: Kiss, *Floods*, 385–386.	AO, vol. 5. 118–120 no. 55.
8	May 21 and June 14, 1348	MNL OL DL 64 031.	AOklt, vol. 32. 169 no. 291 and 194 no. 355.	–
9	August 31, 1348	MNL OL DL 64 661, 68 880, and 68 881.	AOklt, vol. 32. 285–286 no. 573, 611. 303–304 no. 639, and 340 no. 694.	–
10	October 5, 1349	DL 68 895, 68 894, and 68 917.	AOklt, vol. 33. 254–255 no. 504, 255 no. 505, 335–336 no. 684., and 364 no. 745. See: Kiss, *Foods*, 391–392.	–
11	November 6, 1354	MNL OL DF 200 928, 200 929. and MNL OL DL 15 680.	AOklt, vol. 38. 338 no. 462.	AO, vol. 6. 244–249 no. 160.

	Date	*Archival registry number*	*Summary / Reference*	*Edition*
12	January 11, 1357	MNL OL DL 31 283.	AOklt, vol. 40. 250–251 no. 529. See: Kiss, *Floods*, 398.	AO, vol. 6. 525–529 no. 331.
13	May 19, 1359	MNL OL DL 68 916.	MNL OL database	-
14	October 28, 1374	MNL OL DL 70 662, 71 545, and 71 550.	Tringli, *A Perényi*, 89 no. 131.	*Oklevéltár a gróf Csáky család történetéhez*, 2 vols [Cartulary to the history of the Csáky family], ed. László Bártfai Szabó (Budapest: Stephaneum, 1919), vol. 1. 135–148. (with wrong dating)
15	May 10, 1382	MNL OL DL 52 420.	*A nagykállói Kállay-család levéltára*, 2 vols [Cartulary of the Kállay family of Nagykálló] (Budapest: [no publisher], 1943), 219 no. 1917.	Frigyes Pesthy, *Krassó vármegye története*, vol. 3 [History of Krassó County] (Budapest: Athenaeum, 1882), 159 no. 103.
16	June 11, 1382	MNL OL DL 91 893.	Degré, *Magyar halászati jog*, 139, and MNL OL database	*Zala vármegye története. Oklevéltár*, 2 vols [History of Zala County. Cartulary]. Eds. Imre Nagy, Dezső Véghely, and Gyula Nagy (Budapest: [no publisher], 1886), vol. 2. 186 no. 61.
17	July 6, 1387	MNL OL DF 207 457.	ZsO, vol. 1. 15 no. 138.	Dreska, "A pannonhalmi konvent hiteleshelyének," 13 no. 1.
18	November 27, 1393	MNL OL DF 237 891.	ZsO, vol. 1. 357–358 no 3227.Kiss, *Floods*, 628.	-
19	March 25, 1401	MNL OL DL 65 287.	MNL OL database	-
20	March 17, 1405	MNL OL DL 92 239.	ZsO, vol. 2/1. 446 no. 3726.	App. no. 2.

	Date	*Archival registry number*	*Summary / Reference*	*Edition*
21	November 5, 1406	MNL OL DL 68 950.	ZsO, vol. 2/1. 641 no. 5091. sz.	App. no. 3.
22	January 16, 1407	MNL OL DL 34 054.	ZsO, vol. 2/2. 5 no. 5239.	–
23	May 31, 1408	MNL OL DL 9410.	ZsO, vol. II/2. 138 no. 6118.	–
24	June 3, 1408	MNL OL DL 42 946.	ZsO, vol. 2/2. 139 no. 6125.	–
25	January 18, 1411	MNL OL DL 92 385.	ZsO, vol. 3. 84 no. 53.	–
26	February 21, 1411	MNL OL DF 279 140.	ZsO, vol. 3. 106–108 no. 173.	–
27	February 23, 1415	MNL OL DL 53822 and 53 823.	ZsO, vol. 5. 113 no. 218.	*Oklevelek Temesvármegye*, vol. 1. 508–509 no. 310.
28	May 1, 1415	MNL OL DL 53 871.	ZsO, vol. 5. 191 no. 571.	*Oklevelek Temesvármegye*, vol. 1. 511–512 no. 312.
29	May 14, 1416	MNL OL DL 10 465.	ZsO, vol., vol. 5. 509–510 no. 1889.	–
30	January 15, 1418	MNL OL DL 53 984.	ZsO, vol. 6. 380 no. 1377.	*Oklevelek Temesvármegye*, vol. 1. 543–544 no. 343 (with wrong dating).
31	February 15, 1418	MNL OL DL 53 990.	ZsO, vol. 6. 405 no. 1499.	*Oklevelek Temesvármegye*, vol. 1. 546–547 no. 346.
32	April 3, 1419	MNL OL DL 43 628.	ZsO, vol. 7. 99 no. 265.	–
33	February 11, 1420	MNL OL DL 54 110.	ZsO, vol. 7. 314 no. 1345.	*Oklevelek Temesvármegye*, vol. 1. 558–559 no. 361.
34	June 21 to 25, 1420	MNL OL DL 86 495.	ZsO, vol. 7. 432–433 no. 1871.	–
35	August 10, 1421	MNL OL DL 98 381.	ZsO, vol. 8. 251–252 no. 859. See: Kiss, *Floods*, 453.	–
36	March 31, 1425	MNL OL DL 43 628.	ZsO, vol. 12. 160–161 no. 381.	–
37	July 20, 1433	MNL OL DL 106 963.	MNL OL database	–
38	April 12, 1454	MNL OL DL 66 938.	MNL OL database	–

	Date	Archival registry number	Summary / Reference	Edition
39	March 13, 1467	MNL OL DL 16 498.	MNL OL database	–
40	April 12, 1469	MNL OL DL 39 456.	Fekete Nagy, "A Petróczy levéltár," 261–262 no. 202.	–
41	June 1, 1471	MNL OL DL 36 393. 87–88, no 2.	Jakó, *A kolozsmonostori konvent*, vol. 1. 717–718 no. 2022.	–
42	September 18, 1472	MNL OL DL 17 372.	Olexik, "Középkori levéltártörténeti adatok," 270–271 no. 9.	–
43	December 4, 1479	MNL OL DL 65 632.	MNL OL database	–
44	February 24, 1487	MNL OL DL 83 932.	MNL OL database	–
45	May 3, 1496	MNL OL DL 95 726.	MNL OL database	–
46	February 7, 1515	MNL OL DL 106 744.	MNL OL database, K. Németh, "Vizek és vízgazdálkodás, III.," 9. (with wrong dating)	–
47	March 6, 1520	MNL OL DL 30 554.	MNL OL database	–
48	September 26, 1520	MNL OL DL 67 529.	MNL OL database	–
49	December 23, 1520	MNL OL DL 29 981.	MNL OL database	
50	September 1, 1524	MNL OL DL 36 400. 257–259, no. 1.	Jakó, *A kolozsmonostori konvent*, vol. 2. 444 no. 4041.	–
51	February 17, 1525	MNL OL DL 63 037.	MNL OL database	–

Appendix D List of the Kings of Medieval Hungary

Árpádian dynasty
Stephen I (Saint) 1000–1038
Peter Orseolo 1038–1041 and 1044–1046
Samuel Aba 1041–1044
Andrew I 1046–1060
Béla I 1060–1063
Salamon 1063–1074
Géza I 1074–1077
Ladislas I (Saint) 1077–1095
Coloman (the Learned) 1095–1116
Stephen II 1116–1131
Béla II (the Blind) 1131–1141
Géza II 1141–1162
Stephen III 1162–1172
Béla III 1172–1196
Emeric 1196–1204
Ladislas III 1204–1205
Andrew II 1205–1235
Béla IV 1235–1270
Stephen V 1270–1272
Ladislas IV (the Cuman) 1272–1290
Andrew III 1290–1301

Přemysl dynasty
Wenceslas 1301–1305

Wittelsbach dynasty
Otto 1305–1307

Angevin dynasty
Charles I (Robert) 1301–1342
Louis I (the Great) 1342–1382
Mary 1382–1395
Charles II (the Short) 1385–1386

Luxembourg dynasty
Sigismund 1387–1437

Habsburg dynasty
Albert 1437–1439
Ladislas V (Postumus) 1440–1457

Jagiellonian dynasty
Wladislas I (Warnenczyk) 1440–1444

Hunyadi dynasty
Matthias Corvinus 1458–1490

Jagiellonian dynasty
Wladislas II 1490–1516
Louis II 1516–1526

Szapolyai dynasty
John I 1526–1540
John II Sigismund 1540–1551 and 1556–1570; Prince of Transylvania, 1570–1571

Habsburg dynasty
Ferdinand I 1526–1564

BIBLIOGRAPHY

Archival Sources

Arhivele Naţionale ale României [Romanian National Archive]
Serviciul Judeţean Maramureş, Baia Mare [Maramureş County Branch Arhive]
Fond Primăria oraşului Baia Mare [Arhive of the town of Baia Mare]
Seria 2 – Documente privilegiale [Privilegial documents]
Budapesti Történeti Múzeum [Budapest History Museum]
Metszettár [Collection of prints]
Hessisches Staatsarchiv, Marburg
Wilhelmshöher Kriegskarten
Magyar Nemzeti Levéltár [Hungarian National Archive]
Győr-Moson-Sopron Vármegyei Levéltár Soproni Levéltára, Sopron [Sopron City Archive of the Archive of Győr-Moson-Sopron County]
IV. 6. Sopron Város Kamarási Hivatalának iratai [Files of the chamberlain of the town of Sopron]
Kamarási számadások [Account books]
IV.1003.a.
Tanácsülési jegyzőkönyvek (Rathsprotocoll, Protocollum Senatorium) 1625.
Országos Levéltár, Budapest [State Archive]
E Magyar Kamara Archívuma [Archive of the Hungarian Chamber]
41 Litterae ad cameram exaratae
156 Magyar Kamara Archívuma. Urbaria et Conscriptiones
N Regnicolaris Levéltár [Regnicolaris archive]
80 Limitanea Hungarico-Styriaca 1753–1756
P Családi levéltárak [Family Archives]
507 Nádasdy család nádasdladányi levéltára [Nádasladány archive of the Nádasdy family]
Levelezés [Letters]

1314 A herceg Batthyány család levéltára. Missiles [Archive of the Batthyány family. Letters]
1322 A herceg Batthyány család levéltára. Instrukciók [Archive of the Batthyány family. Instructions]
Q Diplomatikai Levéltár Mohács előtti gyűjtemény [Archive of diplomatics. Collection Ante-Mohácsiana]
DL Diplomatikai Levéltár [Archive of Diplomatics]
S Térképtár [Map collection]
11 Kamarai térképek (1747–1882) [Maps of the Hungarian Chamber (1747–1882)]
73 Új szerzeményű térképek (1701–1940) [Newly acquired maps (1701–1940)]
X Másolatok gyűjteménye [Collection of copies]
DF Diplomatikai Fényképgyűjtemény [Archive of photo copies of diplomatics]
Országos Széchényi Könyvtár, Budapest [National Széchenyi Library]
Kéziratár [Manuscript collection]
Cod. Lat.
TK Térképtár [Map collection]
Österreichische Nationalbibliothek, Vienna
Handschriftensammlung
Codex 8609
Österreichische Staatsarchiv, Vienna
Kriegsarchiv
Kartensammlung KPS B
IVa 527. Ungarn
IXa 715. Siebenbürgen
Prímási Levéltár Esztergom [Archiepiscopal Archive, Esztergom]
Acta Radicalia
Soproni Múzeum [Museum of Sopron]
KP 54.509.1
Štátny archív v Prešov [Town Archive of Prešov]
Magistrát Prešov [Magistrate of Prešov]
Zbierka máp a plánov [Collection of Maps and Plans]
Stiftsbibliothek St. Gallen [Library of Saint Gall Abbey]
Codex 1092

Printed Sources

1504–1566, memoria rerum. A Magyarországon legutóbbi László király fiának legutóbbi Lajos királynak születése óta esett dolgok emlékezete. Verancsics-évkönyv [Memory of the things that happened since the birth of Louis II, the son of Wladislas II. Verancsics memorial book] (Magyar Helikon), ed. József Bessenyei. Budapest: Magyar Helikon, 1981.

A nagykállói Kállay-család levéltára, 2 vols *[Cartulary of the Kállay family of Nagykálló].* Budapest: [no publisher], 1943.

A nagy-károlyi gróf Károlyi család oklevéltára, 5 vols *[Cartulary of the Károly family of Carei],* ed. Kálmán Géresi. Budapest: Nyomatott Kocsi Sándornál, 1882–1897.

A pannonhalmi Szent-Benedek-Rend története, I–XII/B *[History of the Benedictine order of Pannonhalma]*, eds. László Erdélyi, and Pongrácz Sörös. Budapest: Stephaneum, 1902–1916.

A Perényi család levéltára 1222–1526 [Cartulary of the Perényi family, 1222–1526] (Magyar Országos Levéltár kiadványai, II. Forráskiadványok, 44), ed. István Tringli. Budapest: Magyar Országos Levéltár, 2008.

A Szapolyai család oklevéltára / Documenta Szapolyaiana I. Levelek és oklevelek / Epistulae et litterae (1458–1526), ed. Tibor Neumann. Budapest: MTA Bölcsészettudományi Kutatóközpont Történettudományi Intézet, 2012.

A zichi és vásonkeői gróf Zichy-család idősb ágának okmánytára / Codex diplomaticus domus senioris comitum Zichy de Zich et Vasonkeő, 12 vols, eds. Imre Nagy et al. Pest – Budapest: Magyar Történelmi Társulat, 1871–1931.

Agricola, Georgius. *De re metallica libri XII*. Basileae: Hieronymus Froben & Nicolaus Episcopius, 1556.

Agricola, Georgius. *De re metallica*, eds. Herbert Clark Hoover, and Lou Henry Hoover. New York: [no publisher], 1950.

Akta Grodzkie i Ziemskie z czasów Rzeczypospolitej Polskiej z Archiwum tak zwanego bernardyńskiego we Lwowie, 25 vols *[Town and land deeds from the times of the Polish Republic in the so-called Bernardine Archives in Lviv]*. Lviv: Towarzystwo Naukowe, 1868–1914.

Anjou-kori oklevéltár, 43 vols *[Angevin cartulary]*, ed. in chief, Gyula Kristó, and Tibor Almási. Szeged and Budapest: JATE – MNL OL, 1990–2018.

Anjoukori okmánytár / Codex diplomaticus Hungaricus Andegavensis, 7 vols, eds. Imre Nagy, and Gyula Tasnádi Nagy. Budapest: Magyar Tudományos Akadémia, 1878–1920.

Anon., "A Charter to Re-establish Kraków." In *Epistolae: Medieval Women's Latin Letters*, ed. Joan Ferrante. New York: Columbia University Libraries, 2014. https://epistolae.ctl.columbia.edu/letter/26164.html (last accessed: May 22, 2025)

Árpádkori új okmánytár, 12 vols *[New Árpádian-age cartulary]*, ed. Gusztáv Wenzel. Pest and Budapest: Eggenberger, 1860–1874.

Az Árpád-házi királyok okleveleinek kritikai jegyzéke / Regesta regum stirpis Arpadianae critico-diplomatica, 2 vols, eds. Imre Szentpétery, and Iván Borsa. Budapest: Akadémiai, 1923–1987.

Az Árpád-kori nádorok és helyetteseik okleveleinek kritikai jegyzéke [Critical register of the charters of Árpádian-age palatines and their deputies] (A Magyar Országos Levéltár kiadványai, II. Forráskiadványok, 51), ed. Tibor Szőcs. Budapest: Magyar Országos Levéltár, 2012.

Bándi, Zsuzsanna. "Északkelet-magyarországi pálos kolostorok oklevelei." [Charters of the Pauline abbeys of northeastern Hungary] *Borsodi Levéltári Évkönyv* 5 (1985): 557–725.

Bártfai Szabó, László. *Pest megye történetének okleveles emlékei 1002–1599-ig [Cartulary of the history of Pest County]*. Budapest: [Author's edition], 1938.

Bél, Mátyás. *Bányavárosok a legkegyelmesebb királyok uralkodása alatt. Válogatás Bél Mátyás leírásaiból* (Központi Bányászati Múzeum Közleményei, 3), ed. Erzsébet Bircher. Sopron: Központi Bányászati Múzeum Alapítvány, 2004.

Bél, Mátyás. *Sopron vármegye leírása / Descriptio Comitatus Semproniensi*, 3 vols (Sopron Város Történeti Forrásai, C/4), ed. Katalin Mária Kincses. Sopron: Soproni Levéltár, 2006.

Borsa, Iván. *A Szent-Iványi család levéltára 1230–1525 [Cartulary of the Szentiványi family]* (Magyar Országos Levéltár kiadványai, II. Forráskiadványok, 14). Budapest: MOL, 1988.

Bredekow, H. *Geometrischer Grundt-Riß der Neunmahl Bekriegten Undt Blutig Besiegten Königlichen Ungarischen Resitends Stadt Buda vel Offen.* [Vienna?]: [no publisher], 1686.

Budapest történetének okleveles emlékei/Monumenta diplomatica civitatis Budapest, I, ed. Albert Gárdonyi. Budapest: Budapest Székesfőváros, 1936.

Chaucer, Geoffrey. "The Canterbury Tales." In *The Complete Poetry and Prose of Geoffrey Chaucer*, ed. John H. Fisher. New York: Holt, Rinehart and Winston, 1977.

Codex diplomaticus et epistolaris regni Bohemiae, vol. 1, ed. Gustavus Friedrich. Prague: Sumptibus Comitiorum Regni Bohemiae, 1904/1907.

Codex diplomaticus et epistolaris Slovaciae, 2 vols, ed. Richard Marsina. Bratislava: Academiae Scientiarum Slovacae, 1971–1987.

Codex diplomaticus Hungariae ecclesiasticus ac civilis, 11 vols, ed. Georgius Fejér. Buda: Typis Regiae Universitatis Hungaricae, 1829–1844.

Codex diplomaticus Regni Croatiae, Dalmatiae et Slavoniae. Diplomatički sbornik Kraljevine Hrvatske s Dalmacijom i Slavonijom, 2 vols (Monumenta historica Slavorum meridionalium, 2–3), ed. Ivan Kukuljević Sakcinski. Zagreb: Ex officina societatis typographicae, 1874–1875.

Cosmas Pragensis. "Chronicon Boemorum." In *MGH Scriptores rerum Germanicarum Nova Series*, vol. 2, ed. Bertold Bretholz. Berlin: Wedimann, 1923.

Cosmas of Prague. *The Chronicle of the Czechs*, eds. János M. Bak, and Pavlína Rychterová. (Central European Medieval Texts, 10) New York: Central European University Press, 2020. https://doi.org/10.7829/j.ctv1z7kj6j

D. Szakács, Anita. *Sopron város tanácsa bírósági jegyzőkönyveinek regesztái III. 1570–1579. [Summaries of the minutes of the court of law of the town of Sopron, 1]*. Sopron: Győr-Moson-Sopron Megye Soproni Levéltára, 2011.

Das Ofner Stadtrecht: Eine deutschsprachige Rechtssammlung des 15. Jahrhunderts aus Ungarn (Monumenta historica Budapestinensia, 1), ed. Karl Mollay. Budapest: Akadémiai, 1959.

de La Vigne, Nicolas Marcel. *Bvdae ac Pesti ichonographica descriptio tvm etia(m) earv(m) vicinarv(m), castrorv(m), aggerorv(m), fossarv(m), prospectvv(m), nvda, ac recta scenographica delineatio, aliarvmve dictae Bvdae obsidionis particvlaritatv(m) à sac(ratissimi) caes(arei) mai(estatis) exercitv, eivsqve praecelsorvm foederatorvm copiis factae.* [Vienna] [?]: [no publisher], 1686.

de Núrsia, Sant Benet [Saint Benedict of Nursia]. *Regula monachorum* (Subsidia Monastica, 21), ed. Ignasi M. Fossas. Barcelona: Publicacions de l'Abadia de Montserrat, 1997.

de Saxoferrato, Bartolus. *Tractatus de fluminibus seu Tyberiadis*. Bologna: Johannes Roscius, 1576.

de Saxoferrato, Bartolus. *Tractatus de fluminibus seu Tyberiadis*, ed. Guido Astuti. Turin: Bottega d'Erasmo, 1964.

Diplomata Hungariae Antiquissima, vol. 1. 1000–1131, ed. György Györffy. Budapest: Akadémiai, 1992.

Documenta Artis Paulinorum, 3 vols. Eds. Béla Gyéressy, and Ferenc Hervay. Budapest: MTA Művészettörténeti Kutató Csoportja, 1975–1978.

Documente privind istoria României. C. Transilvania. Veacul XIV, 12 vols (Documenta Romaniae historica). Bucharest: Acad. Republicii Populare Române, 1953–1985.

du Plessis, Paul Jacobus. "An Annotated Translation of Bartolus' Tractatus de Fluminibus seu Tyberiadis (Book 1)." MA-Th., Potchefstroomse Universiteit, 1999.

Duru, Louis M. *Bibliothèque historique de l'Yonne; ou, Collection de légendes, chroniques et documents divers pour servir à l'histoire des différentes contrées qui forment aujord'hui ce département*, vol. 1. Auxerre: Perriquet, 1801.

Elenchus fontium historiae urbanae, vol. 3/2, ed. in chief András Kubinyi. Budapest: Balassi, 1997.

Első telekkönyv. Erstes Grundbuch (1480–1553), ed. Károly Mollay. Sopron: Soproni Levéltár, 1993.

Eperjes szabad királyi város levéltára, 1245–1526, 2 vols *[Archive of the free royal town of Prešov*], ed. Béla Iványi. Szeged: Szegedi Városi Nyomda és Könyvkiadó, 1931–1932.

Fejérpataky, László. *Magyarországi városok régi számadáskönyvei (Selmeczbánya, Pozsony, Beszterczebánya, Nagyszombat, Sopron, Bártfa és Körmöczbánya városok levéltáraiból) [Old account books of Hungarian towns (from the archives of Banská Štiavnica, Bratislava, Banská Bystrica, Trnava, Sopron, Bardejov and Kremnica)]*. Budapest: Magyar Tudományos Akadémia, 1885.

Fekete Nagy, Antal. "A Petróczy levéltár középkori oklevelei: első közlemény." [Medieval documents of the Petróczy Archive, Part 1] *Levéltári Közlemények* 8 (1930): 190–264.

Filarete's Treatise on Architecture: Being the Treatise by Antonio di Piero Averlino, Known as Filarete, 2 vols, ed. John R. Spencer. New Haven, CT: Yale University Press, 1965.

Gedenkbuch. Feljegyzési könyv 1492–1543 (Sopron Város Történeti Forrásai, A/3). Eds. Károly Mollay, and Károly Goda. Sopron: Győr-Moson-Sopron Megyei Levéltár Soproni Levéltára, 2006.

Gerichtsbuch. Bírósági Könyv 1423–1531 (Sopron Város Történeti Forrásai, A/2). Eds. Jenő Házi, and János Német. Sopron: Győr-Moson-Sopron Megye Soproni Levéltára, 2005.

Górecki, Piotr. *A Local Society in Transition. The Henryków Book and Related Documents.* (PIMS Studies and Texts, 155). Toronto: Pontifical Institute of Mediaeval Studies, 2007.

Gruber, Elisabeth. *"Raittung und außgab zum gepew" Kommunale Rechnungspraxis im oberösterreichischen Freistadt Edition und Kommentar der Stadtgrabenrechnung (1389–1392)* (Quelleneditionen des Instituts für Österreichische Geschichtsforschung, 14). Vienna: Böhlau, 2015.

Hájek, Václav. *Kronika česká: Podle originálu z r. 1541*, 4 vols [*Czech Chronicle.* From the beginnings to 1541], ed. Václav Flajšhans. Prague: Česká Akademie věd a umění, 1918–1933.

Harmatta, János, ed. *Lexicon Latinitatis medii aevi Hungariae*, 6 vols. Budapest: Akadémiai – Argumentum, 1987–2017.

Hazai okmánytár / Codex diplomaticus patrius, 8 vols, ed. Imre Nagy. Győr and Budapest: Sauervein Géza – Franklin, 1865–1891.

Házi, Jenő. *Sopron szabad királyi város története,* 13 vols *[The history of the free royal town of Sopron]*. Sopron: Székely, Szabó és Társa Könyvnyomdája, 1921–1943.

Hermann, István. "Egy Pápa belvárosi ház felújítása az 1670-es években. (forrásközlés)." [Rebuilding of a house in the downtown of Pápa in the 1670s (source edition)] *Soproni Szemle* 70 (2016): 222–236.

Historia critica regum Hungariae. vols. 8–19. Stirpis mixtae: 1301–1526, 12 vols. Budae: Weingand & Koepf, 1788–1794.

Horváth, Vladimir. *Bratislavské mestské privilégium [The urban privilege of Bratislava].* Bratislava: Archív hlavného mesta SR: Obzor, 1991.

"Ius regale montanorum." In *Codex juris Bohemici,* ed. Hermenegildus Jireček, 266–435. Prague: Kober, 1867.

Jakó, Zsigmond. *A kolozsmonostori konvent jegyzőkönyvei, 1289–1556,* 2 vols *[The protocol of the abbey of Kolozsmonostor]* (Magyar Országos Levéltár kiadványai, II, Forráskiadványok, 17). Budapest: Magyar Országos Levéltár, 1990.

Keglevich, Kristóf. *A garamszentbenedeki apátság 13. és 14. századi oklevelei (1225–1403) [13th- and 14th-century documents of the abbey of Hronský Beňadik]* (Capitulum, 9). Szeged: Szegedi Tudományegyetem Történti Intézete, 2014.

Knauz, Nándor. "A budai káptalan regestái 1148–1649." [Summary of the documents of the chapter of Buda] *Magyar Történelmi Tár* 12 (1863): 3–156.

Kodeks dyplomatyczny miasta Krakowa 1257–1506, vol. 1 *[Cartulary of the town of Kraków],* ed. Franciszek Piekosiński. Kraków: Akad. Umiejętności, 1879.

Kolosvári, Sándor, and Kelemen Óvári, eds. *A magyar törvényhatóságok jogszabályainak gyűjteménye / Corpus Statutorum Hungariae Municipalium,* 5 vols. Budapest: Magyar Tudományos Akadémia, 1885–1904.

Kőfalvi, Tamás. *A pécsváradi konvent hiteleshelyi oklevéltára, 1254–1526 [Cartulary of the credible place of the convent of Pécsvárad, 1254–1526].* Szeged: [SZTE], 2006.

Kredics, László, and László Solymosi. *A veszprémi püspökség 1524. évi urbáriuma. Urbarium episcopatus Vesprimiensis anno MDXXIV* (Fontes Minores ad Historiam Hungariae Spectantes – Új Történelmi tár, 4). Budapest: Akadémiai, 1993.

Kubinyi, Ferencz, Jr., *Árpádkori oklevelek 1095–1301 / Codex Diplomaticus Arpadianus* (Monumenta Hungariae historica, 1). Pest: Osterlamm, 1867.

Kühnel, Harry. "Die materielle Kultur Wiens im Mittelalter." In *Wien im Mittelalter. Katalog,* 29–45. Vienna: Museum der Stadt Wien, 1975.

Lackner Kristófnak, mindkét jog doktorának rövid önéletrajza. Vitae Christophori Lackhner I. U. D. Hominis, brevis consignatio [Short autobiography of Kristóf Lackner] (Sopron Város Történeti Forrásai, C/5), ed. Gergely Tóth. Sopron: Soproni Levéltár, 2008.

Les gestes des évéques d'Auxerre, vol. 2 (Les classiques de l'histoire de France au Moyen Âge, 43), eds. Michel Sot, and Pierre Bonnerue. Paris: Belles Lettres, 2006.

Liber fundationis claustri Sancte Marie Virginis in Heinrichau (Księga henrykowska), ed. Roman Grodecki. Wrocław: Muzeum Archidiecezjalne we Wrocławiu, 1991.

Lossai, Petri. *Notationes et Delineationes 1498. Az Országos Széchényi Könyvtár Cod. Lat. 197 jelű kódexe [Notationes et Delineationes 1498. The Cod. Lat. 197 of the Széchényi National Library],* eds. Zoltán Poronyi, and Alajos Fleck. Pécs: Pécsi Geodéziai és Térképészeti Vállalat, 1969.

Magyar-Zsidó oklevéltár, 18 vols *[Hungarian-Jewish cartulary],* eds. Ármin Friss et al. Budapest: Magyar Izraeliták Országos Képviselete, 1901–1980.

Matunák, Mihály. "Eredeti oklevelek a körmöczbányai püspöki malomról." [Original charters on the bishop's mill at Kremnica] *Magyar Történelmi Tár* 8 (1907): 321–324.

MGH Capitularia regum Francorum [Legum Sectio 2.1], ed. Alfred Boretius. Hanover: Hahn, 1883.

Monumenta Ecclesiae Strigoniensis , 4 vols, eds. Ferdinandus Knauz et al. Esztergom and Budapest: Ae. Horák – Argumentum – Esztergomi Prímási Levéltár, 1874–1999.

Monumenta Germaniae Historica. Diplomata regum et imperatorum Germaniae, vol. 3 *[Henrici II et Arduini diplomata]*. Hannover: Hahn, 1900–1903.

Monumenta Hungariae Heraldica. Magyar czimeres emlékek, 3 vols. eds. Antal Áldásy, and László Fejérpataky. Budapest: Franklin, 1901–1926.

Oklevelek Hontvármegyei magán-levéltárakból I. 1256–1399 [Charters from private archives of Hont County, 1256–1399], ed. Ferenc Kubinyi, Jr. Budapest: [no publisher], 1888.

Oklevéltár a gróf Csáky család történetéhez, 2 vols *[Cartulary to the history of the Csáky family]*, ed. László Bártfai Szabó. Budapest: Stephaneum, 1919.

Ortvay, Tivadar, and Frigyes Pesty. *Oklevelek Temesvármegye és Temesvár város történetéhez. I. 1183–1430 [Documents to the history of Temes County and the town of Timişoara]* (Temesvármegye és Temesvár város története, 4). Bratislava: Eder István Könyvnyomdája, 1896.

"Passio beatissimi Gerardi (Legenda minor)." [Passion of the Most Blessed Gerard (Minor legend)] [translated by Cristian Gaşpar]. In *Sanctitas principum. Sancti reges, duces, episcopi et abbates Europae Centralis (saec. XI–XIII). The Sanctity of the Leaders. Holy Kings, Princes, Bishops, and Abbots from Central Europe (Eleventh to Thirteenth Centuries)*, eds. Gábor Klaniczay, and Ildikó Csepregi (Central European Medieval Texts, 7), 239–265. Budapest, New York, and Vienna: CEU Press, 2022.

Paulus [Paul the Deacon]. "Historia Langobardorum." In *Monumenta Germaniae Historica Scriptores Rerum Germanicarum,* vol. 48, ed. Georg Waitz, 49–242. Hannover: Hahn, 1878.

Perényi család levéltára 1222–1526 [Cartulary of the Perényi family, 1222 to 1526] (Magyar Országos Levéltár kiadványai, II. Forráskiadványok, 44), ed. István Tringli. Budapest: Magyar Országos Levéltár, 2008.

Pest megye történetének okleveles emlékei 1002–1599-ig [Diplomatic sources on the history of Pest County, 1002–1599], ed. László Bártfai Szabó. Budapest: [no publisher], 1938.

Pesthy, Frigyes. *Krassó vármegye története,* vol. 3 *[History of Krassó County]*. Budapest: Athenaeum, 1882.

Prokopios. *The Wars of Justinian.* Trans. by H. B. Dewing, revised by Anthony Kaldellis. Indianapolis, IN: Hackett, 2014.

Registrum proventuum regni Hungariae: A Magyar Királyság kincstartójának számadáskönyve (1494–1495), ed. Tibor Neumann. Budapest: MTA BTK Történettudományi Intézet, 2019.

"Sachsenspiegel Landrecht." In *Monumenta Germaniae Historica. Fontes iuris Germanici antiqui, Nova series Fontes iuris N.S.*, vol. 1.1, ed. Karl August von Eckhardt. 2nd ed. Göttingen: Musterschmidt, 1955.

Szabó, Dénes. "A dömösi prépostság adománylevele (1138/1329). I–III." [Donation charter of the provostry of Dömös (1138/1329)] *Magyar Nyelv* 32 (1936): 54–57, 130–135, and 203–206.

Szádeczky, Lajos. *Székely oklevéltár V. kötet (1296–1603.) [Székely cartulary, vol. 5]*. Cluj: Ajtai K. Albert Könyvnyomdája, 1896.

Szovák, Kornél. *Supplementum ad Lexicon Latinitatis Medii Aevi Hungariae A–I.* Budapest: Argumentum, 2013.

Teleki, József. *A Hunyadiak kora Magyarországon,* vol. 11 *[History of the age of the Hunyadis in Hungary].* Pest: Emich, 1855.

Thallóczy, Lajos. *Jajca (bánság, vár és város) története [History of (the banate, castle and town of] Jajce* (Magyarország melléktartományainak oklevéltára. Codex diplomaticus partium regno Hungariae adnexarum, 4). Budapest: MTA, 1915.

The Digest of Justinian, eds. Alan Watson, Theodor Mommsen, and Paul Krueger. Philadelphia, PA: University of Pennsylvania Press, 1985.

The History of the King's Works. The Middle Ages, 2 vols, ed. Howard M. Colvin. London: H. M. Stationery Office, 1963.

The Ledger Book of Vale Royal Abbey, ed. John Brownbill. Manchester: [no publisher], 1914.

Tirnitz, József, and Anita Szakács. *Sopron város tanácsa bírósági jegyzőkönyveinek regesztái I. 1533–1554 [Summaries of the minutes of the court of law of the town of Sopron, 1*]. Sopron: Győr-Moson-Sopron Megye Soproni Levéltára, 1996.

Tkalčić, Ivan Krstitelj. *Povjestni spomenici slob.kralj. grad. Zagreba. Monumenta historica liberae regiae civitatis Zagrabiae,* 11 vols. Zagreb: Brzotisk Albrechta, 1889–1905.

Tóth, Péter. *Vas vármegye közgyűlési jegyzőkönyveinek regesztái I. 1595–1600 [Summaries of the minutes of the gatherings of Vas County, I]* (Vas megyei levéltári füzetek, 2). Szombathely: Vas Megyei Levéltár, [N. d.].

Treaties of Peace with Italy, Bulgaria, Hungary, Roumania, and Finland. Washington D.C.: U.S. G.P.O., 1947.

Urkundenbuch des Burgenlandes und der angrenzenden Gebiete der Komitate Wieselburg, Ödenburg und Eisenburg, 5 vols (Publikationen des Instituts für Österreichische Geschichtsforschung), eds. Hans Wagner, and Irmtraut Lindeck-Pozza. Graz, Cologne, and Vienna: Böhlau, 1955–2000.

Urkundenbuch zur Geschichte der Deutschen in Siebenbürgen, 7 vols, eds. Franz Zimmermann et al. Sibiu and Bucharest: Michaelis and Verlag der Rumänischen Akademie, 1892–1991.

Veress, Endre. *Gyula város oklevéltára. 1313–1800 [Cartulary of the town of Gyula].* Budapest: Stephaneum, 1938.

Virágh, Elek. "A csicsva-vári udvarbíráknak, Eszterházy és Homonnay részrül adott utasítása." [The instructions to court judges of the castle of Čičva given by the Eszterházy and the Homonnay families] *Magyar Gazdaságtörténelmi Szemle* 4 (1897): 331–343.

"Vita s. Hedwigis (legenda maior)." In *Monumenta Poloniae Historica,* vol. 4, ed. Aleksander Semkowicz, 510–633. Lwów: Gubrynowicz i Schmidt, 1884.

Výsady miest a mestečiek na Slovensku. I. 1238–1350 [Privileges of towns and cities in Slovakia, 1238–1350], ed. Ľubomír Juck. Bratislava: Veda – Vydavateľstvo Slovenskej akadéémie vied, 1984.

Wagner, Carolus. *Diplomatarium Comitatus Sarosiensis.* Bratislava and Košice: Sumtibus Joannis Michaelis Landerer, 1780.

Wenzel, Gusztáv. *Magyarország bányászatának kritikai története [Critical history of mining in Hungary*]. Budapest: M. Tud. Akadémia Könyvkiadó Hivatala, 1880.

Werbőczy, Stephen. *The Customary Law of the Renowned Kingdom of Hungary: A Work in Three Parts Rendered by Stephen Werbőczy (The "Tripartium") (1517) / Tripartitum opus iuris consuetudinarii inclyti regni Hungariae per Stephanum de*

Werbewcz editum (1517) (The Laws of Hungary Series / Decreta Regni Mediaevalis Hungariae, I/5), eds. János M. Bak, Péter Banyó, and Martyn Rady. Budapest and Idyllwild, CA: Charles Schlacks, Jr., 2005.

Zala vármegye története. Oklevéltár, 2 vols *[History of Zala County. Cartulary]*, eds. Imre Nagy, Dezső Véghely, and Gyula Nagy. Budapest: [no publisher], 1886.

Zbior praw litewskich od roku 1389 do roku 1529 tudzież rozprawy sejmowe o tychże prawach od roku 1544 do roku 1563 [Collection of Lithuanian laws from 1389 to 1529 and parliamentary hearings on these rights from 1544 to 1563], ed. Adam Tytus Działyński. Poznań: W drukarni na Garbarach, 1841.

Secondary Literature

Adamska, Anna. "'From Memory to Written Record' in the Periphery of Medieval Latinitas: The Case of Poland in the Eleventh and Twelfth Centuries." In *Charters and the Use of the Written Word in Medieval Society* (Utrecht Studies in Medieval Literacy, 5), ed. Karl Heidecker, 83–100. Turnhout: Brepols Publishers, 2000. https://doi.org/10.1484/M.USML-EB.6.09070802050003050007070102

Adamska, Anna, and Marco Mostert. *The Development of Literate Mentalities in East Central Europe* (Utrecht Studies in Medieval Literacy, 9). Turnhout: Brepols, 2004. https://doi.org/10.1484/M.USML-EB.6.09070802050003050104040909

Almási, Tibor. "Megjegyzések Gölnicbánya Kun László királytól elnyert privilégiumához és megerősítéseihez." [Notes on Ladislas the Cuman's privilege to Gelnica and its confirmation] *Acta Universitatis Szegediensis de Attila József nominatae. Acta Historica* 102 (1995): 43–49.

Andrásfalvy, Bertalan. "A paraszti halászati jog a Duna mentén Tolna és Baranya megyében." [Peasant fishing rights on the Danube in Tolna and Baranya counties] *Ethnographia* 81 (1970): 316–326.

Andrásfalvy, Bertalan. *Duna mente népének ártéri gazdálkodása Tolna és Baranya megyében az ármentesítés befejezéséig [Flood plain economy of the peoples of the Duna valley in Tolna and Baranya Counties until the completion of the water regulation works]* (Tanulmányok Tolna megye történetéből, 7). Szekszárd: Tolna Megyei Levéltár, 1975.

Andrásfalvy, Bertalan. *A Sárköz és a környező Duna menti települések ősi ártéri gazdálkodása és vízhasználatai a szabályozás előtt [The Sárköz and the ancient flood plain and water management of the settlements along the Danube] (Vízügyi történeti füzetek, 6)*. Budapest: Vízdok, 1973.

Andreška, Jiří. "Development of Fish-Pond Culture in Bohemia." In *The Fishing Culture of the World. Studies in Ethnology, Cultural Ecology and Folklore.* I, ed. Béla Gunda, 77–89. Budapest: Akadémiai, 1984.

Anon. "Hajómalmok." [Ship mills] In *Magyar néprajz III. (Anyagi kultúra 2.) [Hungarian ethnography, III. Material culture, 2]*, ed in chief. Ottó Domonkos. Budapest: Akadémiai, 1991.

Anon. "Vízimalmok." [Watermills] In *Magyar néprajz III. (Anyagi kultúra 2.) [Hungarian ethnography, III. Material culture, 2]*, ed in chief. Ottó Domonkos. Budapest: Akadémiai, 1991.

Arandjelović, Biljana, and Ana Momčilović-Petronijević. "Arhitektura vodnih mlinov v južni Srbiji." [The architecture of water mills in Southern Serbia] *AR Arhitektura, raziskave* 2 (2010): 59–62.

Arnold, Ellen F. "The Plan of St Gall and the Power of Water: An Environmental Reconsideration." *Journal of Medieval History* 51 (2025): 515–529.

Arnoux, Mathieu. "Les moulins à eau en Europe occidentale (IXe–XIIe siècle): aux origines d'une économie institutionelle de l'énergie hydraulique." In *L'acqua nei secoli altomedievali: Spoleto, 12–17 aprile 2007*, vol. 1. 693–746. Spoleto: Fondazione Centro italiano di studi sull'alto Medioevo, 2008.

Arnoux, Mathieu. "Moulins à fer et procédé indirect. Innovation technique et conditions géographiques dans la sidérurgie européenne (XIIIe–XVIe siècles)." In *I mulini nell'Europa medievale: Atti del Convegno di San Quirico d'Orcia, 21–23 settembre 2000 (Biblioteca di storia agraria medievale, 21)*, eds. Paola Galetti, and Pierre Racine, 317–328. Bologna: Clueb, 2003.

Aszt, Ágnes. "A mosonmagyaróvári Bástya utcában végzett régészeti feltárás történeti tanulságai." [Historical lessons of the archaeological excavations at the Bástya Street at Mosonmagyaróvár] *Moson Megyei Műhely* no. 3 (2001): 29–32.

Augustyn, Beatry. "De evolutie van het duinecosysteem in Vlaanderen in de Middeleeuwen: antropogene factoren versus zeespiegelrijzingstheorie." [Evolution of the dune ecosystem in Flanders during the Middle Ages: anthropogenic factors versus sea level change theory] *Historisch-Geografisch Tijdschrift* 13, no. 1 (1995): 9–19.

Augustyn, Beatry. "Evolution of the Dune Ecosystem in Flanders During the Middle Ages: Anthropogenic Factors versus Sea Level Change Theory." Online document: www.vliz.be/imisdocs/publications/100070.pdf (last accessed: April 8, 2018)

Bachmann, Christoph. "Altbayerns erste bisher bekannte Mühlordnung. Die Ordnung der Ingolstädter Müller aus dem Jahr 1437." *Sammelblatt des Historischen Vereins Ingolstadt* 102/103 (1993): 309–316.

Bader, Karl S. stat. "Kollektaneen zu Geschichte und Streuung eines rechtstopographischen Begriffs." *Blätter für deutsche Landesgeschichte* 101 (1965): 8–66.

Balázs, György. "Vízimalmok, szárazmalmok, szélmalmok a 18–19. században." [Water mills, tread mills, wind mills in the 18th–19th centuries]. *A Magyar Mezőgazdasági Múzeum Közleményei* (1995–1997): 83–100.

Bankoff, Greg. "The 'English Lowlands' and the North Sea Basin System: A History of Shared Risk." *Environment and History* 19, no. 1 (2013): 3–37. https://doi.org/10.3197/096734013X1352832843899

Baraczka, István. "Az 1535/36. évi Sopron városi kamarai számadáskönyv néhány tanulsága." [Some lessons of the account book of Sopron from 1535/1536] *Soproni Szemle* 23 (1969): 207–215.

Barber, Peter. "Mapmaking in England, ca. 1470–1650." In *The History of Cartography, Vol.3. Cartography in the European Renaissance,* ed. David Woodward, 1589–1669. Chicago: Chicago University Press, 2007. Online: www.press.uchicago.edu/books/HOC/HOC_V3_Pt2/HOC_VOLUME3_Part2_chapter54.pdf (last accessed: October 5, 2019)

Bártfai Szabó, László. *Óbuda egyházi intézményei a középkorban [Church institutions of Óbuda in the Middle Ages]*. Budapest: [Author's edition], 1935.

Bartha, Antal. "Gazdaságtörténet és szavak." [Economic history and the words] *Magyar Nyelv* 65 (1969): 14–25.

Bartosiewicz, László et al. "Animal Exploitation." In *The Economy of Medieval Hungary* (East Central and Eastern Europe in the Middle Ages, 450–1450, 49), eds. József Laszlovszky et al., 113–165. Leiden and Boston: Brill, 2018. https://doi.org/10.1163/9789004363908_007

Batta, István. "Középkori bányászatunk és kohászatunk a Metercián." [Medieval Hungarian mining and metallurgy on the Metercia] *Bányászati és kohászati lapok. Kohászat* 121 (1988): 275–285.

Béarez, Philippe, Sandrine Grouard, and Benoit Clavel, eds. *Archéologie du Poisson. 30 ans d'archéo-ichtyologie au CNRS*. Antibes: Éditions APDCA, 2008.

Bedy, Vince. *A győri székeskáptalan története [The history of the chapter of Győr]* (Győregyházmegye múltjából, 3). Győr: Győregyházmegyei Alap Nyomdája, 1938.

Bencze, Ünige. "A Medieval Pauline Landscape in the Szekler Land." *Transsylvania Nostra* 9, no. 34 (2015): 10–17.

Bencze, Ünige. "Reconstructing a Monastic Landscape: The Example of the Cistercian Abbey Cârţa (Kerz, Kerc)." In *Monastic Life, Art, and Technology in 11th-16th Centuries* [Special issue of: Annales Universitatis Apulensis. Series Historica], ed. Ileana Burnichioiu, 29–44. Alba Iulia: Universitatea "1 Decembrie 1918" din Alba Iulia, 2015.

Benda, Borbála. "Táplálkozási változások a főúri udvarok és uradalmak étkezésében a 17. században." [Changes in the diets of aristocratic courts and manors in the 17th century] In *Környezetváltozás, termelés, fogyasztás: a történeti ökológia kérdései [Environmental change, production and consumption: the questions of historical ecology]*, (Annales Tataienses, 5), eds. János Fatuska, Éva Mária Fülöp, and László Gyüszi, Jr., 83–98. Tata: Mecénás Kulturális Közalapítvány, 2006.

Benda, Judit. "A csendes háttér. A nők munkalehetőségei a középkori Budán." [The silent background. Possibilities of female work in medieval Buda] *Korall* no. 60 (2015): 11–40.

Benda, Judit. "A kereskedelem épületei a középkori Budán. 3. Malmok, pékek és kenyérszékek a későközépkori Budán." [Trade buildings in medieval Buda, 3. Mills, bakeries, bread shops at late medieval Buda] *Tanulmányok Budapest Múltjából* 38 (2013): 7–30.

Benkő, Elek. *Kolozsvár magyar külvárosa a középkorban [The Hungarian suburb of Cluj in the Middle Ages]* (Erdélyi Tudományos Füzetek, 248). Cluj: Erdélyi Múzeum Egyesület, 2004.

Benkő, Elek. "The Medieval Water Supply System of Pilis Abbey." *Anteaeus* 37 (2021): 193–224. https://doi.org/10.62149/Antaeus.37.2021_08

Bennett, Richard, and John Elton. *History of Corn Milling*, 4 vols. London: Simpkin, Marshall and Co., 1898–1904.

Berend, Nora, ed. *Christianization and the Rise of Christian Monarchy: Scandinavia, Central Europe and Rus' c. 900–1200*. Cambridge: Cambridge University Press, 2007. https://doi.org/10.1017/CBO9780511496400

Berend, Nora, Przemyslaw Urbanczyk, Przemyslaw Wiszewski. *Central Europe in the High Middle Ages: Bohemia, Hungary and Poland, c. 900 – c. 1300*. Cambridge: Cambridge University Press, 2013. https://doi.org/10.1017/CBO9780511813795

Berg-Hobohm, Stefanie and Peter Ettel, eds. *Großbaustelle 793: das Kanalprojekt Karls des Großen zwischen Rhein und Donau; Begleitbuch zur Ausstellung im Museum für Antike Schiffahrt des Römisch-Germanischen Zentralmuseums in Mainz, 30. April 2014 bis 10. August 2014* (Mosaiksteine, 11). Mainz: Verl. des Röm.-German. Zentralmuseums, 2014.

Bergier, Jean-François, ed. *Montagnes, fleuves, forêts dans l'histoire. Barrières ou lignes de convergence? (Berge, Plusse, Walder in der Geschichte. Hindernisse oder Begegnungsräume?)*. St. Katharienen: Scripta Mercaturae, 1989.

Berking, Jonas et al. "Problems Concerning Ancient Water Management in the Mediterranean." *Journal for Ancient Studies* 6 (2016): 74–101. http://doi.org/10.17169/FUDOCS_document_000000025987

Bernardi, Philippe. "Apprentissage et transmission du savoir dans les métiers du bâtiment à Aix-en-Provence à la fin du Moyen Age (1400–1550)." In *Actes du 1er colloque du CRISIMA. Apprentissages, éducation, initiation, éducation au Moyen Age (Les cahiers du CRISIMA*, 1), 69–79. Montpellier: [Univ. Paul-Valéry CRISIMA], 1993.

Bertalan, Herta. "Das Klarissinnenkloster von Obuda aus dem 14. Jahrhundert." *Acta archaeologica Academiae Scientiarum Hungaricae* 34 (1982): 151–176.

Berthold, Jens "Mühlen im Befund – Eine Übersicht zu archäologischen Erscheinungsformen von Wassermühlen." In *Wassermühlen und Wassernutzung im mittelalterlichen Ostmitteleuropa* (Forschungen zur Geschichte und Kultur des östlichen Mitteleuropa, 50), eds. Martina Maríková, and Christian Zschieschang, 235–268. Stuttgart: Franz Steiner, 2015.

Bianca, Concetta, and Francesco Salvestrini, eds. *L'acqua nemica: fiumi, inondazioni e città storiche dall'antichità al contemporaneo: atti del Convegno di studio a cinquant'anni dall'alluvione di Firenze (1966–2016), Firenze, 29–30 gennaio 2015* (Biblioteca del Centro per il collegamento degli studi medievali e umanistici in Umbria, 37). Spoleto: Fondazione Centro italiano di studi sull'alto Medioevo, 2017.

Bloch, Marc. "Avènement et conquêtes du moulin à eau." *Annales d'histoire économique et sociale* 7 (1935): 538–563. https://doi.org/10.1017/S0003441X00017556

Bonan, Giacomo. "Riflessi sull'acqua. Ricerca storica e biografie fluviali." *Contemporanea: Rivista di storia dell'800 e del '900* 22, no. 2 (2019): 317–328. https://doi.org/10.1409/93534

Bond, James. "Canal Construction in the Early Middle Ages: An Introductory Review." In *Waterways and Canal-Building in Medieval England*, ed. John Blair, 153–206. Oxford: Oxford University Press, 2007. https://doi.org/10.1093/acprof:osobl/9780198723134.003.0008

Bónis, György. *Középkori jogunk elemei. Római jog, kánonjog, szokásjog [Elements of Hungarian medieval law. Roman law, canon low, customary law]*. Budapest: Közgazdasági és Jogi Könyvkiadó, 1972.

Boreczky, Anna. "Mikor és hol készült Lossai Péter kódexe?." [When and where was Peter Lossai's codex produced?] *Magyar Könyvszemle* 113 (1997): 249–265.

Bork, Hans-Rudolf et al. *Landschaftsentwicklung in Mitteleuropa: Wirkungen des Menschen auf Landschaften*. Gotha and Stuttgart: Klett-Perthes, 1998.

Boucheron, Patrick. "Water and Power in Milan, c. 1200–1500." *Urban History* 28 (2001): 180–193. https://doi.org/10.1017/S0963926801002024

Brandstätter, Klaus. "Schwerarbeitende Frauen im Mittelalter: Bauarbeiterinnen in Hall in Tirol." In *Margarete "Maultasch": zur Lebenswelt einer Landesfürstin und anderen Tiroler Frauen des Mittelalters; Vorträge der wissenschaftlichen Tagung im Südtiroler Landesmuseum für Kultur- und Landesgeschichte Schloss Tirol, Schloss Tirol, 3. bis 4. November 2006* (Schlern-Schriften, 339), ed. Julia Hörmann-Thurn und Taxis, 277–281. Innsbruck: Wagner, 2007.

Bráz, Béla. *A nagycsepcsényi és muthnai Vladár-család története és leszármazása [The history and genealogy of the Vladár family of Veľký Čepčín and Mutna]*. Turócszentmárton: Magyar Ny., 1907.

Brogiolo, Gian Pietro, ed. *Este, l'Adige e i Colli Euganei: storie di paesaggi*. Sesto Fiorentino: All'Insegna del Giglio, 2017.

Brunner, Otto. *Die Finanzen der Stadt Wien: von den Anfängen bis ins 16. Jahrhundert*. Vienna: Dt. Verl. für Jugend u. Volk., 1929.

Budak, Neven. "Budući da smo htjeli u Zagrebu na brdu Gradecu sagraditi slobodni grad...." [As we wanted to build a free city in Zagreb on the Hill Gradec...]. In *Zlatna bula 1242–1992 [Golden bull]*, ed. Zlatko Stublić, 21–32. Zagreb: Muzej grada Zagreba, 1992.

Burnouf, Joëlle and Philippe Leveau, eds. *Fleuves et marais, une histoire au croisement de la nature et de la culture: sociétés préindustrielles et milieux fluviaux, lacustres et palustres: pratiques sociales et hydrosystèmes* (Archéologie et histoire de l'art, 19). Paris: Ministère de l'éducation nationale, de l'enseignement supérieur et de la recherche, 2004.

Bütow, Sascha. *Straßen im Fluss. Schifffahrt, Flussnutzung und der lange Wandel der Verkehrsinfrastruktur in der Mark Brandenburg und der Niederlausitz vom 13. bis zum 16. Jahrhundert* (Studien zur brandenburgischen und vergleichenden Landesgeschichte, 18). Berlin: Lukas, 2015.

Buzogány, Aron, Szabina Kerényi, and Gergely Olt. "Back to the Grassroots? The Shrinking Space of Environmental Activism in Illiberal Hungary." *Environmental Politics* 31 (2022): 1267–1288. http://doi.org/10.1080/09644016.2022.2113607

C. Tóth, Norbert. *Az esztergomi székes- és társaskáptalanok archontológiája 1100–1543 [Archontology of the cathedral and collegiate chapters of Esztergom, 1100–1543]* (Subsidia ad historiam medii aevi Hungariae inquirendam, 9). Budapest: Magyar Tudományos Akadémia Magyar Medievisztikai Kutatócsoport, 2019.

Campopiano, Michele. "Cooperation and Private Enterprise in Water Management in Iraq: Continuity and Change between the Sasanian and Early Islamic Periods (Sixth to Tenth Centuries)." *Environment and History* 23 (2017): 385–407. https://doi.org/10.3197/096734017X14979473873867

Campopiano, Michele. "Rural Communities, Land Clearance and Water Management in the Po Valley in the Central and Late Middle Ages." *Journal of Medieval History* 39 (2013): 377-393. https://doi.org/10.1080/03044181.2013.834644

Campopiano, Michele, and Gerrit Jasper Schenk, eds. *Conflicts over Water Management and Water Rights from the End of Antiquity to Industrialisation* (Vierteljahrschrift für Sozial- und Wirtschaftsgeschichte – Beihefte, 260). Stuttgart: Franz Steiner, 2024. https://doi.org/10.25162/9783515137539

Carneiro, Robert L. "A Theory of the Origin of the State." *Science* 169, no. 3947 (1970): 733–738. https://doi.org/10.1126/science.169.3947.733

Carozza, Jean-Michel et al. "Lower Mediterranean Plain Accelerated Evolution during the Little Ice Age: Geoarchaeological Insight in the Tech Basin (Roussillon, Gulf of Lion, Western Mediterranean)." *Quaternary International* 266 (2011): 94–104. https://doi.org/10.1016/j.quaint.2011.06.049

Carr-Riegel, Leslie. "Waste Management in Medieval Krakow: 1257–1500." MA thesis, Central European University, 2016.

Carter, F[rancis] W. *Trade and Urban Development in Poland. An Economic Geography of Cracow from its Origins to 1795* (Cambridge Studies in Historical Geography, 20). Cambridge: Cambridge University Press, 1994. https://doi.org/10.1017/CBO9780511522222

Carus-Wilson, E[leonora] M[ay]. "An Industrial Revolution of the Thirteenth Century." *The Economic History Review* 11, no. 1 (1941): 39–60. https://doi.org/10.1111/j.1468-0289.1941.tb01589.x. 2nd ed. *Internal Colonization in Medieval Europe* (The expansion of Latin Europe, 1000–1500, 2), eds. Felipe Fernández-Armesto, and James Muldoon, 301–322. Farnham: Ashgate, 2008.

Cavaciocchi, Simonetta, ed. *Economia e energia, secc. XIII–XVIII: atti della trentaquattresima Settimana di studi, 15–19 aprile 2002* (Istituto Internazionale di Storia Economica F. Datini, Prato. Serie 2. Atti delle settimane di studio e altri convegni, 34). [Florence]: Le Monnier, 2003.

Cavallar, Osvaldo. "River of Law: Bartolus's Tiberiadis (De alluvione)." In *A Renaissance of Conflicts: Visions and Revisions of Law and Society in Italy and Spain,* eds. John A. Marino, and Thomas Kuehn, 31–129. Toronto: Centre for Reformation and Renaissance Studies, 2004.

Cavallar, Osvaldo. "The Wheels of Watermills and the Wheel of Fortune A consilium of Donatus Ricchi de Aldighieris." *Rechtsgeschichte* 13 (2008): 80–116. http://doi.org/10.12946/rg13/080-116

Cembrzyński, Paweł. *Zaopatrzenie w wode i usuwanie nieczystosci w miastach stref baltyckiej i sudecko-karpackiej w XIII–XVI wieku [Water supply and waste disposal in cities of the Baltic and Sudeten-Carpatians zones in the 13th–16th centuries]* (Wratislavia Antiqua, 14). Wrocław: Uniwersytet Wrocławski, 2011.

Cernea, Emil. "Dreptul asupra vadului de moară." [Legal rights on mill dams] *Analele Universității București* 40 (1991): 93–103.

Chapelot, Jean, and Robert Fossier. *Le village et la maison au moyen âge.* Paris: Hachette, 1985. Published in English: *The Village and House in the Middle Ages.* Berkeley, CA: University of California Press, 1986.

Chira, Alexandra. "Documentary Evidence on the Uses of Water-Mills in Medieval Transylvania. A Case Study of the River Someş." In *Usus aquarum: Interdisziplinäre Studien zur Nutzung und Bedeutung von Gewässern im Mittelalter* (Forschungen zur Geschichte und Kultur des östlichen Mitteleuropa, 54), eds. Christoph Mielzarek, and Christian Zschieschang, 51–70. Vienna: Böhlau, 2019. https://doi.org/10.7788/9783412500894.51

Cioc, Mark, *The Rhine: An Eco-Biography, 1815–2000.* Seattle, WA: University of Washington Press, 2002.

Coulson, Charles L. H. *Castles in Medieval Society: Fortresses in England, France, and Ireland in the Central Middle Ages.* Oxford: Oxford University Press, 2003. https://doi.org/10.1093/acprof:oso/9780198208242.001.0001

Creighton, Oliver, and Robert Higham, *Medieval Town Walls. An Archaeology and Social History of Urban Defence.* Stroud: Tempus, 2005.

Csánki, Dezső. *Magyarország történelmi földrajza a Hunyadiak korában, 5 vols [Historical geography of Hungary in the age of the Hunyadi family].* Budapest: MTA, 1890–1914.

Csánki, Dezső, and Néda Relković, "Újbánya város levéltára és rövid története." [Short history of Nová Baňa and its archive] *Levéltári Közlemények* 2 (1924): 86–107.

Csiffáry, Gergely. "A gyöngyössolymosi malomkőbányászat története." [History of mill stone quarrying at Gyöngyössolymos] In *Tanulmányok a kézműipar történetéből [Studies in the history of craftsmanship]*, eds. Gergely Csiffáry , and Klára Dóka, 161–189. Veszprém: Veszprémi Akadémiai Bizottság, 1999.

Curta, Florin. *Eastern Europe in the Middle Ages (500–1300),* 2 vols (Brill's Companions to European History, 19). Leiden and Boston: Brill, 2019. https://doi.org/10.1163/9789004395190

Curtis, Daniel R., and Michele Campopiano, "Medieval Land Reclamation and the Creation of New Societies: Comparing Holland and the Po Valley, c.800–c.1500." *Journal of Historical Geography* 44 (2014): 93–108. https://doi.org/10.1016/j.jhg.2013.10.004

Czeguhn, Ignacio et al., eds. *Wasser – Wege – Wissen auf der iberischen Halbinsel: Vom Römischen Imperium bis zur islamischen Herrschaft* (Berliner Schriften Zur Rechtsgeschichte, 6). Baden-Baden: Nomos Verlagsgesellschaft mbH & Co. KG, 2016.

Czigány, Béla. "Adatok a Győr megyei hajósmolnárok életéhez, I–V." [Data to the life of ship millers in Győr County] *Arrabona* 2 (1960): 133–139, 4 (1962): 97–116, 5 (1963): 207–228, 7 (1965): 413–428, 9 (1967): 149–178, 10 (1968): 123–140, and 12 (1970): 187–211.

Czysz, Wolfgang. *Die ältesten Wassermühlen. Archäologische Entdeckungen im Paartal bei Dasing*. Thierhaupten: Klostermühlenmuseum, 1998.

Czysz, Wolfgang, ed. *Römische und frühmittelalterliche Wassermühlen im Paartal bei Dasing Studien zur Landwirtschaft des 1. Jahrtausends* (Materialhefte zur bayerischen Vorgeschichte, 103). Kallmünz: Verlag Michael Laßleben, 2016.

Danielewski, Marcin. "Korabniki, Niewody i Rybitwy z terenu Kujaw jako osady służebne związane z książęcym regale wodnym." [Korabniki, Niewody and Rybitwy in the Kujawy region as service settlements connected with the ius ducale water system] *Roczniki Dziejów Społecznych i Gospodarczych* 71 (2011): 33–53.

Dányi, Dezső, and Vera Zimányi. *Soproni árak és bérek a középkortól 1750-ig [Prices and wages from Sopron from the Middle Ages to 1750*]. Budapest: Akadémiai, 1989.

Dávidházy, István. "A soproni posztókallók." [The fulling mills of Sopron] *Soproni Szemle* 43 (1989): 193–205.

Dávidházy, István. "18. és 19. századi posztósműhelyek egy városon belüli területi megoszlása" [Location of clothmakters' workshops in 18th–19th-century Sopron]. In *IV. Nemzetközi Kézművesipartörténeti Szimpózium*, 270–275. Veszprém: Magyar Tudományos Akadémia Veszprémi Akadémiai Bizottsága, 1994.

Davis, R. H. C. "The Ford, the River and the City." *Oxoniensis* 38 (1973) 258–267.

de la Moriniere, Simon Barthélemy Joseph Noel. *Histoire generale des peches anciennes et modernes, dans les mers et les fleuves des deux continens,* vol. 1. Paris: Impr. royale, 1815.

Degré, Alajos. *Magyar halászati jog a középkorban [Fishing rights in medieval Hungary]* (A Budapesti Királyi Magyar Pázmány Péter Tudományegyetem Jogtörténeti Szemináriuma Illés József Szemináriumának kiadványai, 5). Budapest: Sárkány Nyomda, 1939.

del Val, María Isabel. "Water: Object of Desire and Source of Conflict in Castile in the Late Middle Ages." *Imago Temporis. Medium Aevum* 8 (2014): 239–262. https://doi.org/10.21001/imagotemporis.v0i0.299267

del Val Valdivieso, María Isabel, ed. *Usos sociales del agua en las ciudades hispánicas a fines de la Edad Media* (Estudios de historia medieval, 6). Valladolid: Universidad de Valladolid, 2002.

Dell' Adami, Rezső. *Az anyagi magyar magánjog codifikátiója. I. A nemzeti eredet problémája. Külön lenyomat a «Magyar Themis»-ből [Codification of the Hungarian common law I. The problem of the national origin.* Separatum from the Magyar Themis] Budapest: Eggenberger, 1877.

Dembińska, Maria, *Przetwórstwo zbożowe w Polsce średniowiecznej (X–XIV wiek) [Grain processing in medieval Poland (10th–14th centuries)]*. Wrocław: Zakład Narodowy im. Ossolńskich, 1973.

Derwich, Marek. "The Earliest Monasteries in Poland." In *Europe's Centre around AD 1000*, eds. Alfried Wieczorek, and Hans-Martin Hinz, 332–334. Stuttgart: Theiss, 2000.

Détshy, Mihály. "Munkások és mesterek az egri vár építkezésein 1493 és 1596 között. 1. közlemény." [Workers and masters at the building works of the castle of Eger between 1493 and 1596, I] *Az Egri Múzeum Évkönyve* 1 (1963): 173–203.

Deutsches Wörterbuch von Jacob und Wilhelm Grimm. Online document: http:// woerterbuchnetz.de/DWB> (last accessed January 10, 2023).

Divald, Kornél, and Béla Iványi. "Az eperjesi Szent Miklós-templom." [The Saint Church of Prešov] In *Az Országos Magyar Szépművészeti Múzeum Évkönyvei, IV [Annual of the Hungarian Museum of Fine Arts, IV]*, ed. Elek Petrovics, 30–73. Budapest: Országos Magyar Szépművészeti Múzeum, 1924–1926.

Długoborski, Wacław, ed. *Dzieje Wrocławia do roku 1807 [The history of Wrocław until 1807*]. Warsaw: Państwowe Wydawnictwo Naukowe, 1958.

Dobrovolný, Petr et al. "Monthly, Seasonal and Annual Temperature Reconstructions for Central Europe Derived from Documentary Evidence and Instrumental Records since AD 1500." *Climatic Change* 101 (2010): 69–107. https://doi.org/10.1007/ s10584-009-9724-x

Domenová, Marcela. "Daňové písomnosti a sociálno-ekonomické pomery mesta Prešov v období stredoveku." [Tax lists and socio-economic processes in the town of Prešov in the Middle Ages] In *Zborník vedeckých prác katedry ekonómie a ekonomiky Anno 2009 [Proceedings of scientific works of the Department of Economics, University of Prešov in 2009]*, 231–255. Prešov: Prešovská univerzita, 2009.

Domenová, Marcela. "K hospodáreniu mesta Prešov na základe daňovej-účtovnej knihy z rokov 1528–1537." [To the economy of Prešov in the light of the book of income and expenditure from the years 1528–1537] *Annales Presovienses* 11 (2011): 25–59.

Domenová, Marcela. "Prešov." In *Lexikón stredovekých miest na Slovensku [Historical lexicon of Slovakian towns]*, eds. Martin Štefánik, and Ján Lukačka, 331–351. Bratislava: Historický ústav SAV, 2010.

Domenová, Marcela. "Prešovské testamenty do roku 1526. (Diplomaticko-obsahový rozbor)." [The last will from Prešovs until 1526 (diplomatic content analysis)] *Historica Carpatica* 33 (2002): 29–68.

Dominkovits, Péter. "Folyóvizek és a XVII. századi vármegyei közigazgatás, bíráskodás (A nyugat-dunántúli törvényhatóságok példáján)." [Waterways and 17th-century county administration, jurisdiction (on the example of Western Transdanubian courts of laws] In *Víz és társadalom Magyarországon a középkortól a XX. századig [Water and Society in the Hungary from the Middle Ages to the 20th century]*, ed. Gergely Krisztián Horváth, 155–188. Budapest: Balassi, 2014.

Domonkos, Ottó. *Ár- és bérlimitációk Sopron városban és Sopron megyében, XVI–XIX. század [Price and and wage limitations at Sopron and in Sopron County, 16th–19th centuries]* (Előmunkálatok a magyarság néprajzához, 8). Budapest: MTA Néprajzi Kutató Csoport, 1980.

Dorondel, Stefan, Stelu Serban, and Daniel Cain, "The Play of Islands: Emerging Borders and Danube Dynamics in Modern Southeast Europe (1830–1900)." *Environment and History* 25 (2019): 521–547. https://doi.org/10.3197/096734018X15254461646413

Draskóczy, István. "Az erdélyi sókamarák ispánjai, 1529–1535: az erdélyi sóbányák sorsa a Szapolyai korszakban." [Counts of the Salt Chambers of Transylvania, from 1529 to 1535: the fate of the Transylvanian salt mines in the Szapolyai era] *Levéltári Közlemények* 75, no. 1 (2004): 27–45.

Draskóczy, István. "Highways between Buda and Kraków." In *On Common Path. Budapest and Kraków in the Middle Ages*, eds. Judit Benda et al., 38–41. Budapest: Budapest History Museum, 2016.

Dreska, Gábor. "A pannonhalmi konvent hiteleshelyének működése a Zsigmond-korban: a pannonhalmi konvent hiteleshelyének 1387 és 1437 között készült kiadványai a Pannonhalmi Bencés Főapátsági Levéltárban." [The functioning of the place of authentication of the abbey of Pannonhalma in the Sigismundian period: the documents issued by the place of authentication of Pannonhalma between 1387 and 1437 in the Archive of the Benedictine Archabbey of Pannonhalma] *Levéltári Közlemények* 68 (1997): 3–61.

Droste, Peter Johannes. *Wasserbau und Wassermühlen an der Mittleren Rur: Die Kernlande des Herzogtums Jülich, 8.–18. Jahrhundert* (Aachener Studien zur alteren Energiegeschichte, 9). Aachen: Shaker Verlag, 2003.

Duby, Georges. *L'économie rurale et la vie des campagnes dans l'Occident médiéval,* 2 vols. Paris: Aubier, 1962. Published in English as *Rural Economy and Country Life in the Medieval West*. London: E. Arnold, 1962.

Durand, Robert. "Les moulins hydrauliques dans le Portugal medieval." In *I mulini nell'Europa medievale: Atti del convegno di San Quirico d'Orcia, 21–23 settembre 2000* (Biblioteca di storia agraria medievale, 21), ed. Paola Galetti, 87–102. Bologna: CLUEB, 2003.

Dyer, Christopher. *Everyday Life in Medieval England*. London: Hambledon, 1994.

Ecsedi, István. "Rekesztő halászat a Tiszán és az alföldi kisvizeken." [Kiddle fishing in the Tisza and the smaller waters in the Middle Ages] *Néprajzi Értesítő* 18 (1926): 61–75.

Edwards, J., and B. Hindle. "Comment: Inland Water Transportation in Medieval England." *Journal of Historical Geography* 19 (1993): 12–14. https://doi.org/10.1006/jhge.1993.1002

Edwards, J., and B. Hindle. "The Transportation System in Medieval England and Wales." *Journal of Historical Geography* 17 (1991): 123–134. https://doi.org/10.1016/S0305-7488(05)80070-2

Eichhorn, Ernst. "Ein Kaiser an der Altmühl. Zwischen Solagrab und Fossa Carolina." In *Karolingisches Franken*, ed. Wolfgang Buhl, 107–143. Würzburg: Echter, 1973.

Endrei, Walter. "Malom." [Mill] In *Korai magyar történeti lexikon (9–14. század) [Early Hungarian historical dictionary, 9th to 14th centuries]*, ed. Gyula Kristó, 441. Budapest: Akadémiai, 1994.

Endrei, Walter. "Malom." [Mill] In *Magyar művelődéstörténeti lexikon,* 14 vols *[Lexicon of cultural history in Hungary]*, ed. Péter Kőszeghy, vol. 7. 257–258. Budapest: Balassi, 2003–2014.

Endrei, Walter. "A magyar hajómalom." [Ship mills in Hungary] *Technikatörténeti Szemle* 21 (1995): 9–21.

Endrei, Walter. "A magyar malomipar, 1550–1800." [Hungarian milling industry, 1550–1800] In *Műszaki innovációk sorsa Magyarországon: malomipar, vaskohászat, textilipar [The fate of technological innovations in Hungary: milling industry, metallurgy, textile industry]*, ed. Walter Endrei, 48–54. Budapest: Akadémiai, 1995.

Endrei, Walter, and György Balázs, "Vízimalom." [Water mill]. In *Magyar művelődéstörténeti lexikon,* 14 vols *[Lexicon of cultural history in Hungary]*, ed. Péter Kőszeghy, vol. 12. 485–488. Budapest: Balassi, 2003–2014.

Engel, Pál, and Norbert C. Tóth, *Királyok és királynék itineráriumai (1382–1438), Zsigmond király és császár (1382–1437) [Itinerary of kings and queens, 1382–1438; King and Emperor Sigismund (1382– 1437)]* (Segédletek a középkori magyar történelem tanulmányozásához, 1). Budapest: Magyar Tudományos Akadémia Támogatott Kutatóhelyek Irodája, 2005.

Engels, Jens Ivo and Gerrit Jasper Schenk. "Infrastrukturen der Macht – Macht der Infrastrukturen: Überlegungen zu einem Forschungsfeld." In *Wasserinfrastrukturen und Macht. Politisch-soziale Dimensionen technischer Systeme von der Antike bis zur Gegenwart* (Historische Zeitschrift, Beiheft 63), eds. Birte Förster, and Martin Bauch, 22–58. Berlin, Munich, and Boston: de Gruyter Oldenbourg, 2015. https://doi.org/10.1515/9783486781052.22

Engels, Johannes. "Der verklagte Flußgott Mäander." *Historia: Zeitschrift für Alte Geschichte* 51, no. 2 (2002): 192–205.

Erdélyi, László. "Szent István-kori bencések hatása a föld- és kertmívelésre meg az iparra." [The role of the Benedictines in the age of St Stephen on gardening and industry] In *Emlékkönyv Szent István király halálának kilencszázadik évfordulóján,* 3 vols *[Memorial volume on the 900th anniversary of Saint Stephen's death]*, ed. Jusztinián Serédi, vol. 1. 479–491. Budapest: Szent István Társulat, 1938.

Éri, István, Lajos Nagy, and Péter Nagybákay, eds. *A magyarországi céhes kézművesipar forrásanyagának katasztere*, 2 vols [Cadastre of the sources of *[Cadastre of the sources of Hungarian guilds]*, Budapest: [no publisher], 1975–1976.

Érszegi, Géza. "A leleszi monostor alapítóleveléről, 1214." [On the foundation charter of the abbey of Leles] *Szabolcs-Szatmár-Beregi Levéltári Évkönyv* 16 (2003): 13–28.

Evenden, Matthew. "Beyond the Organic Machine? New Approaches in River Historiography." *Environmental History* 23, no. 4 (2018): 698–720. https://doi.org/10.1093/envhis/emy054

F. Romhányi, Beatrix. *Kolduló barátok, gazdálkodó szerzetesek: koldulórendi gazdálkodás a késő középkori Magyarországon [Mendicant brothers, farming friars. Mendicant economy in late medieval Hungary]*. Budapest: Martin Opitz, 2018.

F. Romhányi, Beatrix. *Kolostorok és társaskáptalanok a középkori Magyarországon [Monasteries and collegiate chapters in medieval Hungary*]. Budapest: Pytheas, 2000.

F. Romhányi, Beatrix. "A középkori magyar plébániák és a 14. századi pápai tizedjegyzék." [The medieval parishes in Hungary and the 14th-century papal tithe records] *Történelmi Szemle* 61 (2019): 339–360.

F. Romhányi, Beatrix. *Pauline Economy in the Middle Ages* (East Central and Eastern Europe in the Middle Ages, 450–1450, 62). Leiden and Boston: Brill, 2020. https://doi.org/10.1163/9789004424760

F. Romhányi, Beatrix. "Spatial Transformations and Regional Differences in the Medieval Kingdom of Hungary (1000–1500)." *Hungarian Historical Review* 13 (2024): 339–360.

F. Romhányi, Beatrix. "Szempontok a Kárpát-medence térszervezésének változásaihoz (5–14. század)." [Considerations to the changes of the spatial organization of the Carpathian Basin (5th–14th centuries)] In *Hatalmi központok az Avar Kaganátusban – Power Centres of the Avar Khaganate*, eds. Csilla Balogh, József Szentpéteri, and Erika Wicker, 399–420. Kecskemét: Katona József Múzeum – MTA BTK MŐT, 2019.

F. Romhányi, Beatrix, and Katalin Szende. "Cities, Towns and Market Towns in the Context of the Regional Disparities of the Kingdom of Hungary around 1500." *Mesto a dejiny* 14, no. 1 (2025): 6–32. https://doi.org/10.33542/CAH2025-1-01

Fantoni, Giuliana. *L'acqua a Milano: uso e gestione nel basso medioevo (1385–1535)*. Bologna: Cappelli, 1990.

Fehértói, Katalin. "RATA és TILO." [RATA and TILO] *Magyar Nyelvőr* 122 (1998): 64–67.

Fényes, Gabriella. "Középkori vízmű maradványai Budapest Lánchíd utca 19–21. alól: 1/A lelőhely." [Remains of a medieval waterworks from Lánchíd Street 19–21, Budapest. Site 1/A] *Budapest Régiségei* 41 (2008): 193–227.

Ferenczi, László. "Management of Monastic Landscapes. A Spatial Analysis of the Economy of Cistercian Monasteries in Medieval Hungary." PhD-diss., Central European University, 2018.

Ferenczi, László. "'Molendinum ad aquas calidas.' A pilisi ciszterciek az állítólagos 'Fehéregyházán'. Történeti topográfiai és tájrégészeti kutatás a pilisi apátság birtokán." [The Cistercian of Pilis at the alleged 'Fehéregyháza'. Historical topographic and landscape archaeological research at the estates of the Pilis abbey] *Studia Comitatensia* [NS] 1 (2014): 145–160.

Fischer, Theodor, and Markus Strobel. "Von der Fossa Carolina zur Main-Donau-Überleitung." In *Historische Wassernutzung an Donau und Hochrhein sowie zwischen Schwarzwald und Vogesen* (Schriften der Deutschen Wasserhistorischen Gesellschaft, 10), ed. Christoph Ohlig, 61–71. Norderstedt: Books on demand, 2008.

Fish Exploitation in the Past: Proceedings of the 7th Meeting of the ICAZ Fish Remains Working Group. Tervuren: Koninklijk Museum voor Midden-Afrika, 1994.

Fitzmaurice, John. *Damming the Danube: Gabčikovo and Post-Communist Politics in Europe*. Boulder, CO, and Oxford: Westview, 1996.

Fontana, Alessandro, Mattseo Frassine, and Livio Ronchi. "Geomorphological and Geoarchaeological Evidence of the Medieval Deluge in the Tagliamento River (NE Italy)." In *Palaeohydrology. Geography of the Physical Environment,* eds. Jürgen Herget and Alessandro Fontana, 9–116. Cham: Springer, 2020. https://doi.org/10.1007/978-3-030-23315-0_5

Forbes, Robert J. *Studies in Ancient Technology, Volume 2 Irrigation and Drainage, Power [Water- and Windmills], Land Transport and Road Building, The Coming of the Camel.* Leiden: Brill, 1993. https://doi.org/10.1163/9789004450547

Fournier, Patrick, and Sandrine Lavaud, eds. *Eaux et conflits dans l'Europe médiévale et moderne: actes des XXXIIes journées internationales d'histoire de l'Abbaye de Flaran, 8 et 9 octobre 2010.* Toulouse: Presses Univ. du Mirail, 2012.

Fournier, Patrick, and Sandrine Lavaud. "Rapport Historiographique." In *Eaux et conflits dans l'Europe médiévale et moderne: Actes des XXXIIes journées internationales d'histoire de l'Abbaye de Flaran, 8 et 9 octobre 2010,* eds. Patrick Fournier and Lavaud Sandrine , 7–31. Toulouse: Presses Univ. du Mirail, 2012.

Földesi, Szilvia. "Remarks on the Medieval Millstones at Kána." In *Archaeological Investigations in Hungary in 2004*, ed. Júlia Kisfaludi, 54–55. Budapest: Kulturális Örökségvédelmi Hivatal – Magyar Nemzeti Múzeum, 2005.

Förster, Birte and Martin Bauch, eds. *Wasserinfrastrukturen und Macht von der Antike bis zur Gegenwart* (Historische Zeitschrift. Beiheft, N.F., 63). Berlin: de Gruyter – Oldenbourg, 2015. https://doi.org/10.1515/9783486781052

Freedman, Paul. "The Miller." In *Historians on Chaucer: The General Prologue to the Canterbury Tales*, eds. Stephen H. Rigby, and Alastair J. Minnis, 368–385. Oxford: Oxford University Press, 2014. https://doi.org/10.1093/acprof:oso/9780199689545.003.0021

Frova, Carla. "Le traité de fluminibus de Bartolo da Sassoferrato (1355)." *Médiévales* 18 (1999): 81–89.

Fügedi, Erik. "Az esztergomi érsekség gazdálkodása a XV. század végén. I–II. rész." [Farming of the archbishopric of Esztergom at the end of the fifteenth century, I–II] *Századok* 94 (1960): 82–124, and 505–555.

Fügedi, Erik. "Das mittelalterliche Königreich Ungarn als Gastland." In *Die Deutsche Ostsiedlung des Mittelalters als Problem der europäischen Geschichte: Reichenau-Vorträge, 1970–1972* (Vorträge und Forschungen / Konstanzer Arbeitskreis für Mittelalterliche Geschichte, 18), ed. Walter Schlesinger, 471–507. Sigmaringen: J. Thorbecke, 1975.

Fügedi, Erik. "Győr városának 1271. évi kiváltságlevele." [The 1271-privilege charter of the town of Győr] In *Győr. Várostörténeti tanulmányok [Győr. Studies in the history of the town]*, eds. Lajos Dávid, Alfréd Lengyel, and László Z. Szabó, 109–117. Győr: Győr Megyei Jogú Város Tanácsa, 1971.

Fügedi, Erik. "Középkori magyar városprivilégiumok." [Medieval Hungarian urban privileges] *Tanulmányok Budapest Múltjából* 14 (1961): 17–107.

Gácsová, Alžbeta. "Spoločenská štruktúra mesta Prešova v 15. storočí a v prvej polovici 16. storočia." [The social structure of Prešov in the 15th century and in the first half of the 16th century] *Historický časopis* 18 (1970): 347–378.

Galambos, Judit. "An International Environmental Conflict on the Danube: the Gabcicovo – Nagymaros Dams." In *Environment and Democratic Transition in Central and Eastern Europe. Policy and Politics in Central and Eastern Europe* (Technology, Risk and Society, 7), eds. Anna Vári, and Pál Tamás, 146–156. Dordrecht, Boston, and London: Springer, 1993.

Galetti, Paola, ed. *I mulini nell'Europa medievale: atti del convegno di San Quirico d'Orcia, 21–23 settembre 2000* (Biblioteca di storia agraria medievale, 21). Bologna: CLUEB, 2003.

Gálffy, László. "Dunai társadalmak. Változások a városi folyamtérben Bécs és Buda között (XIII. század vége–XV. század közepe)." [Danube societies: changes in the urban riverspace between Vienna and Buda (from end of the 12th to the mid-15th century)] In *Víz és társadalom Magyarországon a középkortól a XX. század végéig [Water and society in Hungary from the Middle Ages to the end of the 20th century]*, ed. Gergely Krisztián Horváth, 111–155. Budapest: Balassi, 2014.

Galloway, James A. "Metropolitan Food and Fuel Supply in Medieval England: Regional and International Contexts." In *Food Supply, Demand and Trade: Aspects of the Economic Relationship between Town and Countryside (Middle Ages – 19th Century)* (Comparative Rural History Network – Publications, 14), ed. E. Thoen, and P. Van Cruyningen, 7–18. Turnhout: Brepols, 2012. https://doi.org/10.1484/M.CORN-EB.4.00053

Galloway, James A., Derek Keene, and Margaret Murphy. "Fuelling the City: Production and Distribution of Firewood and Fuel in London's Region,

1290–1400." *The Economic History Review* NS 49 (1996): 447–472. https://doi.org/10.1111/j.1468-0289.1996.tb00577.x

Galusová, Lucie. "Vodní mlýn jako objekt archeologického výzkumu." [Water mills as objects of archaeological research] *Archaeologia historica* 40 (2015): 267–293. https://doi.org/10.5817/AH2015-1-15

Galusová, Lucie, and Martina Maríková. "Die Baugestalt der Wassermühlen im mittelalterlichen Böhmen und Mähren." In *Wassermühlen und Wassernutzung im mittelalterlichen Ostmitteleuropa* (Forschungen zur Geschichte und Kultur des östlichen Mitteleuropa, 50), eds. Martina Maríková, and Christian Zschieschang, 309–324. Stuttgart: Franz Steiner, 2015.

Garády, Sándor. "Buda középkori és újkori vízművei, különös tekintettel a Svábhegy-Városkút kutatására." [The medieval and modern water works of Buda with special regard to Svábhegy-Városkút] *Budapest Régiségei* 32 (1998): 245–265.

Garády, Sándor. "Jelentés az 1936–1942. évben végzett ásatásokról." [Report of the excavations carried out between 1936 and 1942] *Budapest Régiségei* 13 (1943): 418–427.

Gardiner, Mark F. "Bateliers, pêcheurs et habitants des ports: les conflits sur les droits des rivières dans l'Angleterre medieval." In *Eaux et conflits dans l'Europe médiévale et moderne: Actes des XXXII*[es] *Journées Internationales d'Histoire de l'Abbaye de Flaran, 8 et 9 octobre 2010,* eds. Patrick Fournier, and Sandrine Lavaud, 59–70. Toulouse: Presses Univ. du Mirail, 2012.

Gardiner, Mark. "Hythes, Small Ports, and Other Landing Places in Later Medieval England." In *Waterways and Canal-Building in Medieval England,* ed. John Blair, 85–109. Oxford and New York: Oxford University Press, 2007. https://doi.org/10.1093/acprof:osobl/9780198723134.003.0005

Geffcken, Heinrich. "Zur Geschichte des deutschen Wasserrechts." *Zeitschrift der Savigny-Stiftung für Rechtsgeschichte. Germanistische Abteilung* 21, no. 1 (1900): 173–217. https://doi.org/10.7767/zrgga.1900.21.1.173

Germanidou, Sophia. "Watermills in Byzantine Greece (Fifth–Twelfth Centuries). A Preliminary Approach to the Archaeology of Byzantine Hydraulic Milling Technology." *Byzantion* 84 (2014): 185–201.

Gestione dell'acqua in Europa (XII–XVIII secc.): selezione di ricerche = Water Management in Europe (12th*–18th Centuries): Selection of Essays)* (Serie II, Atti delle 'settimane de studio' e altri convegni, 49). Florence: Firenze University Press, 2018.

Getzler, J. *A History of Water Rights at Common Law.* Oxford: Oxford University Press, 2004. https://doi.org/10.1093/acprof:oso/9780199207602.001.0001

Gheorgiu, Teodor Octavian. *Cetatile oraselor: apararea urbana în centrul si estul Europei în Evul Mediu [Fortified towns. Urban Defenses in Central and Eastern Europe in the Middle Ages].* Bucureşti: Editura Simetria, 2000.

Ghinzoni, P[ietro]. "Maestro Giacomo Arribotti e il Naviglio Grande reso navigabile." *Archivio Storico Lombardo* 20 (1893): 200–203.

Gille, B. "Le moulin à eau, une révolution médiévale." *Techniques et civilizations* 3 (1954): 1–15.

Gimpel, Jean. *La Révolution industrielle du Moyen âge* (Paris: Seuil, 1975). Published in English as: *The Medieval Machine: The Industrial Revolution of the Middle Ages.* New York: Holt, Rinehart & Winston, 1976.

Ginzburg, Carlo. *The Cheese and the Worms. The Cosmos of a Sixteenth-Century Miller.* Baltimore: Johns Hopkins University Press, 1992.

Göbel, Ilka. *Die Mühle in der Stadt: Müllerhandwerk in Göttingen, Hameln und Hildesheim vom Mittelalter bis ins 18. Jahrhundert* (Veröffentlichungen des Instituts für historische Landesforschung der Universität Göttingen, 31). Bielefeld: Verl. für Regionalgeschichte, 1993.

Goda, Károly. "A soproni városvezető réteg a 15–16. században. A polgármesteri és városbírói tisztség összehasonlító igazgatási és társadalomtörténete." [The leading elite of Sopron in the 15th–16th centuries. The administrative and social history of the positions of majors and town judges] PhD-diss., Eötvös Loránd Tudományegyetem, 2011.

Goliński, Mateusz. *Wokół socjotopografii późnośredniowiecznej Świdnicy*, 2 vols *[The sociotopography of late medieval Swidnica]*. Wrocław: Wydawn. Uniw. Wrocławskiego, 2000.

Gräf, Daniela. *Boat Mills in Europe from Early Medieval to Modern Times* (Veröffentlichungen des Landesamtes für Archäologie mit Landesmuseum für Vorgeschichte, 51 = Bibliotheca molinologica, 19). Dresden: Landesamt für Archäologie mit Landesmuseum für Vorgeschichte, 2006.

Graus, Frantisek. *Dějiny venkovského lidu v Čechách v době* předhusitské, 2 vols *[History of the rural population in Bohemia in the pre-Hussite period]* (Prague: Státnínakladatelstvi politické literatury, 1953–1957.

Graus, Igor. "Vývoj fortifikácie Banskej Bystrice do konca 16. storočia v podmienkach špecifík pasívnej obrany stredoslovenských banských miest." [Development of the fortifications of the town of Banská Bystrica until the end of the 16th century under the specific conditions of passive defense of the Middle-Slovakian mining towns] *Vojenská história* 4 (2000): 3–21.

Grélois, Emmanuel. "Les usages concurrents de la rivière: l'Allier en Basse-Auvergne (XIIIee-XVIIIe siècle)." In *Eaux et conflits dans l'Europe médiévale et moderne: Actes des XXXIIes Journées Internationales d'Histoire de l'Abbaye de Flaran, 8 et 9 octobre 2010*, eds. Patrick Fournier, and Sandrine Lavaud, 107–118. Toulouse: Presses Univ. du Mirail, 2012.

Grewe, Klaus. "Der Karlsgraben bei Weißenburg i.B. und der Fulbert-Stollen von Maria Laach: zwei große Wasserbauprojekte des Mittelalters in Deutschland." In *Europäische Technik im Mittelalter, 800 bis 1400: Tradition und Innovation*, ed. Uta Lindgren, 111–115. Berlin: Gebr. Mann, 1996.

Grewe, Klaus, ed. *Die Wasserversorgung im Mittelalter*. Mainz am Rhein: von Zabern, 1991.

Guillerme, André. *Les temps de l'eau: la cité, l'eau et les techniques: nord de la France, fin IIIe–début XIXe siècle* (Seyssel: Champ Vallon, 1983). Published in English as: *The Age of Water: The Urban Environment in the North of France, A. D. 300–1800* (Environmental History Series, 9). College Station, TX: A & M University Press, 1988.

Guittonneau, Pierre-Henri. "Petites villes et espace fluvial: pratiques sociales et conflits d'usage autour de Paris au XVe siècle." In *The Power of Space in Late Medieval and Early Modern Europe: The Cities of Italy, Northern France and the Low Countries* (Studies in European Urban History, 30), eds. Marc Boone, and Martha C. Howell, 57–76. Turnhout: Brepols, 2013. https://doi.org/10.1484/M.SEUH.1.101407

Gulyás, László Szabolcs. "A középkori magyar városfejlődés migráció- és ipartörténeti vonatkozásai a történeti személynévtan tükrében II. Személynévadás és kézműipar." [Migration and industrial historical aspects of medieval Hungarian urban development in light of historical onomastics, II. Personal naming and crafsmanship] *Névtani Értesítő* 37 (2015): 137–156.

Gulyás, László Szabolcs. "Középkori mezővárosi foglalkozásneveink forrásértékéről." [On the source value of profession names from medieval Hungarian market towns] *Századok* 142 (2008): 437–462.

Gulyás, László Szabolcs. "A mezővárosi önkormányzat funkciói és társadalmi háttere a középkori Hegyalján." [Functions of the autonomy of market towns and its social background in the medieval Hegyalja region] PhD-diss., Debreceni Egyetem, 2008.

Gulyás, László Szabolcs. "Városfalépítés a középkori Eperjesen." [Town wall building at medieval Prešov] In *Falak és választóvonalak a történelemben [Walls and division line sin history]* (A Nyíregyházi Főiskola Történettudományi és Filozófia Intézete Kiadványai. Terminus Könyvek, 1), eds. Attila Buhály, Gábor Reszler, and György Szoboszlay, 127–142. Nyíregyháza: Nyíregyházi Főiskola Történettudományi és Filozófia Intézet, 2014.

Gulyás, László Szabolcs. *Városfejlődés a középkori Máramarosban [Urban development in medieval Maramureș]* (Erdélyi Tudományos Füzetek, 280). Cluj: Erdélyi Múzeum-Egyesület, 2014.

Gündisch, Gustav. "Winz und die Sächsische Nationsuniversität." In *Emlékkönyv Kelemen Lajos születésének nyolcvanadik évfordulójára [Memorial book in honor of the 80th birthday of Lajos Kelemen]* (A Bolyai Tudományegyetem Kiadványai, I-1), eds. András Bodor et al., 311–325. Bucharest and Cluj: Tudományos Könyvkiadó, 1957.

Györffy, György. *István király és műve [King Stephen and his legacy]*. Budapest: Akadémiai, 1977.

Györffy, György. *König Stephan der Heilige*. Budapest: Corvina, 1988.

Gyulai, Éva. "Miskolc középkori topográfiája." [Medieval topography of Miskolc] In *Miskolc története I. A kezdetektől 1526-ig [History of Miskolc, I. From the beginnings to 1526]*, ed. István Dobrossy, 175–254. Miskolc: Borsod-Abaúj-Zemplén Megyei Levéltár, 1996.

Gyulai, Éva. "Termelés és kereskedelem a középkori Miskolcon." [Production and trade at medieval Miskolc] In *Miskolc története I. A kezdetektől 1526-ig [History of Miskolc, I. From the beginnings to 1526]*, ed. István Dobrossy, 255–336. Miskolc: Borsod-Abaúj-Zemplén Megyei Levéltár, 1996.

Haidvogl, Gertrud. et al., "Historical Ecology of Riverine Fish in Europe." *Aquatic Sciences* 77 (2015): 315–324. https://doi.org/10.1007/s00027-015-0400-0

Hammer, Carl I. "'A Suitable Place for Putting up a Mill.' Water Power Landscapes and Structures in Carolingian Bavaria." *Vierteljahrschrift für Sozial- und Wirtschaftsgeschichte* 95 (2008): 319–334.

Härtel, Reinhard. "Von Wasser- und Mühlenrecht im Hochmittelalter." In *Recht und Geschichte. Festschrift Hermann Baltl zum 70. Geburtstag*, ed. Helfried Valentinitsch, 219–236. Graz: Leykam, 1988.

Hehmeyer, Ingrid. *A History of Water Engineering and Management in Yemen* (Handbook of Oriental Studies. Section 1: The Near and Middle East, 129). Leiden and Boston: Brill, 2019. https://doi.org/10.1163/9789004387713

Heimpel, Hermann. "Die Federschnur. Wasserrecht und Fischrecht in der 'Reformation Kaiser Siegmunds'." *Deutsches Archiv für Erforschung des Mittelalters* 19 (1963): 451–488.

Herényi, István. *A nyugati gyepü erődítményei: erődítmények, várak Nyugat-Magyarországon [Constructions of the Western gyepű: fortifications and castles in Western-Hungary]*. Budapest: Heraldika, 2007.

Hlaváček, Ivan. "The Use of Charters and Other Documents in Přemyslide Bohemia." In *Charters and the Use of the Written Word in Medieval Society* (Utrecht Studies in Medieval Literacy, 5), ed. Karl Heidecker, 133–144. Turnhout: Brepols Publishers, 2000. https://doi.org/10.1484/M.USML-EB.6.09070802050003050007070102

Hoffmann, István. *A Tihanyi alapítólevél mint helynévtörténeti forrás [The foundation charter of Tihany as a source of onomastics]* (A Magyar Névarchívum Kiadványai, 16). Debrecen: Debreceni Egyetemi Kiadó, 2010.

Hoffmann, Richard. "Medieval Fishing." In *Working with Water in Medieval Europe: Technology and Resource-Use* (Technology and Change in History), ed. Paolo Squatriti, 331–393. Leiden, Boston, and Cologne: Brill, 2000. https://doi.org/10.1163/9789047400110_012

Hoffmann, Richard, and Christoph Sonnlechner, "Vom Archivobjekt zum Umweltschutz: Maximilians Patent von 1506." *Jahrbuch des Vereins für Geschichte der Stadt Wien* 62/63 (2006/2007): 71–124.

Hoffmann, Richard C. "A Brief History of Aquatic Resource Use in Medieval Europe." *Helgoland Marine Research* 59 (2005): 22–30. https://doi.org/10.1007/s10152-004-0203-5

Hoffmann, Richard C. "Economic Development and Aquatic Ecosystems in Medieval Europe." *The American Historical Review* 101 (1996): 631–669. https://doi.org/10.1086/ahr/101.3.631

Hoffmann, Richard C. "Footprint Metaphor and Metabolic Realities: Environmental Impacts of Medieval European Cities." In *Natures Past: The Environment and Human History* (Comparative Studies in Society and History), ed. Paolo Squatriti, 288–325. Ann Arbor, MI: University of Michigan Press, 2007.

Hohensinner, Severin et al. "Changes in Water and Land: The Reconstructed Viennese Riverscape from 1500 to the Present." *Water History* 5 (2013): 145–172. https://doi.org/10.1007/s12685-013-0074-2

Hohensinner, Severin et al. "Two Steps Back, One Step Forward: Reconstructing the Dynamic Danube Riverscape Under Human Influence in Vienna." *Water History* 5 (2013): 121–143. https://doi.org/10.1007/s12685-013-0076-0

Holl, Imre. "Feuerwaffen und Stadtmauern. Angaben zur Entwicklung der Wehrarchitektur des 15. Jahrhunderts." *Acta Archaeologica Academiae Scientiarum Hungaricae* 33 (1981): 201–243.

Holl, Imre. "Középkori városi élet – városi építészet." [Medieval urban life – urban architecture] *Archaeologiai Értesítő* 116 (1989): 52–76.

Holl, Imre. "Sopron középkori városfalai." [Medieval town walls of Sopron] *Archaeologiai Értesítő* 94 (1967): 155–183, 95 (1968): 188–205, 98 (1971): 24–44, and 100 (1973): 180–207.

Holl, Imre. "Sopron (Ödenburg) im Mittelalter." *Acta Archaeologica Academiae Scientiarum Hungaricae* 31 (1979): 105–145.

Holl-Gyürki, Katalin. "The Use of Glass in Medieval Hungary." *Journal of Glass Studies* 28 (1986): 70–81.

Holt, Richard. "Medieval England's Water-Related Technologies." In *Working with Water in Medieval Europe: Technology and Resource-Use* (Technology and Change in History), ed. Paolo Squatriti, 51–100. Leiden, Boston, and Cologne: Brill, 2000. https://doi.org/10.1163/9789047400110_006

Holt, Richard. *The Mills of Medieval England*. Oxford and New York: Basil Blackwell, 1988.

Holub, József. *Zala megye középkori vízrajza [Medieval hydrography of Zala County]* (A Göcseji Múzeum közleményei, 23). Zalaegerszeg: Göcseji Múzeum, 1963.

Horn, Walter, and Ernest Born. "Water Power and the Plan of St. Gall." *Journal of Medieval History* 1 (1975): 219–258.

Horváth, Gergely Krisztián, ed. *Víz és társadalom Magyarországon a középkortól a XX. századig [Water and Society in Hungary from the Middle Ages to the 20th century]*. Budapest: Balassi, 2014.

Horváth, Richárd. "Mitől vár a vár? Márkusfalvi Máriássy István várépítési kísérlete 1507-ben." [What makes a castle a castle? The castle-building attempt of István Márkusfalvi Máriássy in 1507] In *Tiszteletkör. Történeti tanulmányok Draskóczy István egyetemi tanár 60. születésnapjára [Lap of honor. Studies in honor of the 60th birthday of Professor István Draskóczy]*, eds. Gábor Mikó, Bence Péterfi, and András Vadas, 421–430. Budapest: ELTE Eötvös Kiadó, 2012.

Horváth, Richárd. "Vár volt-e a tari vár?." [Was the castle or Tar a castle?] *Castrum. A Castrum Bene Egyesület Hírlevele* 15, no. 1–2 (2012): 43–48.

Horváth, Richárd. "Várépítés engedélyezése az Árpád-kori Magyarországon." [Permission of castle-building in Árpádian-age Hungary] In *Várak nyomában. Tanulmányok a 60 éves Feld István tiszteletére [In pursuit of castles… Studies in honor of the 60th birthday of István Feld]*, eds. György Terei et al., 79–93. Budapest: Castrum Bene Egyesület, 2011.

Ihrig, Dénes, ed. *A magyar vízszabályozás története [The history of water regulation in Hungary]*. Budapest: Országos Vízügyi Hivatal, 1973.

Ingenhaeff, Wolfgang, ed. *Wasser – Fluch und Segen: Schwazer Silber*. Innsbruck: Berenkamp, 2004.

Irásné Melis, Katalin. "Pest kialakulása és a középkori város építési periódusai." [The formation of Pest and its medieval building periods] *Budapest Régiségei* 47 (2014): 135–238.

Isenberg, Gabriele, and Barbara Scholkmann, eds. *Die Befestigung der mittelalterlichen Stadt* (Städteforschung, A 45). Cologne, Weimar, and Vienna: Böhlau, 1997.

Iványi, Béla. "Eperjes középkori ágyúöntőháza és puskapormalma, I–II." [Medieval cannon foundry and gundpowder mill of Prešov] *Hadtörténeti Közlemények* 15 (1913): 113–122 and 264–272.

Iványi, Béla. "Eperjes szabad királyi város jegyzőkönyve, 1424–1509." [The protocols of the free royal town of Prešov, 1424–1509] *Történelmi Tár* 10 (1909): 121–122.

Jaccottey, Luc, and Gilles Rollier, eds. *Archéologie des moulins hydrauliques, à traction animale et à vent, des origines à l'époque médiévale et moderne en Europe et dans le monde Méditerranéen: Actes du colloque international, Lons-le-Saunier du 2 au 5 novembre* 2011, 2 vols. Besançon: Presses Universitaires de Franche-Comté, 2016.

Jakab, Gusztáv et al. "The Environmental History of a Former Salt Town in Transylvania (Sic, Northern Romania)." *Acta Archaeologica Academiae Scientiarum Hungaricae* 69 (2018): 185–206. https://doi.org/10.1556/072.2018.69.1.8

Jakab, Gusztáv et al. "Relict Anthropogenic Ecosystem from the Middle Ages: History of a Salt Marsh from Transylvania (Sic, N Romania)." *Environmental Archaeology* 25 (2019): 96–113. https://doi.org/10.1080/14614103.2019.1578547

Jankó, Ferenc, József Kücsán, and Katalin Szende. *Sopron* (Magyar Várostörténeti Atlasz, 1). Sopron: Győr-Moson-Sopron Megyei Levéltár Soproni Levéltára, 2010.

Jankovich B., Dénes. "Adatok a Körösvidék középkori vízrajzához és a vizek hasznosításához." [Data to the medieval hydrography and water use of the Körös region] *Békés Megyei Múzeumok Közleményei* 16 (1996): 305–349.

Jankovská, Vlasta, and Jan Pokorný. "Palaeoecology of a Medieval Fishpond System (Vajgar, Czech Republic)." *Folia Geobotanica* 37, no. 3 (2002): 253–273.

Jaritz, Gerhard, and Katalin Szende, eds. *Medieval East Central Europe in a Comparative Perspective: From Frontier Zones to Lands in Focus*. New York: Routledge, 2016. https://doi.org/10.4324/9781315618876

Jásek, Jaroslav. "An Attempt at an Outline of the Historical Development of Water Supply and Sewerage of Medieval Settlements in the Czech Lands." In *Wassermühlen und Wassernutzung im mittelalterlichen Ostmitteleuropa* Forschungen zur Geschichte und Kultur des östlichen Mitteleuropa, 50), eds. Martina Maríková, and Christian Zschieschang, 57–76. Stuttgart: Franz Steiner, 2015.

Jones, E. T. "River Navigation in Medieval England." *Journal of Historical Geography* 26 (2000): 60–75. https://doi.org/10.1006/jhge.1999.0191

Jørgensen, Dolly. "Local Government Responses to Urban River Pollution in Late Medieval England." *Water History* 2, no. 1 (2010): 35–52. https://doi.org/10.1007/s12685-010-0016-1

Juhász, Lajos. "A Vas megyei Farkaserdő a XVII–XVIII. században." [Farkas Forest in Vas County in the seventeenth–eighteenth centuries] *Századok* 71 (1937) [Supplementum]: 553–575 [37–69].

K. Németh, András. "Barátok tava. Tájrégészeti és történeti adatok egy helynév eredetéhez." [The lake of the brothers. Landscape archaeological and historical data to the origin of a placename] *Archaeologiai Értesítő* 138 (2013): 367–374.

K. Németh, András. "Vizek és vízgazdálkodás a középkori Tolna megyében I. Vízimalmok." [Waters and water management in medieval Tolna County, I. Water mills] *A Wosinsky Mór Múzeum Évkönyve* 35 (2013) [2015]: 121–151.

K. Németh, András. "Vizek és vízgazdálkodás a középkori Tolna megyében. II. Halászat." [Waters and water management in medieval Tolna County, II. Fishing] *A Wosinsky Mór Megyei Múzeum Évkönyve* 36 (2014): 223–251.

K. Németh, András. "Vizek és vízgazdálkodás a középkori Tolna megyében III. Vízrajz." [Waters and water management in medieval Tolna County, III. Hydrography] *A Wosinsky Mór Megyei Múzeum Évkönyve* 37 (2015): 7–39.

K. Németh, András, and Gábor Máté. "Szempontok és példák a középkori eredetű malmok és malomhelyek folytonosságának vizsgálatához." [Notes and examples to the study of the continuity of medieval mills and mill sites] In *Középkori elemek a mai magyar anyagi kultúrában [Medieval elements in the present- day Hungarian material culture]* (Tanulmányok a Kárpát-medence anyagi kultúrája köréből, 1), eds. Anikó Báti, and Zsigmond Csoma, 47–68. Budapest: Agroinform, 2014.

Kádas, István. "Sárosi 'reform' Miklós fia Miklós ispánsága idején (1374–1380)?." ['Reform' in Sáros County during the ispánság of Miklós son of Miklós (1374–1380)?] In *Micae mediaevales V. Fiatal történészek dolgozatai a középkori Magyarországról és Európáról [Studies of young historians on medieval Hungary and Europe]* ELTE BTK Történelemtudományok Doktori Iskola Tanulmányok – Konferenciák, 9), eds. Laura Fábián et al., 127–144. Budapest: ELTE BTK Történelemtudományok Doktori Iskola, 2016.

Kain, Roger. J. P. "Maps and Rural Land Management in Early Modern Europe." In *The History of Cartography, Vol. 3. Cartography in the European Renaissance*, ed. David Woodward, 705–718. Chicago: Chicago University Press, 2007.

Online: www.press.uchicago.edu/books/HOC/HOC_V3_Pt1/HOC_VOLUME3_Part1_chapter28.pdf (last accessed: October 5, 2019)

Kaiser, Knut et al., eds. *La città e il fiume, Historische Perspektiven auf Wasserhaushalt und Wassernutzung in Mitteleuropa* (Cottbuser Studien zur Geschichte von Technik, Arbeit und Umwelt, 38). Münster: Waxmann, 2012.

Kalhous, David. *Anatomy of a Duchy. The Political and Ecclesiastical Structures of Early Přemyslid Bohemia* (East Central and Eastern Europe in the Middle Ages, 450–1450, 19). Leiden and Boston: Brill, 2012. https://doi.org/10.1163/9789004229815

Kálmán, Dániel. "A középkori magyar armálisok címerei 1437–1490 között." [Coats of arms of the medieval Hungarian grants of arms between 1437 and 1490] In *Micae mediaevales V. Fiatal történészek dolgozatai a középkori Magyarországról és Európáról [Studies of young historians on medieval Hungary and Europe]* (ELTE BTK Történelemtudományok Doktori Iskola Tanulmányok – Konferenciák, 9), eds. Laura Fábián et al., 145–158. Budapest: ELTE BTK Történelemtudományok Doktori Iskola, 2016.

Kammerer, Odile, and Odile Redon, eds. *Médiévales* no. 36 (1999): Le fleuve [Special issue].

Karácsonyi, János. *A hamis, hibáskeltű és keltezetlen oklevelek jegyzéke 1400-ig: a Történelmi Tár 1908 évi számában megjelent "pótlások..."-kal kiegészítve* [Register of forged, wrongly dated and undtated charters up to 1400, complemented by the additions published in the *Történelmi Tár* in 1908] (*Aetas Könyvek*, 1). Szeged: Aetas, 1988.

Karaman, Igor. *Tisućljetni Zagreb [Zagreb Through a Thousand Years*]. Zagreb: Školska knjiga, 1994.

Karbić, Marija, and Bruno Škreblin. "Grad na rijeci ili pored nje: srednjovjekovni Zagreb i rijeka Sava." [The Town On or Close to the River: Medieval Zagreb and the Sava River] In *Rijeka Sava u povijesti [River Sava in history]* (Bibliotheca Croatica: Slavonica, Sirmiensia et Baranyansia Posebna izdanja, 15), ed. Branko Ostajmer, 163–180. Slavonski Brod: Hrvatski Institut za Povijest, 2015.

Karczag, Ákos, and Tibor Szabó. *Erdély, Partium és a Bánság erődített helyei* [Fortified places in Transylvania, Partium and Banate]. 2nd ed. Budapest: Semmelweis, 2012.

Karlinszky, Balázs Péter. "A veszprémi káptalan a középkorban. A veszprémi székeskáptalan középkori birtokai." [The chapter of Veszprém in the Middle Ages. The medieval estates of the chapter] PhD-diss., Pázmány Péter Katolikus Egyetem, 2013.

Károlyi, Zsigmond. *A vízhasznosítás, vízépítés és vízgazdálkodás története Magyarországon [The history of water use, water construction and water management in Hungary]* (Műszaki tudománytörténeti kiadványok, 13). Budapest: Tankönyvkiadó, 1960.

Katić, Lovre. "Solinski mlinovi u prošlosti." [The ancient mills of Salona] *Starohrvatska prosvjeta* Serija III. 2 (1952): 201–219.

Kázmér, Miklós. *Régi magyar családnevek szótára. XIV–XVII. század [Dictionary of old Hungarian family names, 14th–17th centuries*]. Budapest: Magyar Nyelvtudományi Társaság, 1993.

Keats-Rohan, K[atharine] S. B. *Domesday People: a Prosopography of Persons Occurring in English Documents, 1066–1166. I. Domesday Book.* Woodbridge: Boydell, 1999.

Keene, Derek. "Issues of Water in Medieval London to c. 1300." *Urban History* 28 (2001): 161–179. https://doi.org/10.1017/S0963926801002012

Keglevich, Kristóf. *A garamszentbenedeki apátság története az Árpád- és az Anjou-korban, 1075–1403 [The history of the abbey of Hronský Beňadik in the Árpádian and Angevin period, 1075–1403]* (Capitulum, 8). Szeged: Szegedi Tudományegyetem Történeti Intézete, 2012.

Keglevich, Kristóf. "A garamszentbenedeki bencések és Újbánya polgárai a 14. század közepén." [The Benedictines of Hronský Beňadik and the burghers of Nová Baňa in the mid-14th century] In *Tiszteletkör. Történeti tanulmányok Draskóczy István egyetemi tanár 60. születésnapjára* [Lap of honor. Studies in honor of the 60th birthday of Professor István Draskóczy], eds. Gábor Mikó, Bence Péterfi, and András Vadas, 137–144. Budapest: ELTE Eötvös Kiadó, 2012.

Kenyeres, István. "The Fate of the Medieval Archives of Buda and Pest." In *Medieval Buda in Context* (Brill's Companions to European History, 10), eds. Balázs Nagy et al., 52–68. Leiden and Boston: Brill, 2016. https://doi.org/10.1163/9789004307674_004

Kiersnowski, Ryszard. "Znaki graniczne w Polsce średniowiecznej." [Border signs in medieval Poland]. *Archeologia Polski* 5 (1960): 257–289.

Kisch, Guido. *Das Fischereirecht im Deutschordensgebiet* (Forschungen und Quellen zur Rechts- und Sozialgeschichte des Deutschordenslandes, 3). 2nd ed. Sigmaringen: Thorbecke, 1978.

Kiss, Andrea. *Floods and Long-Term Water-Level Changes in Medieval Hungary* (Springer Water). Cham: Springer, 2019.

Kiss, Andrea. " 'Rivulus namque, qui dicitur Fuk, fluens de prefato lacu' – Fok, Sár, Foksár." In *Antropogén ökológiai változások a Kárpát-medencében [Anthropogenic ecological changes in the Carpathian Basin]*, eds. Bertalan Andrásfalvy, and Gábor Vargyas, 49-64. [Pécs] and Budapest: PTE Néprajz–Kulturális Antropológia Tanszék–L'Harmattan, 2009.

Kiss, Andrea. "Study on the Historical Geography of the First Extent Perambulation Sketch from the Carpathian Basin." *Zbornik Odsjeka za povijesne znanosti ZPDZ HAZU* 19 (2002): 127–141.

Kiss, Lajos. *Földrajzi nevek etimológiai szótára [Etymological dictionary of geographic names]*. Budapest: Akadémiai, 1978.

Kiss, Lajos. "Középkori földrajzi nevek magyarázata." [Explanations of medieval place names] *Magyar Nyelv* 86 (1990): 161–172.

Klaić, Nada. *Povijest Zagreba. 1. Zagreb u srednjem vijeku [History of Zagreb, vol. 1. Zagreb in the Middle Ages]*. Zagreb: SNL, 1982.

Klaić, Nada. *Zagreb u srednjem vijeku [Zagreb in the Middle Ages]*. Zagreb: Liber, 1982.

Klápšte, Jan. *The Czech Lands in Medieval Transformation* (East Central and Eastern Europe in the Middle Ages, 450–1450, 17). Leiden and Boston: Brill, 2012. https://doi.org/10.1163/9789004226463

Klápště, Jan, ed. *Water Management in Medieval Rural Economy* (Ruralia V – Památky archeologické, Supplementum, 17). Prague: Institute of Archaeology, Academy of Sciences of the Czech Republic, 2005.

Klein, Willi. *Zur Geschichte des Mühlenwesens im Main-Kinzig-Kreis* (Hanauer Geschichtsblätter, 40). Hanau: Selbstverl. d. Hanauer Geschichtsvereins u. d. Wetterauischen Gesellschaft für die Gesamte Naturkunde zu Hanau, 2003.

Kleinhans, Maarten G., Henk J. T. Weerts, and Kim M. Cohen. "Avulsion in Action: Reconstruction and Modelling Sedimentation Pace and Upstream Flood

Water Levels Following a Medieval Tidal-River Diversion Catastrophe (Biesbosch, The Netherlands, 1421–1750 AD)." *Geomorphology* 118 (2010): 65–79. https://doi.org/10.1016/j.geomorph.2009.12.009

Klement, Judit. *Hazai vállalkozók a hőskorban: A budapesti gőzmalomipar vállalkozói a 19. század második felében [Hungarian enterpreneurs in the heroic age: the enterpreneurs of steam mill industry in the second hald of the 19th century]*. Budapest: Eötvös Kiadó, 2012.

Kljajić, Ivka, and San Mikulec, "Cartographic-Historical Analyses of the Medveščak Stream Regulation and Quantum GIS." *Kartografija i geoinformacije – Journal Cartography and Geoinformation* 12, no. 19 (2013): 30–43.

Knezović, Marin. "Voda u hrvatskim ranosrednjovjekovnim ispravama." [Water in the Croatian early medieval documents] *Ekonomska i ekohistorija* 3, no. 1 (2007): 35–50.

Kniezsa, István. *A magyar nyelv szláv jövevényszavai,* vol. 1 *[Slavic loanwords in Hungarian]*. Budapest: Akadémiai, 1955.

Knoll, Martin. "Fluss und Flusslandschaft. Eine Kultur- und Umweltgeschichte." *Mitteilungen des Verbandes bayerischer Geschichtsvereine* 26 (2014): 65–86.

Knoll, Martin, Uwe Lübken, and Dieter Schott, eds. *Rivers Lost, Rivers Regained: Rethinking City-River Relations* (History of the Urban Environment). Pittsburgh, PA: University of Pittsburgh Pres, 2017. https://doi.org/10.2307/j.ctt1qnw8gv

Koehne, Carl. *Das Recht der Mühlen bis Zum Ende der Karolingerzeit. Ein Beitrag znr Geschichte des Deutschen Gewerberechts* (Deutschen Staats- und Rechtsgeschichte, 71). Breslau: M. & H. Marcus, 1904.

Koehne, Carl. "Die Mühle im Rechte der Völker." *Beiträge zur Geschichte der Technik* 5 (1913): 27–53.

Kőfalvi, Tamás. "Malmok és malomhelyek a pécsváradi konvent hiteleshelyi gyakorlatában." [Mills and mill sites in the practice of the credible place of the convent of Pécsvárad] In *Studia Professoris – Professor Studiorum. Tanulmányok Érszegi Géza hatvanadik születésnapjára. [Studies in honor of the 60th birthday of Géza Érszegi]*, eds. Tibor Almási, István Draskóczy, and Éva Jancsó, 154–164. Budapest: Magyar Országos Levéltár, 2005.

Kohl, Werner. *Recht und Geschichte der alten Münchner Mühlen* (Miscellanea Bavarica Monacensia, 15 = Neue Schriftenreihe des Stadtarchivs München, 32). Munich: Stadtarchiv, 1969.

Kolláth, Ágnes. "Székesfehérvár vízellátása a középkorban és a kora újkorban." [The water supply of Székesfehérvár in the Middle Ages and the early modern times] In *Határon innen és túl. Gazdaságtörténeti tanulmányok a magyar középkorról [Within and beyond the borders. Studies in the economic history of medieval Hungary]*, eds. Boglárka Weisz, and István Kádas, István, 351–379. Budapest: Bölcsészettudományi Kutatóközpont, Történettudományi Intézet, 2021.

Koppány, Tibor. "A gátai számadáskönyvek. Adatok egy 17. század eleji dunántúli uradalom építkezéseihez." [The account books of Gattendorf. Data to the building works of an early-17th-century manor in the Transdanubia] *Műemlékvédelmi Szemle* 10, no. 1–2 (2000): 145–160.

Kotyza, Oldřich. "K počátkům vodních mlýnů v českých zemích aneb o existenci vltavských jezů a hydraulických mlýnů v Praze 10. století. Poznámky k břevnovskému aktu ze 14. ledna 993." [On the origin of water mills in the Czech lands or on the existence of the weirs on the Vltava and hydraulic mills in Prague in the 10th century. Notes on the Břevnov act of 14 January 993] In *Středověká Evropa v pohybu k poctě Jana Klápště / Medieval Europe in Motion in Honour of Jan Klápště*, eds. Ivana Boháčová, and Petr Sommer, 461–499. Prague: Archeologický ústav AV ČR, Praha, v. v. i., 2014.

Koutná-Karg, Dana. "Die Anfange des Klosters Brevnov." In *Tausend Jahre Benediktiner in den Klöstern Brevnov, Braunau und Rohr*, ed. P. Johannes Hofmann OSB, 219–230. St. Ottilien: EOS Verlag, 1993.

Kovács, Éva. *A Tihanyi összeírás mint helynévtörténeti forrás [The conscription of Tihany as a source of historical toponymy]* (Magyar Névarchívum Kiadványai, 34). Debrecen: Debreceni Egyetemi Kiadó, 2015.

Kovács, Györgyi, and Csilla Zatykó, eds. *"Per sylvam et per lacus nimios" – The Medieval and Ottoman Period in Southern Transdanubia, Southwest Hungary: The Contribution of the Natural Sciences*. Budapest: MTA BTK Régészeti Intézet, 2016.

Körmendy, Adrienne. *Melioratio terrae: Vergleichende Untersuchungen über die Siedlungsbewegung im östlichen Mitteleuropa im 13.–14. Jahrhundert* (Prace Komisji historycznej / Poznańskie Towarzystwo przyjaciół nauk, Wydział historii i nauk społecznych, 48). Poznań: Wydawn. Poznańskiego Towarzystwa przyjaciół nauk, 1995.

Kreamer, David K. "The Past, Present, and Future of Water Conflict and International Security." *Journal of Contemporary Water Research & Education* 149 (2012): 87–95. https://doi.org/10.1111/j.1936-704X.2012.03130.x

Kristó, Gyula. "A korai feudalizmus (1116–1231)." [Age of early feudalism (1116–1231)] In *Magyarország története I/1. Előzmények és magyar történet 1242-ig. [History of Hungary, i/1. Prehistory and Hungarian history until 1242]* (Magyarország története: tíz kötetben), ed. György Székely, 1007–1415. Budapest: Akadémiai, 1978.

Kropač, Ingo Herbert. *Mühlen und Mühlenrecht in der Steiermark während des Mittelalters* (Dissertationen der Karl-Franzens-Universität Graz, 61). Graz: Verlag für d. Techn. Univ. Graz, 1982.

Kubicki, Rafał. *Młynarstwo w państwie zakonu krzyżackiego w Prusach w XIII–XV w. (do 1454 r.) [Milling industry in the state of the Teutonic Order in Prussia in the 13th–15th century (to 1454)]*. Gdańsk: Uniwersytetu Gdańskiego, 2012.

Kubicki, Rafał. "Water Mills and Rural Settlement in the Dominion of the Teutonic Order in Prussia." In *Usus aquarum: Interdisziplinäre Studien zur Nutzung und Bedeutung von Gewässern im Mittelalter* (Forschungen zur Geschichte und Kultur des östlichen Mitteleuropa, 54), eds. Christoph Mielzarek, and Christian Zschieschang, 29–50. Vienna: Böhlau, 2019. https://doi.org/10.7788/9783412500894.29

Kubicki, Rafał. "Windmills in the State of Teutonic Order in Prussia, 14th – First Half of the 15th Centuries." In *Origines et mutationes. Transfer – Exchange – Power*, eds. Aleksandra Girsztowt, Piotr Kitowski, and Andrzej Gierszewski, 215–232. Kraków: Libron, 2017.

Kubinyi, András. "A középkori Pápa." [Medieval Pápa] In *Tanulmányok Pápa város történetéből. A kezdetektől 1970-ig [Studies in the history of the town of Pápa. From the beginnings to 1970]*, ed. András Kubinyi , 75–124. Pápa: Pápa Város Önkormányzata, 1994.

Kubinyi, András. "Buda, Medieval Capital of Hungary." In *Medieval Buda in Context* (Brill's Companions for European History, 10), eds. Balázs Nagy et al., 366–385. Leiden and Boston: Brill, 2016. https://doi.org/10.1163/9789004307674_017

Kubinyi, András. "Budafelhévíz topográfiája és gazdasági fejlődése." [Topography and economic development of Budafelhévíz] *Tanulmányok Budapest Múltjából*

16 (1964): 85–180. Later published in: *Tanulmányok Budapest középkori történetéről*, 2 vols *[Studies in the history of medieval Buda]* (Várostörténeti tanulmányok, 10), eds. István Kenyeres, Péter Kis, and Csaba Sasfi, vol. 1. 115–182. Budapest: Budapest Főváros Levéltára, 2009.

Kubinyi, András. "Budai és pesti polgárok családi összeköttetései a Jagelló-korban." [Family contacts of the burghers of Buda and Pest in the Jagiello period] *Levéltári Közlemények* 37 (1966): 227–291. Later published in: *Tanulmányok Budapest középkori történetéről*, 2 vols *[Studies in the history of medieval Buda]* (Várostörténeti tanulmányok, 10), eds. István Kenyeres, Péter Kis, and Csaba Sasfi, vol. 2. 513–570. Budapest: Budapest Főváros Levéltára, 2009.

Kubinyi, András. "A budai német patriciátus társadalmi helyzete családi összeköttetései tükrében a 13. századtól a 15. század második feléig." [The social position of the German patriciate of Buda in the light of their family ties from the thirteenth to the second half of the fifteenth century] *Levéltári Közlemények* 42 (1971): 203–269. Published again in: András Kubinyi , *Tanulmányok Budapest középkori történetéről*, 2 vols *[Studies in the history of medieval Buda]* (Várostörténeti tanulmányok, 10), eds. István Kenyeres, Péter Kis, and Csaba Sasfi, vol. 2. 457–512. Budapest: Budapest Főváros Levéltára, 2009.

Kubinyi, András. "Budapest története a későbbi középkorban Buda elestéig." [History of Budapest in the later Middle Ages, to the fall of Buda] In *Budapest története*, 5 vols *[History of Budapest]*, ed. László Gerevich, vol. 2. 7–240. Budapest: Budapest Főváros Tanácsa, 1975–1980.

Kubinyi, András. "Királyi kancellária és udvari kápolna Magyarországon a XII. század közepén." [Royal chancery and court chapel in Hungary in the middle of the 12th century] *Levéltári Közlemények* 46 (1975): 63–126.

Kubinyi, András. "Städtische Wasserversorgungsprobleme im mittelalterlichen Ungarn." In *Städtische Versorgung und Entsorgung im Wandel der Geschichte (Stadt in der Geschichte.* Veröffentlichungen des südwestdeutschen Arbeitskreises für Stadtgeschichtsforschung, 8), ed. Jürgen Sydow, 180–190. Sigmaringen: Jan Thorbecke, 1981.

Kubinyi, András. "Városhálózat a késő középkori Kárpát-medencében." [*Urban network in the late medieval Carpathian Basin] Történelmi Szemle* 46 (2004): 1–30.

Kücsán, József. "A Várkerület belső ívének beépítése, 1766–1801." [The building-in of the inner circle of the Castle area of Sopron, 1766–1801] *Soproni Szemle* 69 (2015): 307–336.

Kuffart, Hajnalka. "Az esztergomi érsek *pisetum*-jövedelme Battyányi István udvarbíró idején." [The *pisetum*-income of the archbishop of Eszterom in the period of the provisorship of István Batthyányi] In *Veretek, utak, katonák. Gazdaságtörténeti tanulmányok a magyar középkorról [Mints, routes, soldiers. Studies in the economic history of medieval Hungary]*, eds. István Kádas, Renáta Skorka, and Boglárka Weisz, 93–109. Budapest: MTA BTK Történettudományi Intézet, 2018.

Kumorovitz L., Bernát. "A zselicszentjakabi alapítólevél 1061-ből, 'Pest' legkorábbi említése." [The foundation charter of Zelicsszentjakab from 1061 and the earliest mention of 'Pest'] *Tanulmányok Budapest múltjából* 16 (1964): 43-83.

L'acqua nei secoli altomedievali: Spoleto, 12–17 aprile 2007, 2 vols *(Settimane di Studio del Centro Italiano di Studi sull'Alto Medioevo, 55*). Spoleto: Presso La sede della Fondazione, 2008.

Lakatos, Bálint. "Lossai Péter." [Péter Lossai] In *Magyar művelődéstörténeti lexikon, 14 vols [Lexicon of cultural history in Hungary]*, ed. Péter Kőszeghy, vol. 7. 84–85. Budapest: Balassi, 2003–2014.

Lakatos, Bálint. "Mezővárosi és falusi önkormányzati testületek Magyarországon a késő középkorban." [Councils in late medieval Hungarian market towns and villages] *Századok* 148 (2014): 495–530.

Lakatos, Bálint. *Mezővárosi oklevelek: Települési önkormányzat és írásbeliség a késő középkori Magyarországon, 1301–1526 [Market town charters. Urban autonomy and literacy in late medieval Hungary]*. Budapest: Bölcsészettudományi Kutatóközpont Történettudományi Intézet, 2019.

Lambrecht, Kálmán. *A magyar malmok könyve: történeti anyag [Book of Hungarian mills: the historical material]* (Iparosok olvasótára, 21). Budapest: Lampel, 1915.

Lampen, Angelika. *Fischerei und Fischhandel im Mittelalter: wirtschafts- und sozialgeschichtliche Untersuchungen nach urkundlichen und archäologischen Quellen des 6. bis 14. Jahrhunderts im Gebiet des Deutschen Reiches* (Historische Studien, 461). Husum: Matthiesen, 2000.

Langdon, John. "The Economics of Horses and Oxen in Medieval England." *Agricultural History Review* 30 (1982): 31–40.

Langdon, John. "Inland Water Transport in Medieval England." *Journal of Historical Geography* 19 (1993): 1–11. https://doi.org/10.1006/jhge.1993.1001

Langdon, John. "Inland Water Transport in Medieval England: The View from the Mills: A Response to Jones." *Journal of Historical Geography* 26 (2000): 75–82. https://doi.org/10.1006/jhge.1999.0195

Langdon, John. "Water-Mills and Windmills in the West Midlands, 1086–1500." *Economic History Review* 2nd series 44 (1991) 424–444.

Langdon, John. *Mills in the Medieval Economy: England 1300–1540*. Oxford: Oxford University Press, 2004.

Lászay, Judit, and Adrienn Papp. "Investigations in the Budafelhévíz Baths." In *Régészeti kutatások Magyarországon 2007 / Archaeological Investigation in Hungary 2007*, ed. Júlia Kisfaludi, 63–84. Budapest: Kulturális Örökségvédelmi Hivatal and Magyar Nemzeti Múzeum, 2008.

Lászlóffy, Woldemár, ed. *A Kárpát-medence vízborította es vízjárta területei az ármentesítő es lecsapoló munkálatok megkezdése előtt. [Map] [Water-Covered Areas and Wetlands in the Carpathian Basin Before the Beginning of Flood Protection and Drainage Works]*. Budapest: Vízrajzi Intézet, 1938.

Lee, Lawrence B. "Water Resource History: A New Field of Historiography?." *Pacific Historical Review* 57 (1988): 457–467. https://doi.org/10.2307/3640377

Leguay, Jean-Pierre. *L'eau dans la ville au Moyen Âge*. Rennes: Presses universitaires de Rennes, 2002.

Lenders, H. J. R. et al. "Historical Rise of Waterpower Initiated the Collapse of Salmon Stocks." *Nature. Scientific Reports* 6, art. no. 29269 (2016): 9–18. https://doi.org/10.1038/srep29269

Leonardi, Marco. *Aqua curanda est: le acque e il loro utilizzo nei territori di Friburgo in Brisgovia e Catania dal XIII al XVI secolo* (Aquae. Studi e testi sulle terme, 9). Florence: Leo S. Olschki, 2017.

Leonardi, Marco. "Die Wassernutzung in Freiburg im späten Mitelalter und der frühen Neuzeit." *Zeitschrift des Breisgau-Geschichtsvereins "Schau-ins-Land"* 129 (2010): 55–75. https://doi.org/10.57962/regionalia-17831

Lexer, Matthias. *Mittelhochdeutsches Handwörterbuch*, vol. 3. Leipzig: Hirzel, 1878.

Libor, Jan. *Václav II. a struktury panovnické moci [Venceslas II and the structure of the ruling]*. Brno: Matice moravská, 2006.

Lilley, Keith D., Christopher D. Lloyd, and Steven Trick. "Designs and Designers of Medieval 'New Towns' in Wales." *Antiquity* 81 (2007): 279–293. https://doi.org/10.1017/S0003598X0009517X

Lisac, Andrija-Ljubomir. "Mlinarstvo Zagreba od najstarijih vremena do početka XX stoljeća." [Mills of Zagreb from the earliest times to the beginning of the 20th century] *Zbornik Historijskog zavoda Jugoslavenske akademije* 8 (1977): 217–290.

Lohrmann, Dietrich. "Schiffsmühlen auf dem Tiber in Rom nach Papsturkunden des 10.–11. Jahrhunderts." In *Ex ipsis rerum documentis: Beiträge zur Mediävistik. Festschrift für Harald Zimmermann zum 65. Geburtstag*, eds. Klaus Herbers, Hans-Henning Kortüm, and Carlo Servatius, 277–286. Sigmaringen: Jan Thorbecke, 1991.

Longoni, Raphael, and Oliver Wetter. "Urban Stream Works in Central Europe 1200–1700: Municipal Administration, Hydraulic Engineering and Flood Reconstruction." *Water History* 11 (2019): 31–57. https://doi.org/10.1007/s12685-019-00231-6

Lucas, Adam. *Ecclesiastical Lordship, Seigneurial Power and the Commercialization of Milling in Medieval England*. Farnham: Ashgate, 2014. https://doi.org/10.4324/9781315578620

Lucas, Adam. "Fulling Mills in Medieval Europe. Comparing the Manuscript and Archaeological Evidence." In *Archéologie des moulins hydrauliques, à traction animale et à vent, des origines à l'époque médiévale et moderne en Europe et dans le monde Méditerranéen: Actes du colloque international, Lons-le-Saunier du 2 au 5 novembre 2011*, 2 vols (Annales littéraires de l'Université de Franche-Comté, 959), eds. Luc Jaccottey, and Gilles Rollier, vol. 2. 581–594. Besançon: Presses Universitaires de Franche-Comté, 2016.

Lucas, Adam. "Industrial Milling in the Ancient and Medieval Worlds: A Survey of the Evidence for an Industrial Revolution in Medieval Europe." *Technology and Culture* 46 (2005): 1–30.

Lucas, Adam. "The Role of the Monasteries in the Development of Medieval Milling." In *Wind and Water in the Middle Ages. Fluid Technologies from Antiquity to the Renaissance* (Medieval and Renaissance Texts and Studies, 322 = Penn State Medieval Studies, 2), ed. Steven A. Walton, 89–128. Tempe, AZ: Arizona Center for Medieval and Renaissance Studies, 2006.

Lucas, Adam. *Wind, Water, Work: Ancient and Medieval Milling Technology* (Technology and Change in History, 8). Leiden and Boston: Brill, 2005. https://doi.org/10.1163/9789047417224

Lucas, Adam Robert. "Narratives of Technological Revolution in the Middle Ages." In *Handbook of Medieval Studies: Terms – Methods – Trends*, ed. Albrecht Classen, 967–990. Berlin: De Gruyter, 2010. https://doi.org/10.1515/9783110215588.967

Lukács, László. "Sokféle bírák. A faluközösség önigazgatásának "népi" tisztségviselői." [Many kinds of judges. The "folk" office holders of the self-governance of the village communities] In *Halmok és havasok. Tanulmányok a hatvan esztendős Bárth János tiszteletére [Mounds and mountains. Studies in honor of the 60th birthday of János Bárth]*, eds. Dániel Bárth, and János Laczkó, 395–414. Kecskemét: Bács-Kiskun Megyei Önkormányzat Múzeumi Szervezete, 2004.

Lysá, Žofia. "Privilégium pre Bratislavu z roku 1291." [The privilege of the town of Bratislava from 1291] *Historický časopis* 59 (2011): 189–214.

Lyublyanovics, Kyra. *New Home, New Herds: Cuman Integration and Animal Husbandry in Medieval Hungary from an Archaeozoological Perspective* (Archaeolingua. Central European Archaeological Heritage Series, 10). Oxford: Archaeopress, 2017.

Magina, Livia. " 'Molendinis, Piscinis et Piscaturis': The Utilisation of Water Resources in the Banat in the Medieval and Early Modern Periods." In *Government and Law in Medieval Moldavia, Transylvania and Wallachia (Studies in Russia and Eastern Europe,* 11), eds. Martyn Rady, and Alexandru Simon, 65–70. London: UCL School of Slavonic and East European Studies, 2013.

Magnusson, Roberta, and Paolo Squatriti. "The Technologies of Water in Medieval Italy." In *Working with Water in Medieval Europe: Technology and Resource-Use* (Technology and Change in History), ed. Paolo Squatriti, 217–266. Leiden and Boston: Brill, 2000. https://doi.org/10.1163/9789047400110_010

Magnusson, Roberta J. *Water Technology in the Middle Ages. Cities, Monasteries, and Waterworks after the Roman Empire* (The Johns Hopkins Studies in the History of Technology). Baltimore, MD and London: Johns Hopkins University Press, 2001.

Magnusson, Roberta J. "Water and Wastes in Medieval London." In *A History of Water, series 1. vol. 1: Water Control and River Biographies*, eds. Terje Tvedt, and Eva Jakobsson, 299–313. New York: I. B. Tauris, 2006.

Magyar, Eszter. *A feudalizmus kori erdőgazdálkodás az alsó-magyarországi bányavárosokban, 1255–1747 [Feudal-age forest management in the Lower Hungarian mining towns, 1255–1747]* (Értekezések a Történeti Tudományok Köréből, Új Sorozat, 101). Budapest: Akadémia, 1983.

Majorossy, Judit. "Church in Town. Urban Religious Life in Late Medieval Hungary." PhD-diss., Central European University, 2006.

Majorossy, Judit. "A foglalkozás topográfiája. A társadalmi tértől a személyes térig: a társadalmi mobilitás térbeli elemei a 15. századi Pozsonyban." [Topography of employment. From the social space to the personal space: elements of the spatial mobility in 15th-century Bratislava] *Korall* no. 45 (2011): 102–135.

Majorossy, Judit. "A test és a lélek fürdője. A középkori pozsonyi fürdőkről." [The bath of body and soul. On the medieval bathing houses of Bratislava] In *"Köztes-Európa" vonzásában: Ünnepi tanulmányok Font Márta tiszteletére [In the attraction of "Mitteleuropa" Studies in honor of Márta Font],* eds. Dániel Bagi, Tamás Fedeles, and Gergely Kiss, 343–356. Pécs: Kronosz, 2012.

Majorossy, Judit, and Katalin Szende. "Hospitals in Medieval and Early Modern Hungary." In *Europäisches Spitalwesen. Institutionelle Fürsorge in Mittelalter und Früher* (Neuzeit Quelleneditionen des Instituts für Österreichische Geschichtsforschung, 5), eds. Martin Scheutz et al., 409–454. Vienna and Munich: Oldenbourg, 2008. https://doi.org/10.7767/boehlau.9783205160885.409

Makkai, László. "A malom mint a középkori Európa erő- és munkagépe." [Mill as the power machine of medieval Europe] In *Műszaki innovációk sorsa Magyarországon: malomipar, vaskohászat, textilipar [The fate of technological innovations in Hungary: milling industry, iron smelting, textile industry],* ed. Walter Endrei, 29–35. Budapest: Akadémiai, 1995.

Makkai, László. "Östliches Erbe und westliche Leihe in der ungarischen Landwirtschaft der frühfeudalen Zeit." *Agrártörténeti Szemle* 16 [Supplementum] (1974): 1–53.

Maliniak, Pavol. "Mlynárstvo na strednom Pohroní v stredoveku a na začiatku novoveku." [Milling in central Pohronský region in the Middle Ages and in the beginning of the Modern Age] In *Z histórie technicko-hospodárskeho využitia vodných tokov na strednom Pohroní [From the history of technical and economic use of watercourses in the central Pohronský region]*, eds. Ján Žilák, and Pavel Hronček, 30–45. Banská Bystrica: Centrum vedy a výskumu UMB: Fakulta humanitných vied UMB, 2011.

Mályusz, Elemér. *Turóc megye kialakulása [The formation of Turóc County]*. Máriabesnyő–2nd ed. Gödöllő: Attraktor, 2005.

Maříková, Martina. "Středověké mlýny v českých zemích (Archeologické a písemné prameny)." [Medieval mills in Czech lands (archaeological data and written sources)] *Mediaevalia Historica Bohemica* 10 (2005): 89–148.

Maříková, Martina, and Christian Zschieschang, eds. *Wassermühlen und Wassernutzung im mittelalterlichen Ostmitteleuropa* (Forschungen zur Geschichte und Kultur des östlichen Mitteleuropa, 50). Stuttgart. Steiner, 2015.

Marosi, Endre. *XVI. századi váraink, 1521–1606 [Castles in 16th-century Hungary]*. Miskolc: Hungária-Európa Alapítvány and Borsod-Abaúj-Zemplén Megyei Levéltár, 1991.

Maróti, Egon. "A vízimalom európai elterjedésének történetéhez." [To the history of the spread of water mills in Europe] *Antik Tanulmányok* 22 (1975): 55–74.

Marsina, Richard. "Štúdie k Slovenskému diplomatáru. I. Druhá časť." [Studies in Slovak diplomatics. I. Second part] *Historické štúdie* 18 (1973): 5–119.

Martinez, M. M. "Desarrollo historiografico de la molineria hidraulica en la Espana medieval: perspectivas y resultados." In *I mulini nell'Europa medievale: atti del convegno di San Quirico d'Orcia, 21–23 settembre 2000* (Biblioteca di storia agraria medievale, 21), ed. Paola Galetti, 103–142. Bologna: CLUEB, 2003.

Máthé, Gábor, ed. *A magyar jog fejlődésének fél évezrede. Werbőczy ésésének fésél ésévezrede. Werbésőczy és a Hármaskönyv 500 év múltán [Half a millennium of the development of the Hungarian legal system. Werbőczy and the Tripartitum after 500 years]*. Budapest: Nemzeti Közszolgálati Egyetem, 2014.

Matunák, Michal. *Z dejin Slobodného a Hlavného Banského Mesta Kremnice [History of the privileges and the town of Kremnica]*. Kremnica: Slobodné a hlavné banské mesto, 1928.

Mátyus, Norbert. "Filarete és Bonfini. Egy humanista építészeti tankönyv Itáliában és Magyarországon." [Filarete and Bonini. A humanist architectural treatise in Italy and Hungary] In *Techné. A szakképzés módszerei az ókorban és a középkorban [Techné. Forms of crafts' knowledge transfer in the Ancient Times and the Middle Ages]*, eds. Márta Munding, Kornél Szovák, and László Takács, 345–370. Piliscsaba: Avicenna Közel-Kelet Kutatások Intézete, 2018.

Mauch, Christof, and Thomas Zeller. "Rivers in History and Historiography: An Introduction." In *Rivers in History: Perspectives on Waterways in Europe and North America*, eds. Christof Mauch and Thomas Zeller, 1–10. Pittsburgh, PA: University of Pittsburgh Press, 2008.

Mauri, Luisa Chiappa. *I mulini ad acqua nel milanese (secoli X–XV)*. Milano: Societa' Editrice Dante Alighieri, 1984.

McCormick, Finbar, Thom Kerr, Meriel McClatchie, and Aidan O'Sullivan. *The Archaeology of Livestock and Cereal Production in Early Medieval Ireland, AD 400–1100* (Reconstructing the Early Medieval Irish Economy EMAP Report 5.1). Belfast: [no publisher], 2011.

McEntee, Brian. "Queen Elizabeth of Hungary (1320-1380) and Óbuda: Patronage, Personality and Place." In *La diplomatie des états Angevins: aux XIIIe et XIVe siècles = diplomacy in the countries of the Angevin Dynasty in the Thirteenth–Fourteenth Centuries: Actes du colloque international de Szeged, Visegrád, Budapest: 13–16 septembre 2007*, eds. Zoltán Kordé, and István Petrovics, 209–218. Rome and Szeged: Dipart. di Storia Ungh. del Medioevo e della prima Età Moderna Univ. degli Studi di Szeged and Accademia d'Ungheria in Roma Istituto Storico "Fraknói", 2010.

Melich, János. "Három helynévről: Lehota, Vólya, Ohába." [On three placenames: Lehota, Vólya, Ohába] *Századok* 41 (1907): 321–324.

Melich, János. "Malom." [Mill] *Magyar Nyelv* 16 (1920): 61–66.

Mencl, Václav. *Středověkáá města na Slovensku [Medieval towns in Slovakia]*. Bratislava: Universum, 1938.

Mielzarek, Christoph, and Christian Zschieschang, eds. *Usus aquarum: Interdisziplinäre Studien zur Nutzung und Bedeutung von Gewässern im Mittelalter* (Forschungen zur Geschichte und Kultur des östlichen Mitteleuropa, 54). Vienna: Böhlau, 2019. https://doi.org/10.7788/9783412500894

Mikhail, Alan. "Oriental Democracy." *Global Environment* 7 (2014): 381–404. https://doi.org/10.3197/ge.2014.070206

Mikó, Árpád. "Rész és egész. A magyarországi reneszánsz kőfaragóműhelyek és kutatásuk." [Part and the whole. Hungarian Renaissance stone-carvingworkshops and research into them] In *Koppány Tibor hetvenedik születésnapjára [For the 70th birthday of Tibor Koppány]* (Művészettörténet – műemlékvédelem 10), eds. István Bardoly, and Csaba László, 207–223. Budapest: Országos Műemlékvédelmi Hivatal, 1998.

Millea, Nick, Keith Parry, and Adrian Phillips. "The European Historic Towns Atlas Project." *IMCoS Journal* no. 153 (2018): 26–36.

Minárčic, Miroslav. "Metačná listina svätoantonského panstva z roku 1266 a jej výpovedná hodnota k dejinám banskoštiavnického baníctva." [A document from the year 1266 on the estate of Svätý Anton and its significance in the history of the mining industry at Banská Štiavnica] *Montánna história* 2 (2009): 46–69.

Miňo, Martin. "Niekoľko postrehov k fortifikačným prvkom miest stredoslovenskej banskej oblasti." [Some observations on the fortification of towns in Central Slovakian mining areas] *Archaeologia Historica* 36 (2011): 289–302.

Mira, Giuseppe. *La pesca nel medioevo: nelle acque interne italiane*. Milano: A. Giuffre, 1937.

Mlynarska-Kaletynowa, Mart, and Rafal Eysymontt, eds. *Wrocław – Breslau Atlas* (Historyczny Miast Polskich / Historischer Atlas polnischer Städte, 1). Wrocław: Wydawnictwo Uniwersytetu Wrocławskiego, 2001.

Moldenhauer, Rüdiger. "Mühlen und Mühlenrecht in Mecklenburg." *Zeitschrift der Savigny-Stiftung für Rechtsgeschichte: Germanistische Abteilung* 79 (1962): 195–236.

Mollay, Károly. "Egy évtized soproni külkereskedelmének társadalomtörténeti háttere (1483–1489)." [The social historical background of a decade in the foreign trade of Sopron (1483–1489)] *Soproni Szemle* 49 (1995): 289–316.

Mollay, Károly. "Kőszeg 1532. évi ostroma és Sopron." [The siege of Kőszeg in 1532 and Sopron] *Soproni Szemle* 37 (1983): 193–236.

Mollay, Károly. "A Szélmalom utcai vám 1217–1564." [The customs of the Windmühlgasse, 1217–1564] *Soproni Szemle* 45 (1991): 97–121.

Mollay, Károly. "A Tómalom középkori előzményei (Fejezet a soproni határ történetéből)." [The medieval predecessors of the Tómalom (An episode from the history of the borders of Sopron)] *Soproni Szemle* 46 (1992): 150–167.

Moór, Elemér. "Malom és molnár." [Mill and miller] *Nyelvtudományi Közlemények* 67 (1965): 130–138.

Moór, Elemér. "A szóegyüttesek és a csoportos szóhiányok: művelődéstörténeti tények." [Joint words and joint word omissions: intellectual historical facts] *Magyar Nyelv* 70 (1974): 173–181.

Moore, Scott. "Rethinking Conflict over Water." In *Oxford Research Encyclopedia of Environmental Science*. Oxford: Oxford University Press, 2016. https://doi.org/10.1093/acrefore/9780199389414.013.104

Morgan, John Emrys. "The Micro-Politics of Water Management in Early Modern England: Regulation and Representation in Commissions of Sewers." *Environment and History* 23 (2017): 409–430. https://doi.org/10.3197/096734017X1497947387387

Moździoch, Sławomir, and Krystian Chrzan, eds. *O rzece i wodzie w życiu codziennym człowieka średniowiecza [Rivers and waters in everyday life]* (Spotkania Bytomskie, 8). Wrocław: Wydawnictwo Instytutu Archeologii i Etnologii Polskiej Akademii Nauk – Ośrodek Badań Nad Kulturą Późnego Antyku i Wczesnego Średniowiecza, 2015.

Mumford, Lewis. *Technics and Civilization*. New York: Harcourt, Brace and Co., 1934.

Munro, John H. "Industrial Energy from Water-Mills in the European Economy, 5th to 18th Centuries: The Limitations of Power." In *Economia e energia, secc. XIII–XVIII: atti della trentaquattresima Settimana di studi, 15–19 aprile 2002* (Istituto Internazionale di Storia Economica F. Datini, Prato. Serie 2. Atti delle settimane di studio e altri convegni, 34), ed. Simonetta Cavaciocchi, 223–269. Florence: Le Monnier, 2003.

Muratori, Lodovico Antonio. *Annali d'Italia, dal principio dell'era volgare fino all'anno MDCCL*, vol. 3/2. Rome: Stamperia degli eredi Barbiellini, 1752.

Myśliwski, Grzegorz. "Utilisation of Water in Central Europe (12th–16th cents.)." In *Economia e energia, secc. XIII–XVIII: Atti della trentaquattresima Settimana di studi, 15–19 aprile 2002* (Istituto Internazionale di Storia Economica F. Datini, Prato. Serie 2. Atti delle settimane di studio e altri convegni, 34), ed. Simonetta Cavaciocchi, 321–333. Florence: Le Monnier, 2003.

Nagy, Ágnes. "Brunnen und Zisternen im Mittelalterlichen Ungarn." *Antaeus* 26 (2003): 343–411.

Nagy, Balázs et al., eds. *Medieval Buda in Context* (Brill's Companions for European History, 10). Leiden and Boston: Brill, 2016. https://doi.org/10.1163/9789004307674

Némethy, Béla. "*Vízjogi előzmények* a *magyar Corpus* Jurisban." [Water law antecedents in the Corpus Iuris] *Vízügyi Közlemények* 13 (1931): 126–148.

Nógrády, Árpád. "Az elakadt fejlődés." [Halted development] In *Keresztesekből lázadók. Tanulmányok 1514 Magyarországáról [From crusaders to rebels. Studies on Hungary in 1514]*, eds. Norbert C. Tóth, and Tibor Neumann, 11–30. Budapest: MTA BTK Történettudományi Intézet, 2015.

Nógrády, Árpád. "A középkor végi Magyarország mindennapi kenyere (Beregszász lakosságának gabonavásárlásai 1530-ban és a Jagelló-kori malomvámok)." [The everyday bread of late medieval Hungary (Grain acquisitions of the people of Berehove in the 1530s and the mill customs of the Jagiello period)] *A Nyíregyházi Jósa András Múzeum Évkönyve* 42 (2000): 155–169.

Nógrády, Árpád. *Sáros vármegye [Sáros County]*. Budapest: Bölcsészettudományi Kutatóközpont Történettudományi Intézet, 2022.

Olexik, Ferenc. "Középkori levéltártörténeti adatok." [Medieval archival history data] *Levéltári Közlemények* 13 (1935): 266–274.

Opll, Ferdinand, Heike Krause, and Christoph Sonnlechner. *Wien als Festungsstadt im 16. Jahrhundert. Zum kartografischen Werk der Mailänder Familie Angielini.* Vienna, Cologne, and Weimar: Böhlau, 2017. https://doi.org/110.26530/oapen_626456

Ortvay, Tivadar. *Magyarország régi vízrajza a XIII. század végéig*, 2 vols *[Hydrography of ancient Hungary until the end of the 13th century]*. Budapest: Magyar Tudományos Akadémia, 1892.

Ottaviani, Maria Grazia Nico. *Statuti, territorio e acque nel Medioevo: Perugia e Marsciano, Tevere e Nestóre*. Spoleto: Fondazione Centro italiano di studi sull'alto Medioevo, 2008.

Pais, Dezső. "Pamlény: malom? — és az mn > ml hangváltozás." [Pamlény – mill? – and the mn > ml sonant changes] *Magyar Nyelv* 61 (1965): 455–458

Pálffy, Géza. *Die Anfänge der Militärkartographie in der Habsburgermonarchie. Die regelmäßige kartographische Tätigkeit der Burgbaumeisterfamilie Angielini an den kroatisch-slawonischen und den ungarischen Grenzen in den Jahren 1560–1570 / A haditérképészet kezdetei a Habsburg Monarchiában. Az Angielini várépítész-família rendszeres térképészeti tevékenysége a horvát–szlavón és a magyarországi határvidéken az 1560–1570-es években*. Budapest: Ungarisches Nationalarchiv / Magyar Országos Levéltár, 2011.

Papp, Adrienn. "Fürdők a török kori Budán." [Bathing houses at Ottoman-period Buda] PhD-diss., Eötvös Loránd Tudományegyetem, 2014.

Papp, Adrienn. "Oszmán-török fürdők Budán – A Rudas és Rác fürdő régészeti kutatása (2004–2006)." [Ottoman bathing houses at Buda – archaeological research of th Rudas and Rác bathing houses, 2004–2006] In *A középkor és a kora újkor régészete Magyarországon*, 2 vols *[Archaeology of the Middle Ages and the Early Modern Period in Hungary]*, eds. Elek Benkő, and Gyöngyi Kovács, vol. 1. 207–220. Budapest: MTA Régészeti Intézet, 2010.

Papp, Adrienn. "Török kori lőpormalom Budán." [Ottoman-period gunpowder mill at Buda] *Keletkutatás 2015* (Tavasz [Spring]): 105–113.

Papp, József. "Halászati jog a Hortobágy vizein." [Fishing rights on the waters of the Hortobágy] In *Tanulmányok a Hortobágy néprajzához [Studies in the ethnography of the Hortobágy]*, eds, Béla Gunda, 189–205. Debrecen: Kossuth Lajos Tudományegyetem, 1972–1974.

Pásztor, Lajos. *A magyarság vallásos élete a Jagellók korában [Religious life of the Hungarians in the Jagiello age]*. Budapest: [no publisher], 1940.

Paulinyi, Oszkár. "A bányavíz leküzdése érdekében létesített első tározómedencék és váltókerekek." [The first water reservoirs and water wheels in the fight against mine water] In *Gazdag föld – szegény ország: Tanulmányok a magyarországi bányaművelés múltjából [Rich soil – poor country. Studies in the history of*

Hungarian mining industry] (Gazdaság- és Társadalomtörténeti Kötetek, 3), eds. István Draskóczy, and János Buza, 315–335. Budapest: Corvinus Egyetem, Budapesti Közgazdaságtudományi Egyetem Alapítvány and Miskolci Egyetem, 2005.

Peberdy, R. B. "Navigation on the River Thames between London and Oxford in the Late Middle Ages: A Reconsideration." *Oxoniensis* 61 (1996): 311–340.

Pelham, R[eginald] A[rthur]. *Fulling Mills: A study in the Application of Water Power to the Woollen Industry.* London: Society for the Protection of Ancient Buildings, 1958.

Péterfi, Bence. "Debates Concerning the Regulation of Border Rivers in the Late Middle Ages: The Case of the Mura River." *Hungarian Historical Review* 8 (2019): 313–335.

Peters, Christie. "Industrial Milling and the Prolific Growth of the Cistercian Order in the Twelfth and Thirteenth Centuries." In *Fifty Years of Medieval Technology and Social Change*, ed. Steven A. Walton, 150–164. London: Routledge, 2020.

Petersen, Niels, and Arnd Reitemeier. "Die Mühle und der Fluss. Juristische Wechselwirkungen." In *Wasser in der mittelalterlichen Kultur: Gebrauch – Wahrnehmung – Symbolik / Water in Medieval Culture: Uses, Perceptions, and Symbolism* (Das Mittelalter. Beihefte, 4), eds. Gerlinde Huber-Rebenich, Christian Rohr, and Michael Stolz, 276–290. Berlin: De Gruyter, 2017.

Pető, Zsuzsa Eszter. *Hermits in the Heart of the Hungarian Kingdom: Medieval Monastic Landscape of the Pauline Order in the Pilis* (Archaeolingua. Series Minor, 41). Budapest: Archaeolingua, 2018.

Petráček, Tomáš. "K otázce datace počátků vodních mlýnů v českých zemích v kontextu změny vnímání dynamiky vývoje českých zemích 10.–12. století." [On the issue of dating the beginning of water mills in the Czech lands in the context of changes in the perception of the dynamics of the Czech lands in the 10th–12th centuries] In *Sborník k poctě Jiřího Kalfersta [Studies in honor of George Kalferst]* (Archeologie východních Čech, suppl. 1), eds. Miroslav Novák, and Pavel Burgert, 249–252. Hradec Králové: Muzeum východních Čeché Hradci Králové, 2014.

Pfeifer, Guido. *Ius regale montanorum: ein Beitrag zur spätmittelalterlichen Rezeptionsgeschichte des römischen Rechts in Mitteleuropa.* Ebelsbach: Aktiv Druck & Verlag, 2002.

Piekalski, Jerzy. "Elementy infrastruktury średniowiecznego Wrocławia." [Elements of the infrastructure of medieval Wrocław] In *Wrocław na przełomie średniowiecza i czasów nowożytnych. Materialne przejawy życia codziennego [Wrocław at the turn of the Middle Ages and modern times. Everyday life in light of the material culture]* (Wratislavia Antiqua, 6), eds. Jerzy Piekalski and Krzysztof Wachowski, 9–24. Wrocław: Uniwersytet Wrocławski, 2004.

Piekalski, Jerzy. "The Environment and Living Conditions in Wroclaw (Breslau) in the 16th–18th Centuries." In *Historical Archaeology in Central Europe*, ed. Natascha Mehler, 379–394. Rockville, MD: Society for Historical Archaeology, 2013.

Pinke, Zsolt, and Beatrix Szabó. "Analysis of the Map of the Ministry of Agriculture: Water Covered Areas and Wetlands in the Carpathian Basin Before the Commencement of Flood Protection and Draining." In *8. országos interdiszciplináris Grastyán konferencia előadásai* [Papers presented at the 8th Grastyán interdisciplinary conference], eds. Virág Rab, and Melinda Szappanyos, 207–217. Pécs: PTE Grastyán Endre Szakkollégium, 2010.

Pišút, Peter. "Príspevok historických máp k rekonštrukcii vývoja koryta Dunaja na uhorsko-rakúskej hranici (16.–19. storočie)." [Contribution of historical maps to the reconstruction of the Danube channel evolution near the old Hungarian-Austrian Border (16th–19th centuries)] In *Historické mapy [Historical maps]*, ed. Ján Pravda, 167–181. Bratislava: Kartografická spoločnosť SR– Geografický ústav SAV, 2005.

Pokorná, Adéla et al., "The Oldest Czech Fishpond Discovered? An Interdisciplinary Approach to Reconstruction of Local Vegetation in Mediaeval Prague Suburbs." *Hydrobiologia* 730, no. 1 (2014): 191–213. https://doi.org/10.1007/s10750-014-1837-1

Pražák, Jiří. "Privilegium pervetustum Boleslai." In *Milénium břevnovského kláštera (993–1993). Sborník statí o jeho významu a postavení v českých dějinách [Millennium of the history of the Břevnov monastery (993–1993). Collection of studies on its significance and role in Czech history]*, eds. Ivan Hlaváček, and Marie Bláhová, 13–24. Prague: Karolinum, 1993.

Premerl, Nada, ed. *Potok u srcu Zagreba: uz potok Medveščak-od izvora do ušća / A Brook in the Heart of Zagreb: Along the Medveščak Brook from Source to Mouth*. Zagreb: Muzej grada Zagreba, 2005.

Přikryl, Ivo. "Historický vývoj našeho rybníkářství a rybničních ekosystémů." [Historical fishponds and fishpond ecosystems] *Veronica* 1 (2004): 7–10.

Provansal, Mireille, Georges Pichard, and Edward J. Anthony. "Geomorphic Changes in the Rhône Delta During the LIA: Input from the Analysis of Ancient Maps." In *Sediment Fluxes in Coastal Areas* (Coastal Research Library, 10), eds. Mohamed Maanan, and Marc Robin, 47–72. Dordrecht: Springer Netherlands, 2015. https://doi.org/10.1007/978-94-017-9260-8_3

Rabb, Péter. "Városi védművek a középkori Magyarországon, 2 vols." [Urban fortifications in medieval Hungary]. PhD-diss., Budapesti Műszaki Egyetem, 2005.

Rady, Martyn. *Customary Law in Hungary: Courts, Texts and the Tripartitum*. Oxford: Oxford University Press, 2015. https://doi.org/10.1093/acprof:oso/9780198743910.001.0001

Raptis, K[onstantinos], Th. "Water as Power: Early Christian and Byzantine Watermills in Greece: Typology and Distribution." In *1st IWA International Symposium on Water and Wastewater Technologies in Ancient Civilizations: Symposium Preprint Book*, eds. A. N. Angelakis, and D. Koutsoyiannis, 109–117. Iraklio: National Agricultural Research Foundation, 2006.

Regényi, Kund. "Az eperjesi Szentháromság karmelita konvent története." [The history of the Holy Trinity Carmelite convent of Prešov] In *Tanulmányok a középkori magyar történelemről. Az I. Medievisztikai PhD-konferencia (Szeged, 1999. július 2.) előadásai [Studes in the history of medieval Hungary. Papers presented at the 1st conference in medieval studies, Szeged, 2 July 1999]*, eds. Sarolta Homonnai, Ferenc Piti, and Ildikó Tóth, 103–113. Szeged: Szegedi Középkorász Műhely, 1999.

Reidinger, Erwin. "Stadtplanung im hohen Mittelalter: Wiener Neustadt – Marchegg – Wien." In *Europäische Städte im Mittelalter* (Forschungen und Beiträge zur Wiener Stadtgeschichte, 52), eds. Ferdinand Opll, and Christoph Sonnlechner, 155–176. Innsbruck: Studien Verlag, 2010.

Reynard, Denis. "Conflits et procès autour de l'utilisation de l'eau de la Sionne (Valais, Suisse), du XVe au début du XXe siècle." In *Eaux et conflits dans l'Europe médiévale et moderne: Actes des XXXIIes Journées Internationales d'Histoire de l'Abbaye de Flaran, 8 et 9 octobre 2010*, eds. Patrick Fournier, and Sandrine Lavaud, 189–208. Toulouse: Presses Univ. du Mirail, 2012.

Roff, Shelley E. "'Appropriate to Her Sex?' Women's Participation on the Construction Site in Medieval and Early Modern Europe." In *Women and Wealth in Late Medieval Europe* (The New Middle Ages), ed. Theresa Earenfight, 109–134. New York: *Palgrave*, 2010. https://doi.org/10.1057/9780230106017_7

Rouse, Barbara. "Nuisance Neighbours and Persistent Polluters: The Urban Code of Behaviour in Late Medieval London." In *Medieval Urban Culture* (Studies in European Urban History, 43), eds. Andrew Brown, and Jan Dumolyn, 75–92. Turnhout: Brepols, 2017. https://doi.org/10.1484/M.SEUH-EB.5.114417

Rózsa, György. *Budapest régi látképei. 1493–1800. Alte ansichten von Budapest.* 2nd ed. Budapest: Új Művészet Kiadó, 1999.

Rusu, Adrian Andrei. *Castelarea Carpatică. Fortificaţii şi cetăţi din Transilvania şi teritoriile învecinate (sec. XIII–XIV) [Carpathian castles. Fortifications and fortresses in Transylvania and neighboring territories (13th–14th centuries)]*. Cluj-Napoca: Editura Mega, 2005.

Salvestrini, Francesco. "Le inondazioni a Firenze e nella valle dell'Arno dal XII al XVI secolo." In *L'acqua nemica: fiumi, inondazioni e città storiche dall'antichità al contemporaneo: atti del Convegno di studio a cinquant'anni dall'alluvione di Firenze (1966–2016), Firenze, 29–30 gennaio 2015* (Biblioteca del Centro per il collegamento degli studi medievali e umanistici in Umbria, 37), eds. Concetta Bianca, and Francesco Salvestrini, eds. 31–60. Spoleto: Fondazione Centro italiano di studi sull'alto Medioevo, 2017.

Salvestrini, Francesco. *Libera città su fiume regale. Firenze e l'Arno dall'Antichità al Quattrocento*. Florence: Nardini, 2005.

Sárosi, Edit, Magdolna Szilágyi, and Zsolt Szilágyi. *Kecskemét* (Magyar Várostörténeti Atlasz, 7). Budapest: MTA Történettudományi Intézete, 2018.

Sax, Joseph L. "The Accretion/Avulsion Puzzle: Its Past Revealed, Its Future Proposed." *Tulane Environmental Law Journal* 23 (2010): 305–367.

Schatzki, Theodore R. "Nature and Technology in History." *History and Theory* 42 (2003): 82–93. https://doi.org/10.1046/j.1468-2303.2003.00259.x

Schenk, Gerrit Jasper. "»Knoten« im »Netz«? Überlegungen zur Hafenstadt als »kritischer Infrastruktur«." In *Flusstäler, Flussschifffahrt, Flusshäfen: Befunde aus Antike und Mittelalter Der Tagungsband fasst in 15 Beiträgen die Ergebnisse der gleichnamigen, internationalen Tagung zusammen* (RGZM – Tagungen, 39 = Interdisziplinäre Forschungen zu den Häfen von der Römischen Kaiserzeit bis zum Mittelalter in Europa, 7), eds. Peter Ettel, and Achim Thomas Hack, 207–236. Mainz: Verlag des Römisch-Germanischen Zentralmuseums, 2019.

Scheiber, Sándor. "Középkori zsidó sírkő-töredékek Sopronban." [Medieval Jewish tombstone-fragments at Sopron] *Soproni Szemle* 14 (1960): 357–361.

Schenk, Gerrit Jasper. "»der straßen halb uff der Yll«. Wasserwege im Elsass als »kritische Infrastruktur« für Wirtschaft und Gesellschaft (1350–1550)." In *Weg und Steg. Aspekte des Verkehrswesens von der Spätantike bis zum Ende des Alten Reiches* (Kraichtaler Koll, 11), eds. Kurt Andermann, and Nina Gallion, 121–152. Ostfildern: Thorbecke, 2018.

Schenk, Gerrit Jasper. "Managing Natural Hazards: Environment, Society, and Politics in Tuscany and the Upper Rhine Valley in the Renaissance (1270–1570)."

In *Historical Disasters in Context: Science, Religion, and Politics* (Routledge Studies in Cultural History, 15), eds. Andrea Janku, Gerrit Jasper Schenk, and Franz Mauelshagen, 31–53. New York: Routledge, 2012.

Schenk, Gerrit Jasper, and Stephanie Eifert. "»Kritische Infrastrukturen« als Ergebnisse individueller und kollektiver Kritikalitätszumessungen – ein Ansatz für die Mediävistik?." In *Was heißt Kritikalität? Zu einem Schlüsselbegriff der Debatte um Kritische Infrastrukturen*, eds. Jens Ivo Engels, and Alfred Nordmann, 47–96. Bielefeld: transcript, 2018. https://doi.org/10.14361/9783839442074

Scheutz, Martin, and Herwig Weigl, eds. *Verwaltetes Wasser im Österreich des Spätmittelalters und der Frühen Neuzeit* (Forschungen zur Landeskunde von Niederösterreich, 37). St. Pölten: Verein für Landeskunde von Niederösterreich, 2016.

Schlesinger, Walter. "Die Hufe im Frankenreich." In *Untersuchungen zur eisenzeitlichen und frühmittelalterlichen Flur in Mitteleuropa und ihrer Nutzung, 2 vols* (Abhandlungen der Akademie der Wissenschaften in Göttingen, Phil.- hist. Klasse 3, 115), eds. Heinrich Beck, Dietrich Denecke, and Herbert Jankuhn, vol. 1. 40–70. Göttingen: Vandenhoeck & Ruprecht, 1979.

Schoebel, Martin. "Die Mainzer Rheinmühlen im Mittelalter: Ein Beitrag zur Geschichte der rheinischen Schiffsmühle." *Blätter für deutsche Landesgeschichte* 129 (1993): 79–115.

Schönach, Paula. "River Histories: A Thematic Review." *Water History* 9 (2017): 233–257. https://doi.org/10.1007/s12685-016-0188-4

Schwineköper, Berent. " 'Cum aquis aquarumve decursibus' Zu den Pertinenzformeln der Herrscherurkunden bis sur Zeit Ottos I." In *Festschrift für Helmut Beumann zum 65. Geburtstag*, ed. Kurt-Ulrich Jäschke, 22–56. Sigmaringen: Thorbecke, 1977.

Sedlák, Imrich, ed. *Dejiny Prešova, vol. 1 [The history of Prešov]*. Košice and Prešov: Východoslovenské Vyd. v Košiciach – Muzeum Slovenskej Republiky Rád v Prešove, 1965.

Selmeczi Kovács, Attila. "Kézimalmok. Történeti rétegek – technikai regresszió." [Hand mills. Historical layers – technological regression] *Ethnographia* 92 (1981): 204–232.

Shannon, William D. "Dispute Maps in Tudor Lancashire." *The Local Historian* 42, no. 1 (2012): 2–15.

Sicard, Germain. *Aux origines des sociétés anonymes: les moulins de Toulouse au Moyen âge*. Paris: Édition de l'EHESS, 1953. Published in English as: *The Origins of Corporations: The Mills of Toulouse in the Middle Ages*. New Haven, CT and London: Yale University Press, 2015.

Siklósi, Gyula. "Die Wasserversorgung und das Kanalsystem im mittelalterlichen Székesfehérvár (Stuhlweissenburg)." *Antaeus* 26 (2003): 217–244.

Šikura, Ján Št. *Miestopisné dejiny Turca [Historical geography of Turóc]*. Bratislava: Slovenská akadémia vied a umení, 1944.

Simms, Anngret, and Howard B. Clarke, eds. *Lords and Towns in Medieval Europe. The European Historic Towns Atlas Project*. Farnham: Ashgate, 2015.

Simon, Katalin. "A Duna és a hajómalmok Pest-Budán a 18. században és a 19. század első felében." [The Danube and ship mills at Pest-Buda in the 18th and the first half of the 19th century] *Urbs. Magyar Várostörténeti Évkönyv* 10–11 (2017): 301–321.

Skelton, Leona J. *Tyne after Tyne: An Environmental History of a River's Battle for Protection, 1529–2015*. Winwick: White Horse Press, 2017.

Skelton, R. A., and P. D. A. Harvey. *Local Maps and Plans from Medieval England*. Oxford and New York: Clarendon – Oxford University Press, 1986.

Skorka, Renáta. "On Two Sides of the Border. The Hungarian–Austrian Border Treaty of 1372." *Hungarian Historical Review* 8 (2019): 290–312.

Skorka, Renáta. "A váradi ferences rendi kolostorok topográfiája." [Topography of the Franciscan friaries of Oradea] In *Nagyvárad és Bihar az Anjou-korban [Oradea and Bihor in the Angevin period]* (Tanulmányok Biharország történetéről, 5), ed. Attila Zsoldos, 159–182. Oradea: Varadinum Kulturális Alapítvány, 2018.

Soens, Tim. "The Origins of the Western Scheldt. Environmental Transformation, Storm Surges and Human Agency in the Flemish Coastal Plain (1250–1600)." In *Landscapes or Seascapes? The History of the Coastal Environment in the North Sea Area Reconsidered* (Comparative Rural History of the North Sea Area, 13), eds. Erik Thoen et al., 287–312. Turnhout: Brepols, 2013. https://doi.org/10.1484/M.CORN.1.101557

Sólymos, Szilveszter. "Az első bencés szerzetesek hazánkban." [The first Benedictine monks in Hungary] In *Paradisum plantavit. Bencés monostorok a középkori Magyarországon / Benedictine Monasteries in Medieval Hungary*, ed. Imre Takács, 48–60. Pannonhalma: Pannonhalmi Bencés Főapátság, 2001.

Solymosi, László. "Die glaubwürdigen Orte (loca credibilia) Ungarns im 14.–15. Jahrhundert." *Archiv Für Diplomatik* 55 (2009): 175–190. https://doi.org/10.7788/afd.2009.55.jg.175

Solymosi, László. "Hospeskiváltság 1275-ből." [Hospes-privilege from 1275] In *Tanulmányok Veszprém megye múltjából [Studies in the history of Veszprém County]*, ed. Lajos Madarász, 17–100. Veszprém: Veszprém Megyei Levéltár, 1984.

Sommer, Petr. "The Monastery of Brevnov." In *Europe's Centre Around AD 1000,* 2 vols, eds. Alfried Wieczorek, and Hans-Martin Hinz, vol. 1. 266. Stuttgart: Theiss, 2000.

Sommer, Petr. "The Monastery of Ostrov u Davle." In *Europe's Centre Around AD 1000,* 2 vols, eds. Alfried Wieczorek, and Hans-Martin Hinz, vol. 1. 267. Stuttgart: Theiss, 2000.

Somogyi, Zoltán. *A középkori Magyarország szegényügye [Taking care of poors in medieval Hungary]* (Palaestra Calasanctiana, 37). Budapest: Stephaneum, 1941.

Sonnlechner, Christoph. "Die Etablierung Salzburgs als Netzknoten: karolingische Kirchenstruktur, Raumstrategien und Organisation der Landnutzung um 800." In *Deutsche Königspfalzen* (Beiträge zu ihrer historischen und archäologischen Erforschung, 8: Places of Power – Orte der Herrschaft – Lieux de Pouvoir = Veröffentlichungen des Max-Planck-Instituts für Geschichte, 11/8), ed. Caspar Ehlers, 199–226. Göttingen: Vandenhoeck & Ruprecht, 2007.

Sonnlechner, Christoph. "The Establishment of New Units of Production in Carolingian Times: Making Early Medieval Sources Relevant for Environmental History." *Viator. Medieval and Renaissance Studies* 35 (2004): 21–48. https://doi.org/10.1484/J.VIATOR.2.300191

Sonnlechner, Christoph. "Raumstrategische, agrarökologische und ökonomische Funktionen von Salzburger Stützpunkten im Frühmittelalter." In *Die Kirche im mittelalterlichen Siedlungsraum. Archäologische Aspekte zu Standort, Architektur und Kirchenorganisation* (Beiträge zur Mittelalterarchäologie in Österreich, 21), ed. Sabine Felgenhauer-Schmiedt, 119–130. Vienna: Österreichische Gesellschaft für Mittelalterarchäologie, 2005.

Sowina, Urszula. *Water, Towns and People. Polish Lands against a European Background until the Mid-16th Century* (Polish Studies – Transdisciplinary Perspectives, 17). Frankfurt am Main: Peter Lang, 2016.

Sowina, Urszula, and Wacław Kulczykowski. "The First Urban Water Supply Systems in the State of the Teutonic Knights: The Transfer of Hydrotechnical Knowledge." In *Origines et mutationes. Transfer – Exchange – Power*, eds. Aleksandra Girsztowt, Piotr Kitowski, and Andrzej Gierszewski, 51–62. Kraków: Libron, 2017.

Spitzbart-Glasl, Christina. "Feste Wassermühlen und Schiffsmühlen als Bestandteil der Wiener Gewässerlandschaft." In *Donau-Stadt-Landschaften. Budapest – Wien. Beiträge der Tagungen in Wien (16.04.2014) und Budapest (23.–24.01.2015)*, eds. Máté Tamáska, and Csaba Szabó, 263–278. Berlin: Lit Verlag, 2016.

Squatriti, Paolo. "Digging Ditches in Early Medieval Europe." *Past and Present* 176 (2002): 11–65. https://doi.org/10.1093/past/176.1.11

Squatriti, Paolo. "The Floods of 589 and Climate Change at the Beginning of the Middle Ages: An Italian Microhistory." *Speculum* 85 (2010): 799–826. https://doi.org/10.1017/S0038713410002290

Squatriti, Paolo. *Water and Society in Early Medieval Italy, AD 400–1000.* New York: Cambridge University Press. 1998. https://doi.org/10.1017/CBO9780511583094

Squatriti, Paolo, ed. *Working with Water in Medieval Europe* (Technology and Change in History, 3), ed. Paolo Squatriti. Leiden, Boston, and Cologne: Brill, 2000. https://doi.org/10.1163/9789047400110

Starzyński, Marcin. *Das Mittelalterliche Krakau. der Stadtrat im Herrschaftsgefüge der polnischen Metropole* (Städteforschung, A 92). Cologne: Böhlau, 2015.

Stefánik, Martin. "Italian Involvement in Metal Mining in the Central Slovakian Region from the Thirteenth Century to the Reign of King Sigismund of Hungary." *I Tatti studies* 14/15 (2011–2012): 11–46. https://doi.org/10.1086/its.14_15.41781522

Štefánik, Martin. "Kremnica." In *Lexikon stredovekých miest na Slovensku [Lexicon of medieval towns in Slovakia]*, eds. Martin Štefánik, and Ján Lukačka, 217–241. Bratislava: Historický Ustav SAV, 2010.

Stefánik, Martin. "Nová Baňa." In *Lexikon stredovekých miest na Slovensku [Lexicon of medieval towns in Slovakia]*, eds. Martin Stefánik, and Ján Lukačka, 294–306. Bratislava: Historický Ustav SAV, 2010.

Strätz, Hans-Wolfgang. "Kuttenberger Bergordnung." In *Lexikon des Mittelalters*, vol. 5, eds. Robert Auty et al., 1594. Munich: *Artemis*, 1991.

Sugár, István. "Az egri püspökség és káptalan középkori vizafogó szegyéi a Tiszán." [Sturgeon fishing traps of the bishopric and chapter of Eger on the River Tisza in the Middle Ages] *Archívum. A Heves Megyei Levéltár Közleményei* 9 (1979): 5–33.

Suttor, Marc. "Les conflits pour l'usage et le contrôle de l'eau sur les rivières entre Seine et Meuse à la fin du Moyen Âge et au début de l'Époque modern." In *Eaux et conflits dans l'Europe médiévale et moderne: Actes des XXXIIes Journées Internationales d'Histoire de l'Abbaye de Flaran, 8 et 9 octobre 2010*, eds. Patrick Fournier, and Sandrine Lavaud, 71–86. Toulouse: Presses Univ. du Mirail, 2012.

Sz. Jónás, Ilona. "Női munka a 13. század végi Párizsban." [Female work in late-13h-century Paris] In eadem, *Sokszínű középkor. Természet, társadalom, kultúra a középkori Európában [Colorful Middle Ages. Nature, society, culture in medieval Europe]*, eds. Balázs Nagy, and Veronika Novák, 331–347. Budapest: ELTE Eötvös Kiadó, 2010.

Szabó, Dénes. *A dömösi adománylevél hely- és vízrajza [Topography and hydrography of the donation charter of Dömös]* (A Magyar Nyelvtudományi Társaság Kiadványai, 85). Budapest: Akadémiai Kiadó, 1954.

Szabó, Péter. "Medieval Trees and Modern Ecology: How to Handle Written Sources." *Medium Aevum Quotidianum* 46 (2002): 7–25.

Szabó, Péter. "Sources for the Historian of Medieval Woodland." In *People and Nature in Historical Perspective* (CEU Medievalia, 5), eds. József Laszlovszky, and Péter Szabó, 265–288. Budapest: CEU Press – Archaeolingua, 2003.

Szalai, Béla, and Lajos Szántai. *Magyar várak, városok, falvak metszeteken, 1515–1800,* 3 vols *[Hungarian castles, towns and villages on cuttings].* Budapest: Múzeum Antikvárium, 2006.

Szende, Katalin. "Gyermekek, testvérek, házastársak. Családi viszonyok és örökösödési szokások Eperjesen a középkori végrendeletek tükrében." [Children, siblings, spouses. Family ties and testamentary practices in Prešov in light of medieval last wills] *Történelmi Szemle* 46 (2004): 113–140.

Szende, Katalin. "Iure Theutonico? German Settlers and Legal Frameworks for Immigration to Hungary in an East-Central European Perspective." *Journal of Medieval History* 45 (2019): 360–379. https://doi.org/10.1080/03044181.2019.1612195

Szende, Katalin. "A Kárpát-medence középkori városainak régészeti kutatása az elmúlt két évtizedben." [Archaeological research into medieval towns in the Carpathian Basin since 1990] *A középkor és a kora újkor régészete Magyarországon,* 2 vols *[Archaeology of the Middle Ages and the Early Modern Period in Hungary],* eds. Elek Benkő, and Gyöngyi Kovács, vol. 1. 141–172. Budapest: MTA Régészeti Intézet, 2010.

Szende, Katalin. "Királyi kényszer vagy közösségi akarat? Hivatali írásbeliség a magyarországi bányavárosokban a 13–14. században." [Royal compulsion or common will? Official literacy in the Hungarian mining towns in the 13th and 14th centuries] In *Márvány, tárház, adomány. Gazdaságtörténeti tanulmányok a magyar középkorról [Marble, storage house, donation. Studies in the economic history of the Middle Ages],* eds. Boglárka Weisz, Renáta Skorka, and István Kádas, 507–540. Budapest: MTA BTK Történettudományi Intézet, 2019.

Szende, Katalin. "Kraków and Buda in the Road Network of Medieval Europe." *On Common Path. Budapest and Kraków in the Middle Ages,* eds. Judit Benda et al., 31–37. Budapest: Budapest History Museum, 2016.

Szende, Katalin. "A magyar városok kiváltságolásának kezdetei." [The beginnings of privileging towns in Hungary] In *Debrecen város 650 éves. Várostörténeti tanulmányok [Debrecen is 650 years old. Studies in urban history]* (Speculum Historiae Debreceniense. A Debreceni Egyetem Történelmi Intézet Kiadványai, 7), eds. Attila Pál Bárány, Klára Papp, and Tamás Szálkai, 32–56. Debrecen: DE BTK Történelmi Intézet, 2011.

Szende, Katalin. "Magyar Várostörténeti Atlasz: Feladatok, eredmények, perspektívák." [Hungarian Historic Town Atlas. Tasks, results, perspectives] *Urbs. Magyar Várostörténeti Évkönyv* 8 (2013): 187–199.

Szende, Katalin. "A magyarországi középkori német nyelvű források kiadásának elvei és gyakorlata." [Principles and practice of editing medieval German-language sources in Hungary] *Fons* 7 (2000): 49–62.

Szende, Katalin. "Mennyit ér a kiváltság? Városprivilégiumok kibocsátása és rendelkezéseik betartása I. Károly alatt." [How much privileges worth? The issuing and the implementing of urban privileges during the reign of Charles I] In *Pénz, posztó, piac: Gazdaságtörténeti tanulmányok a magyar középkorról [Money, clorth, market. Studies in the economic history of medieval Hungary]*, ed. Boglárka Weisz, 285–340. Budapest: MTA BTK Történettudományi Intézet, 2016.

Szende, Katalin. "Mills and Towns: Textual Evidence and Cartographic Conjectures from Hungarian Towns in the Pre-Industrial Period." In *Extra muros. Vorstädtische Räume in Spätmittelalter und Früher Neuzeit. Espaces suburbains au bas Moyen Âge et à l'époque moderne* (Städteforschung, A 91), eds. Guy Thewes, and Martin Uhrmacher, 485–516. Cologne and Vienna: Böhlau, 2019. https://doi.org/10.7788/9783412515164.485

Szende, Katalin. "Neighbourhoods, Suburbs and Ethnic Quarters in the Hungarian Towns of the Thirteenth to Fifteenth Centuries." In *Cities–Coins–Commerce: Essays Presented to Ian Blanchard on the Occasion of His 70th Birthday*, ed. Philipp Robinson Rösener, 43–64. Stuttgart: Franz Steiner Verlag, 2012.

Szende, Katalin. "A női munka a középkori gazdaságban." [Female work in medieval economy] *Aetas* 11, no. 1–2 (1995): 179–193.

Szende, Katalin. "Power and Identity. Royal Privileges to the Towns of Medieval Hungary in the Thirteenth Century." In *Urban Liberties and Citizenship from the Middle Ages Up to Now: Libertés et citoyenneté urbaines du moyen âge à nos jours* (Beiträge zur Landes- und Kulturgeschichte, 9), eds. Franz Irsigler, and Michel Pauly, 27–68. Trier: Porta Alba, 2015.

Szende, Katalin. "Scapegoats or Competitors? The Expulsion of Jews from Hungarian Towns in the Aftermath of the Battle of Mohács, 1526." In *Expulsion and Diaspora Formation: Religious and Ethnic Identities in Flux from Antiquity to the Seventeenth Century* (Religion and Law in Medieval Christian and Muslim Societies, 5), ed. John Tolan, 51–83. Turnhout: Brepols, 2015. https://doi.org/10.1484/M.RELMIN-EB.5.109160

Szende, Katalin. "The Sopron Fish Market." In *Genius loci – Laszlovszky 60*, eds. Dóra Mérai et al., 159–164. Budapest: Archaeolingua Alapítvány, 2018.

Szende, Katalin. "Stadt und Naturlandschaft im ungarischen Donauraum des Mittelalters." In *Europäische Städte im Mittelalter* (Forschungen und Beiträge zur Wiener Stadtgeschichte, 52), eds. Ferdinand Opll, and Christoph Sonnlechner, 365–397. Vienna, Bozen, and Innsbruck: Studien Verlag, 2010.

Szende, Katalin. "Town Foundations in East Central Europe and the New World: The Use of the Grid Plan in a Comparative Perspective." In *Medieval East Central Europe in a Comparative Perspective: From Frontier Zones to Lands in Focus*, eds. Gerhard Jaritz, and Katalin Szende, 157–184. New York: Routledge, 2016.

Szende, Katalin. *Trust, Authority, and the Written Word in the Royal Towns of Medieval Hungary* (Utrecht Studies in Medieval Literacy, 41). Turnhout: Brepols, 2018.

Szilágyi, János Ede. "Vízjog: a vizek tulajdonának és használatának főbb magyar előírásai a nemzetközi tendenciák tükrében." [Water rights: the major regulations on the ownership and use of water in Hungary in light of international trends] *Publicationes Universitatis Miskolcinensis Sectio Juridica et Politica* 29 (2011): 595–622.

Szilágyi, Magdolna. *On the Road: The History and Archaeology of Medieval Communication Networks in East-Central Europe* (Series Minor, 35). Budapest: Archaeolingua, 2014.

Szmytka, Rafał. "Pollution in Early Modern Krakow." In *Krakow: An Ecobiography*, eds. Adam Izdebski, and Rafał Szmytka, 108–131. Pittsburgh: University of Pittsburgh Press, 2021. https://doi.org/10.2307/j.ctv201xjkz.8

Szőke, Melinda. *A garamszentbenedeki apátság alapítólevelének nyelvtörténeti vizsgálata [Linguistic historical study of the foundation charter of the abbey of Hronský Beňadik]* (A Magyar Névarchívum Kiadványai, 33). Debrecen: Debreceni Egyetemi Kiadó, 2015.

Szovák, Kornél. "A bencés szerzetesség korai századai Magyarországon." [The early centuries of the Benedictines in Hungary] In *Paradisum plantavit. Bencés monostorok a középkori Magyarországon / Benedictine Monasteries in Medieval Hungary*, ed. Imre Takács, 35–47. Pannonhalma: Pannonhalmi Bencés Főapátság, 2001.

Szűcs, Jenő. "A középkori építészet munkaszervezetének kérdéséhez." [To the question of the labor organization in medieval building industry] *Budapest Régiségei* 18 (1958): 313–363.

Takács, Károly. "Árpád-kori csatornarendszerek kutatása." [Research into Árpádian-period channel systems] In *Táj és történelem [Landscape and history]*, ed. Ágnes R. Várkonyi, 73–106. Budapest: Osiris, 2000.

Takács, Károly. "Árpád-kori csatornarendszerek kutatása a Rábaközben és a Kárpát-medence egyéb területein. I–II." [Research into Árpádian-period channel systems in the Rábaköz region and in other parts of the Carpathian Basin] *Korall* no. 1 (2000): 27–61 and no. 34 (2001): 297–314.

Tagányi, Károly. "Gyepü és gyepüelve, I–IV." [Gyepű and gyepűelve] *Magyar Nyelv* 9 (1913): 97–104, 145–152, 201–206, and 254–266.

Takács, Károly. "Medieval Hydraulic Systems in Hungary: Written Sources, Archaeology and Interpretation." In *People and Nature in Historical Perspective* (CEU Medievalia, 5), eds. József Laszlovszky, and Péter Szabó, 289–312. Budapest: CEU Press and Archaeolingua, 2003.

Takács, Lajos. *Határjelek, határjárás a feudális kor végén Magyarországon [Boundary signs, paermabulation at the end of feudal age in Hungary]*. Budapest: Akadémiai, 1987.

Takács, Levente. "A magyar földmérés(i irodalom) kezdetei és az ókori hagyomány." [The beginnings of geodesy (geodetic literature) in Hungary and the Ancient tradition] In *Hunyadi Mátyás és kora* (Memoria Hungariae, 12 = Studia Folkloristica et Ethnographica, 75), eds. Attila Bárány et al., 24–33. Debrecen: MTA-DE Lendület "Magyarország a középkori Európában" Kutatócsoport, 2019.

Takáts, Sándor. "A komáromi vizahalászat a XVI. században, I–II." [Sturgeon fishing at Komárno in the 16th century] *Magyar Gazdaságtörténelmi Szemle* 4 (1897): 425–445, and 485–509.

Takáts, Sándor. "A magyar faragómolnárok, I–II." [The Hungarian millers] In *Rajzok a török világból*, 4 vols *[Images from the Ottoman Times]*, ed. Sándor Takáts , vol. 2. 422-463. Budapest: Magyar Tudományos Akadémia, 1915-1917.

Takáts, Sándor. "Műveltségtörténeti közlemények: a magyar malom, I–II." [Studies in intellectual history: the Hungarian mill] *Századok* 41 (1907): 143-160, and 236-249.

Takáts, Sándor. "Műveltségtörténeti közlemények: a magyar molnár." [Studies in intellectual history: the Hungarian miller] *Századok* 41 (1907): 52-56.

Takáts, Sándor. *Művelődéstörténeti tanulmányok a XVI–XVII. századból [Studies in the intellectual history of the 16th–17th centuries]*, ed. Kálmán Benda. Budapest: Gondolat, 1961.

Țeicu, Dumitru. *Watermill in the Banat*. Brăila: Museum of Brăila – "Istros" Publishing House, 2012.

Teke, Zsuzsa. "L'energia idraulica nella siderurgia medioevale ungherese." in *Economia e energia sec. XIII–XVIII. Atti della "Trentaquattresima Settimana di Studi", 15–19 aprile 2002*, ed. Simonetta Cavaciocchi, 335–341. Florence: Le Monnier, 2003.

Terei, György. "Az Árpád-kori Kána falu." [The Árpád-age village of Kána] In *A középkor és a kora újkor régészete Mgyarországon,* 2 vols *[Archaeology of the Middle Ages and the Early Modern Period in Hungary]*, eds. Elek Benkő, and Gyöngyi Kovács, vol. 1. 81–111. Budapest: MTA Régészeti Intézet, 2010.

Theissen, Peter. *Mühlen im Münsterland: der Einsatz von Wasser- und Windmühlen im Oberstift Münster vom Ausgang des Mittelalters bis zur Säkularisation (1803)* (Veröffentlichungen der Historischen Kommission für Westfalen., XXII A = Geschichtliche Arbeiten zur westfälischen Landesforschung, wirtschafts- und sozialgeschichtliche Gruppe, 13 = Beiträge zur Volkskultur in Nordwestdeutschland, 101). Münster and New York: Waxmann, 2001.

Thibault, André. "L'analyse des espaces régionaux en France depuis le début du siècle." *Annales de Géographie* 81, no. 444 (1972): 129–170.

Thoen, Erik. et al., eds. *Landscapes or Seascapes? The History of the Coastal Environment in the North Sea Area Reconsidered* (Comparative Rural History of the North Sea Area, 13). Turnhout: Brepols Publishers, 2013.

Thoroczkay, Gábor. "A dömösi prépostság története az alapítástól I. Károly uralkodásának végéig." [The history of the provostry of Dömös from its foundation to the end of the rule of King Charles I] *Fons* 19 (2012): 409–433.

Thoroczkay, Gábor. "Szent István okleveleiről." [On the diplomas of Saint Stephen] *Századok* 143 (2009): 1385–1412.

Timár, Gábor, Pál Sümegi, and Ferenc Horváth. "Late Quaternary Dynamics of the Tisza River: Evidence of Climatic and Tectonic Controls." *Tectonophysics* 410 (2005): 97–110. https://doi.org/10.1016/j.tecto.2005.06.010

Tóth, Gergely. "Előre megírt emlékezet: Lackner Kristóf különös önéletrajza." [The memory written untimely: the curious autobiography of Kristóf Lackner] In *Lackner Kristófnak, mindkét jog doktorának rövid önéletrajza. Vitae Christophori Lackhner I. U. D. Hominis, brevis consignatio [Short autobiography of Kristóf Lackner]* (Sopron Város Történeti Forrásai, C/5), ed. Gergely Tóth, 7–35. Sopron: Soproni Levéltár, 2008.

Tóth, János Attila. "Adatok a dunai hajómalmok régészetéhez." [Data on the archaeology of Danube ship mills] *Archaeologiai Értesítő* 148 (2023): 301–323. https://doi.org/10.1556/0208.2023.00059

Tóth-Szabó, Pál. *A cseh-huszita mozgalmak és uralom története Magyarországon [The history of the Czech-Hussite movements and rule in Hungary].* Budapest: Hornyánszky ny. M. Tud. Akadémia, 1917.

Törnqvist, Torbjörn E. "Middle and late Holocene Avulsion History of the River Rhine (Rhine-Meuse Delta, Netherlands)." *Geology* 22 (1994): 711–714. https://doi.org/10.1130/0091-7613(1994)022<0711:MALHAH>2.3.CO;2

Tracy, James D., ed. *City Walls: The Urban Enceinte in Global Perspective.* New York: Cambridge University Press, 2000.

Travaglini, Carlo M., ed. *La città e il fiume (secoli XIII–XIX)*. Rome: École Française de Rome, 2008.

Tringli, István. "Adalék a magyar vadászati jog középkori történetéhez." [To the medieval history of hunting rights] *Herman Ottó Múzeum Évkönyve* 52 (2013): 213–219.

Tringli, István. "A magyar szokásjog a malomépítésről." [Hungarian customary law of mill construction] In *Tanulmányok a középkorról [Studies on medieval history]* (Analecta Medievalia, 1), ed. Tibor Neumann, 251–267. Budapest and Piliscsaba: PPKE, 2001.

Tringli, István. "A magyar szokásjog és a protestáns természetjog magyarul." [Hungarian Customary Law and Protestant Natural Law in Hungarian] *Történelmi Szemle* 64 (2022): 173–201.

Tringli, István. "Pest megye a késő középkorban." [Pest County in the late medieval period] In *Pest megye monográfiája, vol. 1/2 [Monography of Pest County]*, eds. Domokos Kosáry, Attila Zsoldos, and István Torma, 75–194. Budapest: Pest Megye Monográfia Közalapítvány, 2001.

Tringli, István. "Sátoraljaújhely egyházai a reformáció előtt." [The churches of Sátoraljaújhely before the Reformation] In *Erősségénél fogva várépítésre való. Tanulmányok a 70 éves Németh Péter tiszteletére [Proper for castle building for its strength. Studies in honor of Péter Németh for his 70th birthday]*, eds. Juan Cabello, and Norbert C. Tóth, 377–396. Nyíregyháza: Szabolcs-Szatmár-Bereg Megyei Önkormányzat Múzeumok Igazgatósága, 2011.

Tringli, István. *Sátoraljaújhely* (Magyar Várostörténeti Atlasz, 2). Budapest: MTA Történettudományi Intézet, 2011.

Tringli, István. "Sátoraljaújhely első kiváltságai." In *Tiszteletkör. Történeti tanulmányok Draskóczy István egyetemi tanár 60. születésnapjára [Lap of honor. Studies in history in honor of the 60th birthday of Professor István Draskóczy]*, ed. Gábor Mikó, Bence Péterfi, and András Vadas, 251–259. Budapest: ELTE Eötvös Kiadó, 2012.

Trio, Paul. "The Challenge for a Medieval Center of Industrial Growth: Ypres and the Drinking-Water Problem." In *The Nature and Function of Water, Baths, Bathing, and Hygiene from Antiquity through the Renaissance*, eds. Cynthia Kosso, and Anne Scott, 381–404. Leiden and Boston: Brill, 2009.

Trusen, Winfried. "'Insula in flumine nata.' Ein kanonischer Zivilprozeß aus den Jahren 1357 bis 1363/64 um eine Insel im Mittelrhein." *Zeitschrift der Savigny-Stiftung für Rechtsgeschichte: Kanonistische Abteilung* 68 (1982): 294–338. http://doi.org/10.7767/zrgka.1982.68.1.294

Turner, H. L. *Town Defences in England and Wales*. London: John Baker, 1970.

Tvedt, Terje, ed. *A History of Water*, 9 vols. New York: I. B. Tauris, 2006–2016.

Tvedt, Terje, and Eva Jakobsson. "Introduction: Water History Is World History." In *A History of Water*, series 1. vol. 1: *Water Control and River Biographies*, eds. Terje Tvedt and Eva Jakobsson , ix–xxiii. New York: I. B. Tauris, 2006.

Ulicný, Ferdinand. "Mestské privilégiá Nemeckej Lupce a otázka ich pravosti." [Urban privileges of Partizánska Ľupča and the question of their authenticity] *Historické štúdie* 5 (1959): 302–324.

Uličný, Marián, and Peter Harčar. "Počiatky mesta Prešova na základe výsledkov najnovších archeologických výskumov." [The origins of the town of Prešov deduced from the results of the latest archaeological research] *Archaeologia historica* 35 (2010): 393–403.

Untermann, Matthias. "Erscheinungsformen der Stadtbefestigung." In *Die Befestigung der mittelalterlichen Stadt* (Städteforschung, A 45), eds. Gabriele Isenberg, and Barbara Scholkmann, 3–25. Cologne, Weimar, and Vienna: Böhlau, 1997.

Urbanová, Norma. *Prešov. Pamiatková rezervácia – Prešov [Monument guide: Prešov]* (Pamiatky mestských rezervácií, 8). Bratislava: Tatran, 1986.

Váczy, Péter. "A korai magyar történet néhány kérdéséről." [On some questions of the early Hungarian history] *Századok* 92 (1958): 265–345.

Vadas, András. "Border by the River – But Where is the River? Hydrological Changes and Borders in Medieval Hungary." *Hungarian Historical Review* 8 (2019): 336–360.

Vadas, András. "Economic Exploitation of Urban Moats in Medieval Hungary with Special Regard to the Town of Prešov." *Mesto a Dejiny* 6 no. 2 (2017): 6–21.

Vadas, András. *Egy határfolyó környezettörténete. Háború és vízgazdálkodás a kora újkori Rába-völgyben [The environmental history of a frontier river. War and water-management in the early modern Rába River valley*]. Budapest: Bölcsészettudományi Kutatóközpont, 2021.

Vadas, András. *The Environmental Legacy of War on the Hungarian-Ottoman Frontier, c. 1540–1690* (Environmental Humanities in Pre- modern Cultures). Amsterdam: Amsterdam University Press, 2023.

Vadas, András. "Geography, Natural Resources and Environment." In *The Oxford Handbook of Medieval Central Europe*, eds. Nada Zečević, and Daniel Ziemann, 23–38. Oxford: Oxford University Press, 2022. https://doi.org/10.1093/oxfordhb/9780190920715.013.2

Vadas, András. "Some Remarks on the Legal Regulations and Practice of Mill Constructions in Medieval Hungary." In *Wasser in der mittelalterlichen Kultur: Gebrauch – Wahrnehmung – Symbolik / Water in Medieval Culture: Uses, Perceptions, and Symbolism* (Das Mittelalter. Beihefte, 4), eds. Gerlinde Huber-Rebenich, Christian Rohr, and Michael Stolz, 291–314. Berlin: de Gruyter, 2017.

Vadas, András. "Technologies on the Road between West and East. The Spread of Water Mills and the Christianization of East Central Europe." In *The Medieval Networks in East Central Europe: Commerce, Contacts, Communication*, eds. Balázs Nagy, Felicitas Schmieder, and András Vadas, 123–138. New York: Routledge, 2018.

Vadas, András. "Terminológiai és tartalmi kérdések a középkori malomhelyek körül." [Questions of the terminology and meaning of the term *locus molendini* in the Middle Ages] *Történelmi Szemle* 57 (2015): 619–648.

Vadas, András. "Vízgazdákodás és háborús védekezés: Csákány és a Vas megyei Rába-mente a kora újkorban." [Water management and war defense: Csákány and the course of the Rába in Vas County in the Early Modern period] In *Víz és társadalom Magyarországon a középkortól a XX. századig [Water and Society in the Hungary from the Middle Ages to the 20th century]*, ed. Gergely Krisztián Horváth, 207–245. Budapest: Balassi, 2014.

Vadas, András. "Water-Use in the Medieval Mining Towns of the Kingdom of Hungary." In *Conflicts over Water Management and Water Rights from the End of Antiquity to Industrialisation* (Vierteljahrschrift für Sozial- und Wirtschaftsgeschichte – Beihefte, 260), eds. Michele Campopiano, and Gerrit Jasper Schenk, 71–94. Stuttgart: Franz Steiner, 2024.

Vadas, András. "The 'waters leave their beds frequently': A Western-Hungarian Town and the Flooding of the Rába/Raab River in the Seventeenth Century." *Water History* 5 (2013): 267-286. https://doi.org/10.1007/s12685-013-0087-x

Vadas, András, and János M. Bak. "Diets and Synods in Buda and Its Environs." In *Medieval Buda in Context* (Brill's Companions for European History, 10), eds. Balázs Nagy et al., 322–344. Leiden and Boston: Brill, 2016. https://doi.org/10.1163/9789004307674_015

Vadas, András, and László Ferenczi. "Small Urban Waters and Environmental Pressure before Industrialization: The Case of Hungary." *Journal of Historical Geography* 82 (2023): 98–109. https://doi.org/10.1016/j.jhg.2023.09.008

Vajda, Tamás. "Adatok a Dráva menti középkori fokgazdálkodásról." [On the 'fok' economy along the River Dráva] In *Tanulmányok a középkorról. A II. Medievisztikai PhD-konferencia [Studes in medieval history. Papers presented at the second PhD-conference in medieval studies]*, eds. Boglárka Weisz, László Balogh, and József Szarka, 125–137. Szeged: Szegedi Középkorász Műhely, 2001.

Vajda, Tamás. "Árpád- és Anjou-kori vízimalmaink tájalakító hatása." [Water mills as driving forces of landscape change in the Árpádian and Angevin Periods] In *Micae mediaevales II. Fiatal történészek dolgozatai a középkori Magyarországról és Európáról [Studies of young historians on medieval Hungary and Europe, II]*, eds. Péter Jakab et al., 59–75. Budapest: ELTE BTK Történelemtudományok Doktori Iskola, 2012.

Vajda, Tamás. "Földesúri jogok és kötelezettségek a középkori vízimalmok körül." [Seigniorial right and duties connected to medieval waster mills] In *Víz és társadalom Magyarországon a középkortól a XX. századig [Water and Society in Hungary from the Middle Ages to the 20th century]*, ed. Gergely Krisztián Horváth, 79–110. Budapest: Balassi, 2014.

Vajda, Tamás. "Hazai vízimalmaink 1301–1325 közötti okleveles adatai." [Documentary evidence on water mills in Hungary from 1301 to 1325] In *Középkortörténeti tanulmányok. A III. Medievisztikai PhD-konferencia (Szeged, 2003. május 8–9.) előadásai [Studies in medieval history. Paper presented at the 3rd PhD-conference in medieval studies, Szeged, 8–9 May 2003]*, ed. Boglárka Weisz, 193–213. Szeged: Szegedi Középkorász Műhely, 2003.

Vajda, Tamás. "Korai bencés apátságaink vízimalmaink: a szerzetesek szerepe a technikai fejlődésben." [Water mills of the early Benedictine abbey of Hungary and their role in technological development] In *Episcopus, Archiabbas Benedictinus, Historicus Ecclesiae. Tanulmányok Várszegi Asztrik 70. születésnapjára [Studies in honor of Asztrik Várszegi on the occasion of his 70th birthday]* (METEM Könyvek, 85), eds. Ádám Somorjai OSB, and István Zombori, 25–45. Budapest: Magyar Egyháztörténeti Enciklopédia Munkaközösség – Historia Ecclesiastica Hungarica Alapítvány, 2016.

Vajda, Tamás. "Meghatározható-e a Dél-Alföld késő középkori lélekszáma?." [Is it possible to give an estimate of the late medieval population of the Southern Great Hungarian Plain] In *Műhelyszemináriumi dolgozatok I [Research seminar papers, 1]*, eds. Éva Révész, and Szilvia Kovács, 145–180. Szeged: SZTE BTK Történelemtudományi Doktori Iskola, 2013.

Vajda, Tamás. "Okleveles adatok Árpád-kori vízimalmainkról." [Documentary data on Árpádian-age water mills] In *Medievisztikai tanulmányok. A IV. Medievisztikai PhD-konferencia (Szeged, 2005. június 9–10.) előadásai [Studies in medieval history. Papers read at the 4th conference of medievalists, Szeged, 9– 10 June 2005]*, eds. Szabolcs Marton, and Éva Teiszler, 193–220. Szeged: Szegedi Középkorász Műhely, 2005.

Vajda, Tamás. "Régészeti adatok középkori malmainkról." [Archaeological data on medieval mills in Hungary] In *Középkortörténeti tanulmányok VI. [Studies in medieval history, 6]*, eds. Péter G. Tóth, and Pál Szabó, 307–318. Szeged: Szegedi Középkorász Műhely, 2010.

Vajda, Tamás. "Természeti adottságok szerepe és a környezet-átalakítás mértéke középkori vízimalmainknál." [Role of natural constraints and interventions in the environment at water mills in medieval Hungary] *Paper presented at the Second conference of the Hungarian working group of the European Society for Environmental History, Szeged*, November 5, 2011.

Vajda, Tamás. "Újabb adatok Árpád-kori vízimalmainkról." [New data on Árpádian-age water mills] In *Középkortörténeti tanulmányok 5. Az V. Medievisztikai PhD-konferencia [Studies in medieval history, 5. Papers read at the 5th conference of medievalists]*, eds. Éva Révész Éva, and Miklós Halmágyi, 211–245. Szeged: Szegedi Középkorász Műhely, 2007.

Vajda, Tamás. "Vízimalmok építésével előidézett, mesterséges áradások a középkori Magyarországon." [Floods created by the building of mills in medieval Hungary] In *Középkortörténeti tanulmányok 8. A VIII. Medievisztikai PhD-konferencia (Szeged, 2013. június 17–19.) előadásai [Studies in medieval history, 8. Papers presented at the 8th PhDconference in medieval studies, Szeged, 17– 19 June 2013]*, eds. Márta Tóber, and Ágnes Maléth, 347–362. Szeged: Szegedi Középkorász Műhely, 2015.

Vajkai, Zsófia. "Középkori malmaink." [Mills in medieval Hungary] In *Műszaki innovációk sorsa Magyarországon: malomipar, vaskohászat, textilipar [The fate of technological innovations in Hungary: milling industry, iron smelting, textile industry]*, ed. Walter Endrei, 36–47. Budapest: Akadémiai, 1995.

Vajkai, Zsófia. "Malomtípusok és a molnár mesterség a XIX. századi Magyarországon II." [Mill types and millers in 19th-century Hungary] *A Magyar Mezőgazdasági Múzeum Közleményei* (1981–1983): 355–363.

Vavrušková, Kateřina. "Po stopách mlynářů v Kutné Hoře." [In the footsteps of the millers in Kutna Hora] *Krásné Město* no. 2 (2011): 14–18.

Végh, András. *Buda I. kötet, 1686-ig [Buda, I. Until 1686]* (Magyar várostörténeti atlasz, 4). Budapest: Budapesti Történeti Múzeum, Pázmány Péter Katolikus Egyetem Bölcsészet- és Társadalomtudományi Kar Régészeti Tanszék, and Archaeolingua Alapítvány, 2015.

Végh, András. "Buda-Pest 1300 – Buda-Pest 1400. Two Topographical Snapshots." In *Medieval Buda in Context* (Brill's Companions for European History, 10), eds. Balázs Nagy et al., 169–203. Leiden and Boston: Brill, 2016. https://doi.org/10.1163/9789004307674_009

Viczián, István. "Geomorphological Research in and around Berzence, on the Border of the Drava Valley and Inner Somogy Microregion, Hungary." In *"Per sylvam et per lacus nimios" – The Medieval and Ottoman Period in Southern Transdanubia, Southwest Hungary: The Contribution of the Natural Sciences*, eds. Gyöngyi Kovács, and Csilla Zatykó, 75–91. Budapest: MTA BTK Régészeti Intézet, 2016.

Viczián, István, and Csilla Zatykó. "Geomorphology and Environmental History in the Drava Valley, Near Berzence." *Hungarian Geographical Bulletin* 60 (2011): 357–377.

Vidal de la Blache, Paul. *Tableau géographique de la France*. Paris: Hachette, 1908.

von Gierke, Julius. *Geschichte des deutschen Deichrechts,* 2 vols (Deutschen Staatsund Rechtsgeschichte, 63 and 128). Breslau: M. & H. Marcus, 1901–1917.

von Sternberg, Kaspar. *Umrisse einer Geschichte der böhmischen Bergwerke*. Prague: Haase, 1836–1837.

Vörös, Károly. "Ein Grenzkonflikt zwischen Steiermark und Ungarn im Jahre 1742." *Zeitschrift des Historischen Vereines für Steiermark* 82 (1991): 195–197.

Vručina, Petra. "Kako gospodariti vodom u srednjemu vijeku? Primjer Zagrebačkoga kaptola i njegovih vlastelinstava u 14. i 15. stoljeću / Water Management in the Middle Ages: The Example of the Zagreb Chapter in the 14th and 15th Centuries." *Miscellanea Hadriatica et Mediterranea* 8 (2021): 33–64. https://doi.org/10.15291/misc.3612

Walther, Helmut. "Wasser in Stadt und Contado. Perugias Sorge um Wasser und der Flußtraktat Tyberiadis des Perusiner Juristen Bartolus von Sassoferrato." In *Mensch und Natur im Mittelalter*, 2 vols (Miscellanea Mediaevalia, 21), eds. Albert Zimmermann, and Andreas Speer, vol. 2. 882–897. Berlin: de Gruyter, 1991.

Walton, Steven A., ed. *Fifty Years of Medieval Technology and Social Change*. London and New York: Routledge, 2020. https://doi.org/10.4324/9781315582313

Walton, Steven A., ed. *Wind and Water in the Middle Ages: Fluid Technologies from Antiquity to the Renaissance* (Penn State Medieval Studies, 2 = Medieval and Renaissance Texts and Studies, 322). Tempe, AZ: Arizona Center for Medieval and Renaissance Studies, 2006.

Weczerka, Hugo, ed. *Breslau* (Deutscher Städteatlas IV/5). Münster: Institut für vergleichende Städtegeschichte, 1989.

Weisz, Boglárka. "Az alsó-magyarországi bányavárosok kiváltságai a Zsigmond-korban." [Privileges provided to the Lower Hungarian mining towns in the age of King Sigismund] *Urbs. Magyar Várostörténeti Évkönyv* 12 (2018): 21–48.

Weisz, Boglárka. "A bányaváros, mint önálló várostípus a 14. században." [Mining town as a distinct type of town in the fourteenth century] *Bányászattörténeti Közlemények* 19 (2015): 31–57.

Weisz, Boglárka. "Kassa kereskedelmi életének jogi háttere a középkorban." [The legal background of the trade life of Košice in the Middle Ages] In *Arcana tabularii. Tanulmányok Solymosi László tiszteletére*, 2 vols *[Studies in honor of László Solymosi]*, eds. Attila Bárány, Gábor Dreska, and Kornél Szovák, vol. 2. 899–908. Debrecen and Budapest: ELTE BTK, 2014.

Weisz, Boglárka. "The Legal Background of the Trade Life of Košice in the Middle Ages." In *Košice in the Coordinates of European History*, eds. Mária Hajduová, and Martin Bartoš. 94–111. Košice: The City of Košice, 2013.

Weisz, Boglárka. "Mining Town Privileges in Angevin Hungary." *Hungarian Historical Review* 2 (2013): 288–312.

Weisz, Boglárka. "A nemesércbányászathoz kötődő privilégiumok az Árpád- és az Anjou-korban." [Privileges connected to precious ore mining in the Árpádian and Angevin periods] *Bányászattörténeti Közlemények* 6 (2008): 13–43.

Weisz, Boglárka. "A nemesércbányászathoz kötődő privilégiumok az Árpád- és az Anjou-korban." [Privileges connected to precious ore mining in the Árpádian and Angevin periods] *Történelmi Szemle* 50 (2008): 141–162.

Weisz, Boglárka, and Attila Zsoldos. "Másolatok és kiadások. Győr 1271. évi kiváltságlevelének ma ismert szövegváltozatai." [Copies and editions. The known variants of the text of the 1271 privilege of Győr] In *Márvány, tárház, adomány. Gazdaságtörténeti tanulmányok a magyar középkorról [Marble, storage house, donation. Studies in the economic history of the Middle Ages]*, eds. Boglárka Weisz, Renáta Skorka, and István Kádas, 237–256. Budapest: MTA BTK Történettudományi Intézet, 2019.

Wenzel, Gusztáv. *Az alsó-magyarországi bányavárosok küzdelmei a nagylucsei Dóczyakkal, 1494–1548 [The struggles of the Lower Hungarian mining towns with the Dóczy family of Nagylucse]* (Értekezések a történeti tudományok köréből, VI/6). Budapest: Magyar Tudományos Akadémia, 1876.

Werther, Lukas et al. "On the Way to the Fluvial Anthroposphere—Current Limitations and Perspectives of Multidisciplinary Research." *Water* 13, no. 16 (2021): 2188. https://doi.org/10.3390/w13162188

Wetter, Oliver et al. "The Year-Long Unprecedented European Heat and Drought of 1540 – A Worst Case." *Climatic Change* 125 (2014): 349–363. https://doi.org/10.1007/s10584-014-1184-2

White, Lynn, Jr., *Medieval Technology and Social Change*. London: Oxford University Press, 1962.

Wiemann, Harm. "Beiträge zur Geschichte des Mühlenrechts. Dargestellt an Mühlen der Herrschaft Crimmitschau vom 14.–17. Jahrhundert." *Zeitschrift der Savigny-Stiftung für Rechtsgeschichte, Germanistische Abteilung* 66 (1948): 477–500. http://doi.org/10.7767/zrgga.1948.66.1.477

Wiesiołowski, Jacek. *Socjotopografia późnośredniowiecznego Poznania [The social topography of late medieval Poznan*]. Warszaw: Państw. Wydawn. Nauk, 1982.

Wikander, Örjan. *Exploitation of Water-Power or Technological Stagnation? A Reappraisal of the Productive Forces in the Roman Empire* (Scripta Minora 1983–84, 3). Lund: Gleerup, 1984.

Wikander, Örjan, ed. *Handbook of Ancient Water Technology* (Technology and Change in History, 2). Boston and Leiden: Brill, 2000.

Wikander, Örjan. "The Water-Mill." In *Handbook of Ancient Water Technology* (Technology and Change in History, 2), ed. Örjan Wikander, 371–400. Leiden: Brill, 2000. https://doi.org/10.1163/9789004473829_019

Wilke, Carsten. "Medieval Hebrew Inscriptions: Towards a European Database." *Jewish Studies at the Central European University* 7 (2013): 147–172.

Winiwarter, Verena, Martin Schmid, and Gert Dressel. "Looking at Half a Millennium of Co-Existence: The Danube in Vienna as a Socio-Natural Site." *Water History* 5 (2013): 108. https://doi.org/10.1007/s12685-013-0079-x

Winkler, Elemér. *A soproni céhek története a XV–XIX. században [Life of the guilds of Sopron in the 15th–19th centuries*]. Sopron: Tóth Kny., 1921.

Wittfogel, Karl A. *Oriental Despotism: A Comparative Study of Total Power*. New Haven, CT: Yale University Press, 1957.

Wolfe, Michael. *Walled Towns and the Shaping of France from the Medieval to the Early Modern Era*. New York: Palgrave Macmillan, 2009.

Wöller, István. *A Tapolca vízimalmai Pápán és környékén [The water mills of the River Tapolca at Pápa and its surroundings] (Fejezetek Pápa város történetéből, 1)*. Pápa: Pápa Város Önkormányzata, 1994.

Woodward, Donald. *Men at Work: Labourers and Building Craftsmen in the Towns of Northern England, 1450–1750*. Cambridge: Cambridge University Press, 1995.

Worthen, Shana. "The Influence of Lynn White, Jr's Medieval Technology and Social Change." *History Compass* 7, no. 4 (2009): 1201–1217. https://doi.org/10.1111/j.1478-0542.2009.00615.x

Yim, Laura Lehua. "A Watercourse 'in Variance': Re-situating a Sixteenth-Century Legal Map from Ashbourne, Derbyshire." *Imago Mundi* 68, no. 2 (2016): 147–163. https://doi.org/10.1080/03085694.2016.1171484

Zanoboni, Maria Paola. "L'acqua come spazio economico: attività commerciali e manifatturiere lungo i navigli milanesi (sec. XV)." *Storia economica* 16 (2013): 143–193.

Zatykó, Csilla. "Adalékok egy 1460-as Zala megyei birtokmegosztó oklevelhez." [Contributions to an estate division charter from Zala County from 1460] In *"Es tu scholaris" – ünnepi tanulmányok Kubinyi András 75. születésnapjára* [Studies in honor of the 75th birthday of András Kubinyi] *[Studies in honor of the 75th birthday of András Kubinyi]* (Monumenta historica Budapestinensia, 13), eds. Beatrix F. Romhányi et al., 125–131. Budapest: Budapesti Történeti Múzeum, 2004.

Zatykó, Csilla. "Aspects of Fishing in Medieval Hungary." In *Processing, Storage, Distribution of Food. Food in the Medieval Rural Environment. Ruralia VIII*, ed. Jan Klápště, 399–408. Turnhout: Brepols, 2011.

Zatykó, Csilla. "Eltűnt berzencei malmok nyomában." [In pursuit of the lost mills of Berzence] In *Genius loci – Laszlovszky 60*, eds. Dóra Mérai et al., 108–112. Budapest: Archaeolingua Alapítvány, 2018.

Zatykó, Csilla. "Medieval Villages and their Landscape: Methods of Reconstruction." In *People and Nature in Historical Perspective* (CEU Medievalia, 5), eds. József Laszlovszky, and Péter Szabó, 343-375. Budapest: CEU Press – Archaeolingua, 2003.

Zinkiewicz, Jacek. "Creatio. Foundation of Kraków." In *On Common Path: Budapest and Kraków in the Middle Ages. Exhibition Catalogue, Budapest March 18 – July 24, 2016*, eds. Judit Benda et al., 87–92. Budapest: Budapest History Museum, 2016.

Zschieschang, Christian. "Beobachtungen zur Entstehung von Mühlennamen westlich der Oder." In *Usus aquarum: Interdisziplinäre Studien zur Nutzung und Bedeutung von Gewässern im Mittelalter* (Forschungen zur Geschichte und Kultur des östlichen Mitteleuropa, 54), eds. Christoph Mielzarek, and Christian Zschieschang, 87–162. Vienna: Böhlau, 2019. https://doi.org/10.7788/9783412500894.87

Zschieschang, Christian. "Zur Benennung von Mühlen im Mittelalter." In *Wassermühlen und Wassernutzung im mittelalterlichen Ostmitteleuropa* (Forschungen zur Geschichte und Kultur des östlichen Mitteleuropa, 50), eds. Martina Maríková, and Christian Zschieschang, 193–232. Stuttgart: Franz Steiner, 2015.http://doi.org/10.1515/9783110547023-004

Zsidi, Paula. "A Duna szerepe Aquincum topográfiájában." [The role of the Danube in the topography of Aquincum] *Budapest Régiségei* 41 (2008): 57–83.

Zsoldos, Attila. "Confinium és marchia. Az Árpád-kori határvédelem néhány intézményéről." [Confinium and marchia: on some institutions of Árpádian-age frontier protection] *Századok* 134 (2000): 99–116.

Zsoldos, Attila. "Doncs mester zolyómi ispánsága." [The ispánság of Doncs at Zólyom] *Századok* 158 (2024): 1179–1197.

Zsoldos, Attila. *A Druget-tartomány története, 1315–1342 [The history of the Druget lands, 1315–1342]*. Budapest: MTA Bölcsészettudományi Kutatóközpont Történettudományi Intézet, 2017.

Zsoldos, Attila. "Hűséges oligarchák." [Faithful oligarchs] In *A történettudomány szolgálatában. Tanulmányok a 70 éves Gecsényi Lajos tiszteletére [In the service of historical research. Studies in honor of Lajos Gecsényi on his 70th birthday]*, eds. Magdolna Baráth, and Antal Molnár, 347–354. Budapest and Győr: Magyar Országos Levéltár Győr-Moson-Sopron Megye Győri Levéltára, 2012.

Zsoldos, Attila. "Károly és a városok." [King Charles I and towns] In *Pénz, posztó, piac: Gazdaságtörténeti tanulmányok a magyar középkorról [Money, clorth, market. Studies in the economic history of medieval Hungary]*, ed. Boglárka Weisz, 267–283. Budapest: MTA BTK Történettudományi Intézet, 2016.

Zsoldos, Attila. *Magyarország világi archontológiája, 1000–1301 [Lay archontology of Hungary, 1000–1301]* (História könyvtár. Kronológiák, adattárak, 11). Budapest: História–MTA Történettudományi Intézete, 2011.

Zsoldos, Attila. "The Path of Košice to the Privileges of 1347." In *Košice in the Coordinates of European History*, eds. Mária Hajduová, and Martin Bartoš. 34–49. Košice: The City of Košice, 2013.

Zsoldos, Attila. "Sienai ötvösből szepesi alispán. Olaszok és itáliaiak a Druget-nádorok szolgálatában." [A Sienese Goldsmith as Alispán of Szepes: Italians in the Service of the Druget Palatines] *Világtörténet* 7 [39] (2017): 53–74.

Zsoldos, Attila. "A vasi várispánság felbomlása." [Dissolution of the royal castle district of Vas] *Vasi Szemle* 54, no. 1 (2000): 27-46.

ZUG – Zentrum für Umweltgeschichte Universität für Bodenkultur Wien, ed. *Wasser Stadt Wien – Eine Umweltgeschichte.* Vienna: Universität für Bodenkultur Wien, 2019.

Zupko, Ronald Edward, and Robert Anthony Laures. *Straws in the Wind: Medieval Urban Environmental Law. The Case of Northern Italy.* Boulder, CO: Westview Press, 1996.

Związek, Tomasz. "Kształtowanie sieci młynów wodnych na przykładzie powiatu konińskiego (ok. 1300–1550)." [The formation of a network of water mills on the example of the Konin district (approx. 1300–1550)]." *Studia Geohistorica* no. 2 (2014): 118–142. https://doi.org/10.12775/SG.2014.09

Związek, Tomasz. "The Will, the Mill and the Town. Prolegomena to the Functioning of the Miller Family in the Late Medieval Town." In *Usus aquarum: Interdisziplinäre Studien zur Nutzung und Bedeutung von Gewässern im Mittelalter* (Forschungen zur Geschichte und Kultur des östlichen Mitteleuropa, 54), eds. Christoph Mielzarek, and Christian Zschieschang, 71–86. Vienna: Böhlau, 2019. https://doi.org/10.7788/9783412500894.71

Zycha, Adolf. *Das böhmische Bergrecht des Mittelalters*, 2 vols. Berlin: Vahlen, 1900.

Финдрик, Ранко [Findrik, Ranko]. "Увод у проучавање старих воденица [Uvod u proucavanje starih vodenica]." [Introduction to the study of ancient mills] *Саопштења [Saopstenje]* 15 (1983): 95–117.

Websites and Online Databases

Carolingian Culture at Reichenau and St. Gall database
Online: www.stgallplan.org/ (last accessed: March 16, 2024)

Domesday Book Mapped: Online database
Online: https://opendomesday.org/ (last accessed: July 26, 2024)

Hungarian Cultural History Database – Medieval and Early Modern period
Online: http://mamul.btk.mta.hu/ (last accessed: April 1, 2023)

Hausbuch der Mendelschen Zwölfbrüderstiftung – Online Database
Online: www.nuernberger-hausbuecher.de/ (last accessed: March 16, 2023)

Historic Town Atlas Project – Bibliography of the published volumes
Online: www.historictownsatlas.org.uk/content/bibliography (last accessed: September 26, 2024)

Kremnické dukáty – Kremnica coins

Online: www.corycats.sk/mint/index.php?go=grofi_is (last accessed: November 15, 2023)
Lexicon Latinitatis Medii Aevi Hungariae online
Online: http://clph.elte.hu/llmaeh/llmaeh/index.html (last accessed: October 5, 2024)
Mapire – Historical Maps database
Online: http://mapire.eu (last accessed: June 8, 2025)
Monasterium database
Online: http://monasterium.net/ (last accessed: 16 June 2025)
The Age of the Reformation. Database of the Hungarian National Archives, State Archive (1526–1570)
Online: https://adatbazisokonline.hu/imgview/reformacio-kora-mnl-ol-1526-1570/ (last accessed: December 12, 2024)
The Digital Library of Medieval Hungary
Online: http://mol.arcanum.hu/medieval/ (last accessed: 20 June, 2015). Now integrated to www.hungaricana.hu (last accessed: 20 September, 2024)
Water-Powered Flour Mills in the Czech Republic – Database
Online: http://vodnimlyny.cz (last accessed: November 13, 2024)

INDEX

Abaúj County 138
Abenstorffer, Ulrich 200
account books 24, 99, 110, 129, 198–200, 203, 205, 207, 212, 214–15
Adige (river) 135
advocatus 122, 127
Ágfalva (Agendorf) 201
Albeus 69
Andrew I, king of Hungary 51, 70
Angielini, Nicholas, military engineer, map maker 86–7
Annales School 5
Arduin 50
Arnoux, Mathieu 78
Arnulf, Eastern Frankish king 69
Arribotti, Giacomo, architect 211
Austria 1, 9, 14, 97, 149, 202, 214
Auxerre 97
avulsio 134–6

Baia de Arieş (Aranyosbánya, Offenberg) 114, 242
Baia Mare (Nagybánya, Frauenbach) 116, 119–20, 242
Baia Sprie (Felsőbánya, Mittelstadt) 119–20, 242
Bakonybél 46
Balaton (lake) 70
Balkan (peninsula) 45–6, 54
Baltic Sea 11
Bamberg (Michelsberg) 49
Banate (region) 54
Bánfalvi (river) 99, 101–3
Banská Belá (Bélabánya, Düllen) 242
Banská Bystrica (Besztercebánya, Neusohl) 111, 116–19, 242
Banská Štiavnica (Selmecbánya, Schemnitz) 111, 116, 242
Barátka (part of present-day Levice) 70
Bartolus of Sassoferrato, jurist 132–6, 147
Batthyány, Adam, aristocrat, royal chamberlain 208, 210, 214
Batthyány, Christoph, aristocrat 214
Batthyány, Francis, ban of Croatia, 214
Batthyány, Hungarian family 149–50, 207
Batthyány, Magdalene, aristocrat, wife of László Csáky 149
Bavaria 45–6
Bel, Matthias, Lutheran preacher, scholar 108
Béla II, king of Hungary 70
Béla IV, king of Hungary 71, 116–7, 127
Belisarius, military commander 80–1
Bencze, Ünige 15
Benedict of Nursia 45
Benkő, Elek 15
Bennett, Richard 4, 43
Berencs *see* Rétközberencs
Berhida 154
Bihor Mountains 152

Birmingham 95
Bloch, Marc 4–5, 43–4
Blume, Peter, burgher of Prešov 105, 110
Bodrog (river) 120
Boleslaus, bishop of Vác 70
Bolesław V the Chaste, high duke of Poland 122–3
borders 70, 72, 85, 121, 123, 130–50, 202
Born, Ernest 45
Bratislava (Pozsony, Pressburg) 1–2, 120–1
Břevnov 46, 52
Brezno (Breznóbánya, Bries) 242
British Isles 4, 8
Bruno (Prunward), saint 46
Buda (part of present-day Budapest): Holy Spirit Hospital 91–2; Malom Stream 91–4; Malomtó 90, 93
Budafelhévíz (part of present-day Budapest) 83–4, 89–94
Builth 205
Büksavnica (river) 76–7, 130–1
Bulgaria 45, 229
Byzantine Empire 45, 94

canals 5, 8, 101–2, 135, 140, 211
carpenters 83, 204, 206–9, 214
Caspar, Gaspar, map maker 106–7
channels 1, 4, 8, 13–4, 23, 52, 73, 76, 85–6, 89, 93, 108, 135–6, 146–7, 197, 201, 210, 212, 226
Charlemagne 5
Charles I, king of Hungary 91, 99, 118, 123, 126, 137
Charles IV, Holy Roman emperor 212
Chaucer, Geoffrey, poet 206
Chupor, Hungarian family 89
Cistercians 53, 71, 85–6, 89, 122, 211
cisterns 11, 93
Cluj (Kolozsvár, Klausenburg) 110
Cojocna (Kolozs, Salzgrub) 242
Collner (river) 124–6, 129
Čoltovo (Csoltó) 141–2, 232
Conrad of Erling, master builder 204
Cosmas of Prague, chronicler 53
Constance Lake (Bodensee) 45
Coulson, Charles 95
critical infrastructures 10
Croatia 22, 44
Cenad (Csanád; bishopric) 51
Csepel Island 83
Cucca 135
customary law 14, 23, 64–5, 67–8, 75, 77, 79, 119, 128, 131, 133–4, 136, 139–41, 143, 147, 154
Czechoslovakia 1–2, 229
Czech Republic 2, 16, 95

Dacia (province) 54
Dalmatia 54
Danube (river) 1–2, 5, 7, 12, 14–5, 81, 83–4, 90, 92, 94, 121, 136, 151–2, 206, 229
Danube Bend (region) 2
Danube-Tisza Interfluve (Duna–Tisza köze) 79
Degré, Alajos 12, 66, 118, 140
Dej (Dés, Deesch/Burglos) 242
Dell'Adami, Rezső 140
Digest, legal collection 140, 146, 148
dikes 4, 23, 140, 226, 230
ditches: ditch diggers 64, 89, 95, 97, 99, 102, 104–5, 197–200, 205, 210–4, 228
Dobczyce 117
Dobersdorf (Dobrafalva, part of present-day Rudersdorf) 149–50
Dóczy, Hungarian family 114
Dolný Turček (Alcsóturcsek) 128, 130
Domesday Book 7, 211
Dömös 70–1
Doncs, ispán of Zólyom 118, 140–1
Dörögdi, Nicholas, bishop of Eger 143
Dráva (river) 15, 136
Druget, Nicholas, palatine 232
Druget, Phillip, palatine 137
Duby, Georges 5
Dunacsún (part of present-day Bratislava) 1
Durham 205

ecology 8, 10, 16
Edward I, king of England 211
Eger (bishopric) 143, 208
Egervár 207
Eggenberg, Austrian noble family 150
Egypt 8, 24
Elisabeth of Poland, queen-consort to Charles I 92
Elton, John 4, 43
Endrei, Walter 47
Esztergom (archbishopric) 11, 71, 129
Esztergom (Gran): cathedral chapter 75, 208–9
European Historic Town Atlas 95, 97

Fehérvár *see* Székesfehérvár
Fejér County 48
Fekete-Körös (river) 77
Felhévíz *see* Budafelhévíz
Felsőmutna 127–8
Felső-Petánc (part of present-day Petanjci) 149–50
female work 205
Ferdinand I, king of Hungary 203
Ferdinand III, king of Hungary 148
Ferenczi, László 14
Fertő (lake) 99
Fertő (region) 202
Filarete (Antonio Averlino), architect 96–7
fishing 2, 4, 7–10, 12, 49, 64, 81–3, 89, 114, 116–22, 127, 143, 150–2, 213, 228–9; fishponds 4, 6, 12, 14–7, 23–4, 49, 63, 78, 94, 96–7, 101–2, 136, 199, 201, 211–4, 226, 235
Flaran 8
floods 12–4, 67–8, 135, 138–9, 142–3, 146–9, 207, 209–10, 228–30
Florence 227
Fok *see* Siófok
forests 17, 49, 64, 71, 111, 114, 116, 119, 121–2, 124, 128, 131, 151; deforestation 128; legislation 114, 117; timber harvesting 111, 201
France 4, 8, 95, 135, 227
Francia 78
free royal towns 103, 110, 152, 212, 228
Freistadt 97
Fügedi, Erik 121

Gabčíkovo–Nagymaros Dam 2, 14, 229
Gassino 49
Gattendorf (Lajtakáta) 214
Gelnica (Gölnicbánya, Göllnitz) 111, 117, 242
geomorphology 135–6
George, canon of the chapter of Zagreb 89
Gerlavágása 138
Germany 4, 8, 10, 45–6, 49, 134
Gerse (part of present-day Gersekarát) 46, 233
Géza, prince of Hungary 46
Ginzburg, Carlo 206
Göldner, John, burgher of Kremnica 127
Gombaszög (present-day part of Slavec in Slovakia) 73–4
Gömör County 141
Gračani (part of present-day Zagreb) 89
Great Hungarian Plain (Alföld; region) 13, 48, 152
Guillerme, André 95
Güssing (Németújvár) 214
Gúti Ország, Michael, palatine of Hungary 92
Gutkeled, Stephen, palatine of Hungary 85
Győr (Raab) 69, 120–2, 202, 213; Malomsok 121
Györffy, György 21, 46, 49, 126
Gyula 207

Hahót, Nicholas, ban of Croatia-Dalmatia 85
Hanság (region) 202
Hedwig of Silesia (saint), duchess of Silesia 53
Hejőkeresztúr 209
Henry II, Holy Roman emperor 49
Henryków 53
Hernád (river) 73–4
Heves County 77
Hídvég *see* Sajóhídvég
Holocene 135
Holt, Richard 5, 95
Holub, József 66
Holy Roman Empire 49–50, 53, 68–9, 71, 148, 204
Horn, Walter 45
Horná Štubňa (Újlehotka) 128
Horohalya Stream 76
Horvathjarfalu *see* Jarovce
hospes 117–8, 120–2, 127–8, 130
Hron (river) 76, 114, 117, 208
Hronovce (Lekér) 121
Hronský Beňadik (Garamszentbenedek) 70, 76, 123, 130–1
Hugo of Noyers, bishop of Auxerre 97
Hundred Years' War 95
Hussites 105
Hybe (Hibbe) 117, 242

Iberian Peninsula 9, 44
Ikva (river) 99, 102–3, 213
industries 1–2, 5, 7–8, 12, 15, 17, 22, 63–5, 84, 102–3, 108, 110, 154, 228; gunpowder production 108, 116; industrial mills 5, 12, 79; metallurgy 5; paper production 116; textile industry 5, 102, 116
inland transportation 8

Ipoly (river) 1, 71
Italy 49, 135
Iványi, Béla 104, 108

Jadro (river) 54
Jakab, Gusztáv 15
Jarovce (Horvátjárfalu) 1
Jasov (Jászó) 111
Jedrzejów 122
Jenő (part of present-day Budapest) 83
Jihlava (Iglau) 102
Jobbágy, Jeremiah, provisor of Güssing 214
Johannesberg *see* Kremnické Bane
John I Szapolyai, king of Hungary 203
John of Luxembourg, king of Bohemia 123
John of Zredna (Vitéz), archbishop of Esztergom 76
Justinian, Roman emperor 146, 148
Jutas (part of present-day Veszprém) 78

K. Németh, András 67
Kána (part of present-day Budapest) 58
Karlsgraben (Fossa Carolina; canal) 5
Kékes (part of present-day Pilisszentlászló) 73–4
Keresztúr *see* Hejőkeresztúr
Kiskovácsi (part of present-day Berhida) 154
Klápšte, Jan 52
Klostermarienberg (Borsmonostor) 71
Kniezsa, István 47
Kochuba (part of present-day Cociuba Mare) 76
Kőfalvi, Tamás 66
Koksó (part of present-day Kokšov-Bakša) 72
Kolon (district) 48
Kompa (river) 70
Körmend 207–10
Kóród (part of present-day Coroisânmărtin) 142, 144
Körös (river) 152
Košice (Kassa, Kaschau) 96, 104, 138, 234
Kőszeg (Güns) 96
Kotyza, Oldřich 52
Kraków 15, 84, 104, 122–3
Kremnica (Körmöcbánya, Kremnitz) 111, 113, 116, 119, 123–32, 242
Kristó, Gyula 46
Krupina (Korpona, Karpfen) 127–8, 242
Kubinyi, András 91–2
Kümpf, Urban, miller 201, 206
Kunešov (Szentmihály/Kunosvágása, Kuneschhau) 126–7
Kutná Hora (Kuttenberg) 123–4

Lackner, Christoph, mayor of Sopron 102
Lád *see* Sajólád
Ladislas I, king of Hungary 69
Ladislas IV (the Cuman), king of Hungary 72, 117, 121
Lafnitz (river) 149, 210
Langdon, John 9
language contact 47
language use 22
Langwart, Peter, burgher of Prešov 105, 110
Lekeňa (Lekenye; part of present-day Bohúňovo) 141–2, 232
Lent 12
Lesser Poland 103
Lincoln 205
locatores 126–7
Lombards 135
Lorsch 8
Louis I (the Great), king of Hungary 73, 92, 104, 119–20, 131, 136, 138
Louis II, king of Hungary 21
Louny 128
Low Countries 134–5
Ľubietová (Libetbánya, Libethen) 111, 242
Ľubotín (Lubotény) 136–9
Lucas, Adam 9, 44
Lupoldus (Hyppolit), chamber count of Kremnica 124
Luxembourg dynasty 212

Macedonia 45
Magdeburg Law 122
Main (river) 5
Makkai, László 46–7
Margaret Island 82–3, 91–2
market towns 209, 212
Matthias, king of Hungary 92, 113, 154; library (Bibliotheca Corviniana) 97
Mecklenburg 227
Medvednica Hills 85
Medveščak (Cirkvenik; river) 85–9
Megyehíd 72
Menocchio, miller 206

Michel, Zacharias, map maker 101
Miechów 122
Mihályi 71
Mikhail, Alan 8
Milan 211
millers 6, 25, 51, 53, 83, 93, 127, 198–9, 201–10, 212, 215, 228
mills: hand mills 43–5, 48, 51, 53; horse mills 116; malt mill 108; mill dams 71, 77, 154, 208–10; mill races 8, 14, 43, 71, 73, 85–9, 110, 136, 200–1, 204, 227; millstones 202; ore stamp mills 125; sawmills 111, 114; ship mills 80–4, 89
moats 25, 94–110, 198–204, 213, 215, 228
Moldavia 45, 53
Mongol invasion 48, 85, 123
Monte Cassino 45
Moór, Elemér 47
Moravia 11
Moson (part of present-day Mosonmagyaróvár) 110
Mossóczy, Zakariás, jurist, bishop of Nitra 67
Mumford, Lewis 43–4
Mura (river) 136, 149–50
Muratori, Ludovico Antonio 135
Murska Sobota (Muraszombat) 149
Mutna (*Chernakou*; forest) 127–8

Nádasdy family 149, 154
Nádasdy, Andrew, Hungarian nobleman 150
Nádasdy, Francis, Hungarian aristocrat 154
Nagymaros 2, 14, 229
Nagymartoni, Paul, chief justice of Hungary 143
Nagyszakácsi 143
Naviglio Grande (river) 211
Nemsem, Stephen, provisor of Körmend 207–10
Neolithic 44
Newcastle 205
Nile (river) 147
Nová Baňa (Újbánya, Königsberg) 76–7, 111, 113, 118, 123, 125, 128, 130–2
Nova Ves (part of present-day Zagreb) 89

Óbuda (part of present-day Budapest) 80–4, 89–90, 92–3; Poor Clares 92–3
Ocna Dejului (Désakna) 242

Olpe 124
Ónod 143, 145
Oradea (Várad) 76
Orlov (Orló) 136–9
Ostrov (part of present-day Davle) 46
Ottó, ispán of Somogy 69
Óvár (part of present-day Mosonmagyaróvár) 110

Pannonhalma 46, 69, 71
Pannonia (province) 99
Pápa 202
Pápay, Paul, archdeacon of the cathedral chapter of Győr 213
Paris Peace Treaty 1
Partizánska Ľupča (Németlipcse, Deutschliptsch) 117, 242
Patacs (in the borders of present-day Kamenný Most) 208
Paul the Deacon, chronicler 135
Paulines 86, 89, 120–1, 208–9
Pécsvárad 46, 66, 72, 75–6
Perényi, Emeric, palatine of Hungary 93
Perényi family 14, 23
Perugia 132–3; university 133
Pest (part of present-day Budapest) 80–4, 91, 96; Saint Nicholas Hospital 83
Pető, Hungarian noble family 146–7
Pető, Zsuzsa 14
Pider, Benedict, judge of Buda 92
Plaveč (Palocsa) 138
Po (river) 135, 211
Poland 11, 15–6, 45–6, 53–5, 65, 95, 103, 117, 212
pond masters 199, 205–6, 212–4
Poprad (river) 138–40
Porouathka (river) 141
ports 7, 94–5, 121
Pozsony County 121
Prague 52, 84
Pražák, Jiří 52
Prešov: Carmelites 108, 181; gates 105; mills 105, 108, 126; roads 94, 98, 103–10, 126; town walls 97, 107–10
Procopius, chronicler 80
Pukanec (Bakabánya, Pukantz) 76, 111, 124, 130–1, 242

quarries 200
Quedlinburg 98

Rába (river) 72, 121, 136, 209–10
Rábaköz 13
Rabb, Péter 96
Rábca (river) 71
Rajka 1
Rákos (part of present-day Budapest) 84
Regensburg 46
Reichenau 45–6
Revúca (river) 117, 119
Rhine (river) 5, 81
Rhineland (region) 124
Rhine–Main–Danube Canal 5
Rhuddlan 211
Rimavská Baňa (Rimabánya) 242
Rodna (Óradna, Altrodenau) 242
Roman Empire; water technology 44–5, 80, 99
Roman law; influence in the Middle Ages 133–4, 136, 140, 148, 150, 226
Romania 2, 229
Rome 46, 80
Ronyva (river) 120, 208
Rožňava (Rozsnyó, Rosenau) 111, 114–15
Rudabánya 111
Rum 72
Rusovce (Oroszvár) 1
Ružomberok (Rózsahegy, Rosenberg) 117–20, 122, 242

Šahy (Ság) 71
Sajó (river) 141, 143, 145
Sajóhídvég 143, 145
Salona, Ancient settlement 54
Sármelléke (part of present-day Gersekarát) 146
Sáros County 104, 136–8
Sárosd (river) 68
Sárvíz (river) 146
Sátoraljaújhely 120–1, 208
Sava (river) 84, 89
Saxon Mirror (*Sachsenspiegel*) 140
Scarbantia, Ancient settlement 98
Schönach, Paula 7
Sebesér (river) 76
Séd (river) 154
Serafin, dean at the cathedral chapter of Esztergom 208
Serbia 45, 53
ships 82
Sic (Szék, Secken) 242
Sigismund of Luxembourg, Holy Roman emperor 92, 204
Silesian Beskids 11
Simoncsics, Marc, provisor of Körmend 210
Sió (river) 70
Siófok 70
Skelton, Leona 7
Sklené (Szklenó, Glaserhau) 127
Slovakia 2, 16, 73, 111, 230
Smolník (Szomolnok, Schmölnitz) 111, 242
Soler (valley) 124, 126, 129
Somogy County 68–9
Sopron 11, 79, 94, 96–103, 109–10, 198–207, 210–1, 213–15; fish market 109–10; Foregate 99, 198–201; gallows 207; Jewish cemetery 200; Posztósok árka 102–3; Saint Michael's Church 200; Saint Michael's Gate 200; Saint Sebastian chapel 213
Soviet Union 2
Spišská Nová Ves (Igló, Zipser Neudorf) 111
Squatriti, Paolo 9
Stephen (saint), king of Hungary 46, 48
Stephen V, king of Hungary 117
stonecutters 199–200
stonemasons 200, 205
Strem (Strém) 208
Styria 149–50, 154
Sümeg 214
Szakácsi *see* Nagyszakácsi
Szántó 70
Szapolyai, Stephen, palatine of Hungary 83
Szatmár County 71
Szatóc (Šratovci, Schrottendorf, in the borders of present-day Radenci) 149–50
Szecsőd (part of present-day Molnaszecsőd) 207–8
Székesfehérvár 11, 71; Holy Virgin church 71
Szende, Katalin 14
Szentendrei Island 83
Szentfalva (Szenterzsébetfalva, part of present-day Budapest) 83
Szentmárton (part of present-day Coroisânmărtin) 142, 144
Szepes (Spišská Kapitula) 138
Szigetköz (island) 230
Szomód 71

Takáts, Sándor 12
Tarna (river) 77
Târnava Mică (river) 142–3
Tata 71
Telegdi, Csanád, archbishop of Esztergom 208
Telekes 146–7
Telkibánya 111, 242
Tiber (river) 132–3
Ticino (river) 211
Tihany 51, 70–1
Tisza (river) 136, 152, 197
Torysa (river) 108
trade routes 103
Transylvania 15, 54, 64, 67, 96, 110–11, 116, 151
Tringli, István 14, 67, 77, 139
Trnava (Nagyszombat, Tyrnau) 96
Turda (Torda, Thorenburg) 242
Turin 49
Turóc (Kláštor pod Znievom) 141
Turóc County 127–8
Tyne (river) 7

Újbécs (part of present-day Budapest) 83
Újhely *see* Sátoraljaújhely
Ulpian, jurist 146
Únětice 52
Upper Hungary (region) 64, 111, 116, 119, 151
urbanization 22, 94

Váczy, Péter 46
Váh (river) 117, 119
Vajda, Tamás 15, 47, 66
Vale Royal 211
Varzy 97
Vas County 72, 148–50
Vasvár 120–1; collegiate chapter 70
Veneto (region) 135
Verence Stream 76, 131
Veszprém: bishopric 48–52, 69, 75, 78, 213; cathedral chapter 78, 154
Veszprém County 48, 154
Viczián, István 15
Vienna 6, 98, 136, 202, 206, 209
Vinohradiv (Nagyszőlős) 120–1
Vinţu de Jos (Alvinc, Unterwinz) 209
Visegrád 2
Visegrád County 48
Vistula (river) 11, 122
Vltava (river) 52
Vrančić, Antun, bishop of Pécs, chronicler 84

wages 204–5, 212, 214
Wales 205, 211
Wallachia 45, 53
Warsaw Pact 2, 229
waste 8, 11
wastewater 11, 102–3, 228
water regulation 5, 8, 197
water scarcity 3, 45, 80
Wawel Hill 122
Weilburg in Hesse 49
weirs 8, 25
Weisz, Boglárka 114
Wenceslas IV, king of Bohemia 212
Wenzel, Gusztáv 114
Werbőczy, Stephen, jurist, chief justice of Hungary 24, 68, 75, 139–41, 148
William le Fox 211
William of Boston 211
Winiwarter, Verena 6
Wolfgang, saint 46
Wrocław (Breslau) 15, 84

York 205
Yugoslavia 2

Zagreb 84–89, 91; chapter 86, 89; Cistercians 85–86, 89; Dominicans 86; Gradec 85–89; Kaptol 85–87, 89; Paulines 86, 89; Saint Emeric Church 86
Zala County 66
Zalavár 46
Zatykó, Csilla 15
Žilina (Zsolna) 126–7
Zlatá Idka (Aranyida) 242
Zlatna (Zalatna, Klein-Schlatten) 242
Zólyom County 118, 141
Zselicszentjakab (part of present-day Kaposvár) 69, 71

For Product Safety Concerns and Information please contact our EU representative GPSR@taylorandfrancis.com
Taylor & Francis Verlag GmbH, Kaufingerstraße 24, 80331 München, Germany

www.ingramcontent.com/pod-product-compliance
Lightning Source LLC
LaVergne TN
LVHW010652300926
849161LV00005B/1103